Applied Stochastic Differential Equations

Stochastic differential equations are differential equations whose solutions are stochastic processes. They exhibit appealing mathematical properties that are useful in modeling uncertainties and noisy phenomena in many disciplines. This book is motivated by applications of stochastic differential equations in target tracking and medical technology and, in particular, their use in methodologies such as filtering, smoothing, parameter estimation, and machine learning. It builds an intuitive hands-on understanding of what stochastic differential equations are all about, but also covers the essentials of Itô calculus, the central theorems in the field, and such approximation schemes as stochastic Runge–Kutta. Greater emphasis is given to solution methods than to analysis of theoretical properties of the equations. The book's practical approach assumes only prior understanding of ordinary differential equations. The numerous worked examples and end-of-chapter exercises include application-driven derivations and computational assignments. MATLAB/Octave source code is available for download, promoting hands-on work with the methods.

SIMO SÄRKKÄ is Associate Professor of Electrical Engineering and Automation at Aalto University, Adjunct Professor at Tampere University of Technology and Lappeenranta University of Technology, and Technical Advisor or Founder in multiple startup companies. His research interests are in probabilistic modeling and sensor fusion for location sensing, health technology, and machine learning. He has authored over 100 peer-reviewed scientific articles as well as one book, titled *Bayesian Filtering and Smoothing*.

ARNO SOLIN is Assistant Professor of Computer Science at Aalto University and Technical Advisor in multiple startup companies. His research interests focus on models and applications in sensor fusion for tracking and navigation, brain imaging, and machine learning problems. He has published over 30 peer-reviewed scientific papers, and has won several hackathons and competitions in mathematical modeling, including the 2014 Schizophrenia Classification Challenge on Kaggle.

INSTITUTE OF MATHEMATICAL STATISTICS
TEXTBOOKS

IMS Textbooks give introductory accounts of topics of current concern suitable for advanced courses at master's level, for doctoral students and for individual study. They are typically shorter than a fully developed textbook, often arising from material created for a topical course. Lengths of 100–290 pages are envisaged. The books typically contain exercises.

Other Books in the Series

Applied Stochastic Differential Equations

SIMO SÄRKKÄ

Aalto University, Finland

ARNO SOLIN

Aalto University, Finland

CAMBRIDGE
UNIVERSITY PRESS

CAMBRIDGE
UNIVERSITY PRESS

University Printing House, Cambridge CB2 8BS, United Kingdom

One Liberty Plaza, 20th Floor, New York, NY 10006, USA

477 Williamstown Road, Port Melbourne, VIC 3207, Australia

314-321, 3rd Floor, Plot 3, Splendor Forum, Jasola District Centre, New Delhi - 110025, India

79 Anson Road, #06-04/06, Singapore 079906

Cambridge University Press is part of the University of Cambridge.

It furthers the University's mission by disseminating knowledge in the pursuit of
education, learning and research at the highest international levels of excellence.

www.cambridge.org
Information on this title: www.cambridge.org/9781316510087
DOI: 10.1017/9781108186735

First published 2019

A catalogue record for this publication is available from the British Library

Library of Congress Cataloging in Publication data
Names: Särkkä, Simo, author. | Solin, Arno, author.
Title: Applied stochastic differential equations / Simo Särkkä, Arno Solin.
Description: Cambridge, United Kingdom ; New York, NY : Cambridge University
Press, 2019. | Series: Institute of Mathematical Statistics textbooks ; v
10 | Includes bibliographical references and index.
Identifiers: LCCN 2018026584 | ISBN 9781316510087 (hardback : alk. paper) |
ISBN 9781316649466 (paperback : alk. paper)
Subjects: LCSH: Stochastic differential equations–Textbooks.
Classification: LCC QA274.23 .S23 2019 | DDC 315/.350151923–dc23 LC record available at
https://lccn.loc.gov/2018026584

ISBN 978-1-316-51008-7 Hardback
ISBN 978-1-316-64946-6 Paperback

Contents

Preface

This book is an outgrowth of a set of lecture notes that has been extended with material from the doctoral theses of both authors and with a large amount of completely new material. The main motivation for the book is the application of stochastic differential equations (SDEs) in domains such as target tracking and medical technology and, in particular, their use in methodologies such as filtering, smoothing, parameter estimation, and machine learning. We have also included a wide range of examples of applications of SDEs arising in physics and electrical engineering.

Because we are motivated by applications, much more emphasis is put on solution methods than on analysis of the theoretical properties of equations. From the pedagogical point of view, one goal of this book is to provide an intuitive hands-on understanding of what SDEs are all about, and if the reader wishes to learn the formal theory later, she can read, for example, the brilliant books of Øksendal (2003) and Karatzas and Shreve (1991).

Another pedagogical aim is to overcome a slight disadvantage in many SDE books (e.g., the aforementioned ones), which is that they lean heavily on measure theory, rigorous probability theory, and the theory of martingales. There is nothing wrong in these theories – they are very powerful theories and everyone should indeed master them. However, when these theories are explicitly used in explaining SDEs, they bring a flurry of technical details that tend to obscure the basic ideas and intuition for the first-time reader. In this book, without shame, we trade rigor for readability by treating SDEs completely without measure theory.

The book's low learning curve only assumes prior knowledge of ordinary differential equations and basic concepts of statistics, together with understanding of linear algebra, vector calculus, and Bayesian inference. The book is mainly intended for advanced undergraduate and graduate students in applied mathematics, signal processing, control engineering, statistics, and computer science. However, the book is suitable also for researchers and practitioners who need a concise introduction to the topic at a level that enables them to implement or use the methods.

The worked examples and numerical simulation studies in each chapter illustrate how the theory works in practice and can be implemented for solving the problems. End-of-chapter exercises include application-driven derivations and computational assignments. The MATLAB® source code for reproducing the example results is available for download through the book's web page, promoting hands-on work with the methods.

We have attempted to write the book to be freestanding in the sense that it can be read without consulting other material on the way. We have also attempted to give pointers to work that either can be considered as the original source of an idea or just contains more details on the topic at hand. However, this book is not a survey, but a textbook, and therefore we have preferred citations that serve a pedagogical purpose, which might not always explicitly give credit to all or even the correct inventors of the technical ideas. Therefore, we need to apologize to any authors who have not been cited although their work is clearly related to the topics that we cover. We hope you understand.

The authors would like to thank Aalto University for providing the chance to write this book. We also would like to thank Robert Piché, Petteri Piiroinen, Roland Hostettler, Filip Tronarp, Santiago Cortés, Johan Westö, Joonas Govenius, Ángel García-Fernández, Toni Karvonen, Juha Sarmavuori, and Zheng Zhao for providing valuable comments on early versions of the book.

Simo and Arno

1

Introduction

The topic of this book is stochastic differential equations (SDEs). As their name suggests, they really are differential equations that produce a different "answer" or solution trajectory each time they are solved. This peculiar behaviour gives them properties that are useful in modeling of uncertainties in a wide range of applications, but at the same time it complicates the rigorous mathematical treatment of SDEs.

The emphasis of the book is on applied rather than theoretical aspects of SDEs and, therefore, we have chosen to structure the book in a way that we believe supports learning SDEs from an applied point of view. In the following, we briefly outline the purposes of each of the remaining chapters and explain how the chapters are connected to each other. In the chapters, we have attempted to provide a wide selection of examples of the practical application of theoretical and methodological results. Each chapter (except for the Introduction and Epilogue) also contains a representative set of analytic and hands-on exercises that can be used for testing and deepening understanding of the topics.

Chapter 2 is a brief outline of concepts and solutions methods for deterministic ordinary differential equations (ODEs). We especially emphasize solution methods for linear ODEs, because the methods translate quite easily to SDEs. We also examine commonly used numerical methods such as the Euler method and Runge–Kutta methods, which we extend to SDEs in the later chapters.

Chapter 3 starts with a number of motivating examples of SDEs found in physics, engineering, finance, and other applications. It turns out that in a modeling sense, SDEs can be regarded as noise-driven ODEs, but this notion should not be taken too far. The aim of the rest of the chapter is to show where things start to go wrong. Roughly speaking, with linear SDEs we are quite safe with this kind of thinking, but anything beyond them will not work.

In **Chapter 4**, we reformulate SDEs properly as stochastic integral equations where one of the terms contains a new kind of integral called the Itô integral. We then derive the change of variable formula, that is, the Itô formula for the integral, and use it to find complete solutions to linear SDEs. We also discuss some methods to solve nonlinear SDEs and look briefly at Stratonovich integrals.

The aim of **Chapter 5** is to analyze the statistics of SDEs as stochastic processes. We discuss and derive their generators, the Fokker–Planck–Kolmogorov equations, as well as Markov properties and transition densities of SDEs. We also derive the formal equations of the moments, such as the mean and covariance, for the SDE solutions. It turns out, however, that these equations cannot easily be solved for other than linear SDEs. This challenge will be tackled later in the numerical methods chapters.

As linear SDEs are very important in applications, we have dedicated **Chapter 6** to solution methods for their statistics. Although explicit solutions to linear SDEs and general moment equations for SDEs were already given in Chapters 4 and 5, here we also discuss and derive explicit mean and covariance equations, transition densities, and matrix fraction methods for the numerical treatment of linear SDEs. We also discuss steady-state solutions and Fourier analysis of linear time-invariant (LTI) SDEs as well as temporal covariance functions of general linear SDEs.

In **Chapter 7**, we discuss some useful theorems, formulas, and results that are typically required in more advanced analysis of SDEs as well as in their numerical methods. In addition to the Lamperti transform, Girsanov theorem, and Doob's *h*-transform, we also show how to find solutions to partial differential equations with Feynman–Kac formulas and discuss some connections to path integrals in physics. This chapter is not strictly necessary for understanding the rest of the chapters and can be skipped during a first reading.

Although the Itô stochastic calculus that is derivable from the Itô formula is theoretically enough for defining SDEs, it does not help much in practical solution of nonlinear SDEs. In **Chapter 8**, we present numerical simulation-based solution methods for SDEs. The methods are based primarily on Itô–Taylor series and stochastic Runge–Kutta methods, but we also discuss the Verlet and exact algorithm methods.

In many applications we are interested in the statistics of SDEs rather than their trajectories per se. In **Chapter 9**, we develop methods for approximate computation of statistics such as means and covariances or probability densities of SDEs – however, many of the methods are suitable for numerical simulation of SDEs as well. We start with classical and modern Gaussian "assumed density" approximations and then proceed to other linearization methods. We also discuss Taylor and Hermite series approximations of transition densities and their moments, numerical solutions of Fokker–Planck–Kolmogorov equations, simulation-based approximations, and finally pathwise Wong–Zakai approximations of SDEs.

An important and historically one of the first applications of SDEs is the filtering and smoothing theory. In **Chapter 10**, we describe the basic ideas of filtering and smoothing and then proceed to the classical Kushner–Stratonovich and Zakai equations. We also present the linear and nonlinear Kalman–Bucy and Kalman filters and discuss their modern variants. Finally, we present formal equations and approximation methods for the corresponding smoothing problems.

The aim of **Chapter 11** is to give an overview of parameter estimation methods for SDEs. The emphasis is on statistical likelihood-based methods that aim at computing maximum likelihood (ML) or maximum a posteriori (MAP) estimates or are targeted to full Bayesian inference on the parameters. We start with brief descriptions of the ideas of ML and MAP estimates as well as Markov chain Monte Carlo (MCMC) methods. Parameter estimation in linear SDEs is then discussed, and finally we give approximate likelihood methods for parameter estimation in nonlinear SDEs. We also discuss some parameter estimation methods for indirectly observed SDEs.

Chapter 12 addresses the somewhat less traditional topic of connections between machine learning and SDEs. The aim is to discuss links between Gaussian process regression, Kalman filtering, and SDEs, along with applications of the methods across the fields of signal processing and machine learning.

Finally, **Chapter 13** concludes the book with an overview and gives some hints where to go next. We also discuss additional topics such as fractional Brownian motions, Lévy process driven SDEs, and stochastic control problems.

2

Some Background on Ordinary Differential Equations

The chapter provides background on deterministic (nonstochastic) ordinary differential equations (ODEs) from points of view especially suited to the context of stochastic differential equations (SDEs). As SDEs are inherently inhomogeneous differential equations (i.e., they have an input), we will concentrate on solution methods suitable for them. Furthermore, as linear and especially linear time-invariant (LTI) ODE systems are important in applications, we review the matrix exponential– and transition matrix–based methods of solution. We also discuss Fourier– and Laplace transform–based solution methods for LTI ODEs and for computing matrix exponentials. For more details on ODE methods and theory, the reader is referred to the books of Kreyszig (1993), Tenenbaum and Pollard (1985), and Hairer et al. (2008), although the same information can be found in many other books as well.

2.1 What Is an Ordinary Differential Equation?

An ODE is an equation in which the unknown quantity is a function, and the equation involves derivatives of the unknown function. For example, the second-order differential equation for a forced spring–mass system (or, e.g., a resonator circuit in telecommunications) can be generally expressed as

$$\frac{d^2 x(t)}{dt^2} + \gamma \frac{dx(t)}{dt} + \nu^2 x(t) = w(t), \tag{2.1}$$

where ν and γ are constants that determine the resonant angular velocity and damping of the spring. The *force* $w(t)$ is some given function that may or may not depend on time. In this equation, the position variable x is called the *dependent variable* and time t is the *independent variable*. The equation is of *second order*, because it contains the second derivative and no higher-order terms are present. It is *linear*, because $x(t)$ appears linearly in the equation. The equation is *inhomogeneous*, because it contains the *forcing* term $w(t)$. This inhomogeneous term will become essential in later chapters, because replacing it with a random process leads to a stochastic differential equation.

Here a *solution* to the differential equation is defined as a *particular solution*, a function that satisfies the equation and does not contain any arbitrary constants. A *general solution* on the other hand contains every particular solution of the equation parameterized by some free constants. To actually solve the differential equation, it is necessary to tie down the general solution by some initial conditions. In the preceding case, this means that we need to know the spring–mass position $x(t)$ and velocity $dx(t)/dt$ at some fixed initial time $t = t_0$. Given these initial values, there is a unique solution to the equation (provided that $w(t)$ is continuous). Instead of initial conditions, we could also fix some other (boundary) conditions of the differential equation to get a unique solution, but here we only consider differential equations with given initial conditions.

Note that it is common not to write the dependencies of x and w on t explicitly, and write the equation as

$$\frac{d^2 x}{dt^2} + \gamma \frac{dx}{dt} + v^2 x = w. \tag{2.2}$$

Although it sometimes is misleading, this "ink saving" notation is very commonly used, and we will also employ it here whenever there is no risk of confusion. Furthermore, because in this section and in this whole book we mainly consider ordinary differential equations, we often drop the word "ordinary" and just talk about differential equations.

Time derivatives are also sometimes denoted with dots over the variable, such as $\dot{x} = dx/dt$, $\ddot{x} = d^2x/dt^2$ and so on. In this *Newtonian notation*, the previous differential equation would be written as

$$\ddot{x} + \gamma \dot{x} + v^2 x = w. \tag{2.3}$$

Differential equations of an arbitrary order n can (almost) always be converted into vector differential equations of order one. For example, in the preceding spring model, if we define a *state variable* $\mathbf{x}(t) = (x_1(t), x_2(t)) = (x(t), dx(t)/dt)$, we can rewrite the previous differential equation as a first-order vector differential equation:

$$\underbrace{\begin{pmatrix} dx_1(t)/dt \\ dx_2(t)/dt \end{pmatrix}}_{d\mathbf{x}(t)/dt} = \underbrace{\begin{pmatrix} 0 & 1 \\ -v^2 & -\gamma \end{pmatrix} \begin{pmatrix} x_1(t) \\ x_2(t) \end{pmatrix}}_{\mathbf{f}(\mathbf{x}(t))} + \underbrace{\begin{pmatrix} 0 \\ 1 \end{pmatrix}}_{\mathbf{L}} w(t). \tag{2.4}$$

The preceding equation can be seen to be a special case of models of the form

$$\frac{d\mathbf{x}(t)}{dt} = \mathbf{f}(\mathbf{x}(t), t) + \mathbf{L}(\mathbf{x}(t), t)\,\mathbf{w}(t), \tag{2.5}$$

where the vector-valued function $\mathbf{x}(t) \in \mathbb{R}^D$ is generally called the state of the system, $\mathbf{f}(\bullet, \bullet)$ and $\mathbf{L}(\bullet, \bullet)$ are arbitrary functions, and $\mathbf{w}(t) \in \mathbb{R}^S$ is some (vector-valued) forcing function, driving function, or input to the system. Note that we can absorb the second term on the right into the first term to yield

$$\frac{d\mathbf{x}(t)}{dt} = \mathbf{f}(\mathbf{x}(t), t), \tag{2.6}$$

and in that sense Equation (2.5) is slightly redundant. However, the form (2.5) turns out to be useful in the context of stochastic differential equations, and thus it is useful to consider it explicitly.

The first-order vector differential equation representation of an nth-order differential equation is often called the state-space form of the differential equation. Because nth order differential equations can (almost) always be converted into equivalent n-dimensional vector-valued first-order differential equations, it is convenient to just consider such first-order equations instead of considering nth-order equations explicitly. Thus in this book, we develop the theory and solution methods (mainly) for first-order vector differential equations and assume that nth-order equations are always first converted into equations of this class.

The spring–mass model in Equation (2.4) is also a special case of *linear differential equations* of the form

$$\frac{d\mathbf{x}(t)}{dt} = \mathbf{F}(t)\,\mathbf{x}(t) + \mathbf{L}(t)\,\mathbf{w}(t), \tag{2.7}$$

which is a very useful class of differential equations often arising in applications. The usefulness of linear equations is that we can actually solve these equations, unlike general nonlinear differential equations. This kind of equations will be analyzed in the next sections.

2.2 Solutions of Linear Time-Invariant Differential Equations

Consider the following scalar linear homogeneous differential equation with a fixed initial condition at $t = 0$:

$$\frac{dx}{dt} = F\,x, \qquad x(0) = \text{given}, \tag{2.8}$$

where F is a constant. This equation can now be solved, for example, via separation of variables, which in this case means that we formally multiply by dt and divide by x to yield

$$\frac{dx}{x} = F\, dt. \tag{2.9}$$

If we now integrate the left-hand side from $x(0)$ to $x(t)$ and right-hand side from 0 to t, we get

$$\log x(t) - \log x(0) = F\, t, \tag{2.10}$$

which can be solved for $x(t)$ to give the final solution:

$$x(t) = \exp(F\, t)\, x(0). \tag{2.11}$$

Another way of arriving at the same solution is by integrating both sides of the original differential equation from 0 to t. Because $\int_0^t dx/dt\, dt = x(t) - x(0)$, we can express the solution $x(t)$ as

$$x(t) = x(0) + \int_0^t F\, x(\tau)\, d\tau. \tag{2.12}$$

We can now substitute the right-hand side of the equation for $x(\tau)$ inside the integral, which gives

$$x(t) = x(0) + \int_0^t F\left[x(0) + \int_0^\tau F\, x(\tau')\, d\tau'\right] d\tau$$

$$= x(0) + F\, x(0) \int_0^t d\tau + \int_0^t \left[\int_0^\tau F^2\, x(\tau')\, d\tau'\right] d\tau$$

$$= x(0) + F\, x(0)\, t + \int_0^t \int_0^\tau F^2\, x(\tau')\, d\tau'\, d\tau. \tag{2.13}$$

Doing the same substitution for $x(\tau')$ inside the last integral further yields

$$x(t) = x(0) + F\, x(0)\, t + \int_0^t \int_0^\tau F^2\left[x(0) + \int_0^{\tau'} F\, x(\tau'')\, d\tau''\right] d\tau'\, d\tau$$

$$= x(0) + F\, x(0)\, t + F^2\, x(0) \int_0^t \int_0^\tau d\tau'\, d\tau$$

$$\quad + \int_0^t \int_0^\tau \int_0^{\tau'} F^3\, x(\tau'')\, d\tau''\, d\tau'\, d\tau$$

$$= x(0) + F\, x(0)\, t + F^2\, x(0)\, \frac{t^2}{2} + \int_0^t \int_0^\tau \int_0^{\tau'} F^3\, x(\tau'')\, d\tau''\, d\tau'\, d\tau. \tag{2.14}$$

It is easy to see that repeating this procedure yields the solution of the form

$$x(t) = x(0) + F\,x(0)\,t + F^2\,x(0)\,\frac{t^2}{2} + F^3\,x(0)\,\frac{t^3}{6} + \cdots$$

$$= \left(1 + F\,t + \frac{F^2\,t^2}{2!} + \frac{F^3\,t^3}{3!} + \cdots\right) x(0). \tag{2.15}$$

The series in the parentheses can be recognized to be the Taylor series for $\exp(F\,t)$. Thus, provided that the series actually converges (it does), we again arrive at the solution

$$x(t) = \exp(F\,t)\,x(0). \tag{2.16}$$

The multidimensional generalization of the homogeneous linear differential equation (2.8) is an equation of the form

$$\frac{d\mathbf{x}}{dt} = \mathbf{F}\,\mathbf{x}, \qquad \mathbf{x}(0) = \text{given}, \tag{2.17}$$

where $\mathbf{F}$ is a constant (i.e., time-independent) matrix. For this multidimensional equation, we cannot use the separation of variables method, because it only works for scalar equations. However, the series-based approach works and yields a solution of the form

$$\mathbf{x}(t) = \left(\mathbf{I} + \mathbf{F}\,t + \frac{\mathbf{F}^2\,t^2}{2!} + \frac{\mathbf{F}^3\,t^3}{3!} + \cdots\right) \mathbf{x}(0). \tag{2.18}$$

The series in the parentheses can now be seen as a matrix generalization of the exponential function. This series indeed is the definition of the matrix exponential

$$\exp(\mathbf{F}\,t) = \mathbf{I} + \mathbf{F}\,t + \frac{\mathbf{F}^2\,t^2}{2!} + \frac{\mathbf{F}^3\,t^3}{3!} + \cdots \tag{2.19}$$

and thus the solution to Equation (2.17) can be written as

$$\mathbf{x}(t) = \exp(\mathbf{F}\,t)\,\mathbf{x}(0). \tag{2.20}$$

Note that the matrix exponential cannot be computed by computing scalar exponentials of the individual elements in matrix $\mathbf{F}\,t$. It is a completely different function. Sometimes the matrix exponential is written as $\text{expm}(\mathbf{F}\,t)$ to distinguish it from the elementwise computation, but here we use the common convention to simply write it as $\exp(\mathbf{F}\,t)$. The matrix exponential function can be found as a built-in function in most commercial and open-source mathematical software packages such as MATLAB®and Python. In addition to this kind of numerical solution, the exponential can be evaluated analytically, for example, by directly using the Taylor series expansion, by using the Laplace or Fourier transform, or via the Cayley–Hamilton theorem (Åström and Wittenmark, 1997).

Example 2.1 (Matrix exponential). *To illustrate the difference between the matrix exponential and the elementwise exponential, consider the equation*

$$\frac{d^2 x}{dt^2} = 0, \quad x(0) = given, \quad (dx/dt)(0) = given, \tag{2.21}$$

which in state-space form can be written as

$$\frac{d\mathbf{x}}{dt} = \underbrace{\begin{pmatrix} 0 & 1 \\ 0 & 0 \end{pmatrix}}_{\mathbf{F}} \mathbf{x}, \quad \mathbf{x}(0) = given, \tag{2.22}$$

where $\mathbf{x} = (x, dx/dt)$. Because $\mathbf{F}^n = \mathbf{0}$ for $n > 1$, the matrix exponential is simply

$$\exp(\mathbf{F}\,t) = \mathbf{I} + \mathbf{F}\,t = \begin{pmatrix} 1 & t \\ 0 & 1 \end{pmatrix} \tag{2.23}$$

which is completely different from the elementwise matrix exponential:

$$\begin{pmatrix} 1 & t \\ 0 & 1 \end{pmatrix} \neq \begin{pmatrix} \exp(0) & \exp(t) \\ \exp(0) & \exp(0) \end{pmatrix} = \begin{pmatrix} 1 & e^t \\ 1 & 1 \end{pmatrix}. \tag{2.24}$$

Let us now consider the following linear differential equation with an inhomogeneous term on the right-hand side:

$$\frac{d\mathbf{x}(t)}{dt} = \mathbf{F}\,\mathbf{x}(t) + \mathbf{L}\,\mathbf{w}(t), \tag{2.25}$$

where $\mathbf{x}(t_0)$ is given and the matrices $\mathbf{F}$ and $\mathbf{L}$ are constant. For inhomogeneous equations, the solution methods are numerous, especially if we do not want to restrict ourselves to specific kinds of forcing functions $\mathbf{w}(t)$. However, the following *integrating factor* method can be used for solving general inhomogeneous equations.

If we move the term $\mathbf{F}\,\mathbf{x}(t)$ in Equation (2.25) to the left-hand side and multiply with a term called integrating factor $\exp(-\mathbf{F}\,t)$, we get the following result:

$$\exp(-\mathbf{F}\,t)\frac{d\mathbf{x}(t)}{dt} - \exp(-\mathbf{F}\,t)\,\mathbf{F}\,\mathbf{x}(t) = \exp(-\mathbf{F}\,t)\,\mathbf{L}\,\mathbf{w}(t). \tag{2.26}$$

From the definition of the matrix exponential, we can derive the following property:

$$\frac{d}{dt}[\exp(-\mathbf{F}\,t)] = -\exp(-\mathbf{F}\,t)\,\mathbf{F}. \tag{2.27}$$

The key thing is now to observe that

$$\frac{d}{dt}[\exp(-\mathbf{F}\,t)\,\mathbf{x}(t)] = \exp(-\mathbf{F}\,t)\,\frac{d\mathbf{x}(t)}{dt} - \exp(-\mathbf{F}\,t)\,\mathbf{F}\,\mathbf{x}(t), \quad (2.28)$$

which is exactly the left-hand side of Equation (2.26). Thus we can rewrite the equation as

$$\frac{d}{dt}[\exp(-\mathbf{F}\,t)\,\mathbf{x}(t)] = \exp(-\mathbf{F}\,t)\,\mathbf{L}\,\mathbf{w}(t). \quad (2.29)$$

Integrating from t_0 to t then gives

$$\exp(-\mathbf{F}\,t)\,\mathbf{x}(t) - \exp(-\mathbf{F}\,t_0)\,\mathbf{x}(t_0) = \int_{t_0}^{t} \exp(-\mathbf{F}\,\tau)\,\mathbf{L}\,\mathbf{w}(\tau)\,d\tau, \quad (2.30)$$

which can be further rearranged to give the final solution

$$\mathbf{x}(t) = \exp(\mathbf{F}\,(t - t_0))\,\mathbf{x}(t_0) + \int_{t_0}^{t} \exp(\mathbf{F}\,(t - \tau))\,\mathbf{L}\,\mathbf{w}(\tau)\,d\tau. \quad (2.31)$$

In the preceding solution, we have also used the identity $\exp(\mathbf{F}\,s)\exp(\mathbf{F}\,t) = \exp(\mathbf{F}\,(s + t))$, which is true because the matrices $\mathbf{F}\,s$ and $\mathbf{F}\,t$ commute. The expression (2.31) is the complete solution to Equation (2.25).

2.3 Solutions of General Linear Differential Equations

In this section, we consider solutions to more general, time-varying linear differential equations. The corresponding stochastic equations are a useful class of equations, because they can be solved in (semi)closed form quite analogously to the deterministic case considered in this section.

The solution presented in the previous section in terms of matrix exponential only works if the matrix $\mathbf{F}$ is constant. Thus for the time-varying homogeneous equation of the form

$$\frac{d\mathbf{x}}{dt} = \mathbf{F}(t)\,\mathbf{x}, \quad \mathbf{x}(t_0) = \text{given}, \quad (2.32)$$

the matrix exponential solution does not work. However, we can express the solution in the form

$$\mathbf{x}(t) = \mathbf{\Psi}(t, t_0)\,\mathbf{x}(t_0), \quad (2.33)$$

where $\boldsymbol{\Psi}(t, t_0)$ is the *transition matrix* which is defined via the properties

$$
\begin{aligned}
\frac{\partial \boldsymbol{\Psi}(\tau, t)}{\partial \tau} &= \mathbf{F}(\tau)\, \boldsymbol{\Psi}(\tau, t), \\
\frac{\partial \boldsymbol{\Psi}(\tau, t)}{\partial t} &= -\boldsymbol{\Psi}(\tau, t)\, \mathbf{F}(t), \\
\boldsymbol{\Psi}(\tau, t) &= \boldsymbol{\Psi}(\tau, s)\, \boldsymbol{\Psi}(s, t), \\
\boldsymbol{\Psi}(t, \tau) &= \boldsymbol{\Psi}^{-1}(\tau, t), \\
\boldsymbol{\Psi}(t, t) &= \mathbf{I}.
\end{aligned}
\tag{2.34}
$$

The transition matrix $\boldsymbol{\Psi}(t, t_0)$ does not have a closed-form expression in general. Nevertheless, given the transition matrix we can construct the solution to the inhomogeneous equation

$$
\frac{d\mathbf{x}}{dt} = \mathbf{F}(t)\, \mathbf{x} + \mathbf{L}(t)\, \mathbf{w}(t), \quad \mathbf{x}(t_0) = \text{given}, \tag{2.35}
$$

analogously to the time-invariant case. This time the integrating factor is $\boldsymbol{\Psi}(t_0, t)$, and the resulting solution is

$$
\mathbf{x}(t) = \boldsymbol{\Psi}(t, t_0)\, \mathbf{x}(t_0) + \int_{t_0}^{t} \boldsymbol{\Psi}(t, \tau)\, \mathbf{L}(\tau)\, \mathbf{w}(\tau)\, d\tau. \tag{2.36}
$$

2.4 Fourier Transforms

One very useful method to solve inhomogeneous linear time-invariant differential equations is the Fourier transform. The *Fourier transform* of a function $g(t)$ is defined as

$$
G(i\,\omega) = \mathfrak{F}[g(t)] = \int_{-\infty}^{\infty} g(t)\, \exp(-i\,\omega\, t)\, dt, \tag{2.37}
$$

where i is the imaginary unit. The corresponding *inverse Fourier transform* is

$$
g(t) = \mathfrak{F}^{-1}[G(i\,\omega)] = \frac{1}{2\pi} \int_{-\infty}^{\infty} G(i\,\omega)\, \exp(i\,\omega\, t)\, d\omega. \tag{2.38}
$$

Multidimensional transforms are performed componentwise. Note that there exist many conventions for writing the Fourier transform, but this parameterization will be used throughout in this book. The usefulness of the Fourier transform for solving differential equations arises from the property that for a nonnegative integer n we have

$$
\mathfrak{F}[d^n g(t)/dt^n] = (i\,\omega)^n\, \mathfrak{F}[g(t)], \tag{2.39}
$$

which transforms differentiation into multiplication by $i\omega$. Furthermore, the convolution theorem says that convolution gets transformed into multiplication,

$$\mathfrak{F}[g(t) * h(t)] = \mathfrak{F}[g(t)]\,\mathfrak{F}[h(t)], \tag{2.40}$$

where the convolution is defined as

$$g(t) * h(t) = \int_{-\infty}^{\infty} g(t - \tau)\,h(\tau)\,d\tau. \tag{2.41}$$

To apply the preceding properties to solving ODEs, it is required that the initial conditions are zero. However, this is not a restriction in practice, because it is possible to tweak the inhomogeneous term such that its effect is equivalent to the given initial conditions.

Example 2.2 (Fourier transform solution to spring model). *To demonstrate the usefulness of Fourier transform, we consider the spring model*

$$\frac{d^2 x(t)}{dt^2} + \gamma\,\frac{dx(t)}{dt} + v^2\,x(t) = w(t). \tag{2.42}$$

Taking the Fourier transform of the equation and using the derivative rule, we get

$$(i\omega)^2\,X(i\omega) + \gamma\,(i\omega)\,X(i\omega) + v^2\,X(i\omega) = W(i\omega), \tag{2.43}$$

where $X(i\omega)$ is the Fourier transform of $x(t)$, and $W(i\omega)$ is the Fourier transform of $w(t)$. We can now solve for $X(i\omega)$, which gives

$$X(i\omega) = \frac{W(i\omega)}{(i\omega)^2 + \gamma\,(i\omega) + v^2}. \tag{2.44}$$

The solution to the equation is then given by the inverse Fourier transform

$$x(t) = \mathfrak{F}^{-1}\left[\frac{W(i\omega)}{(i\omega)^2 + \gamma\,(i\omega) + v^2}\right]. \tag{2.45}$$

However, for general $w(t)$, it is useful to note that the term on the right-hand side is actually a product of the (Fourier domain) transfer function

$$H(i\omega) = \frac{1}{(i\omega)^2 + \gamma\,(i\omega) + v^2} \tag{2.46}$$

and $W(i\omega)$. This product can now be converted into a convolution if we start by computing the impulse response function

$$h(t) = \mathfrak{F}^{-1}\left[\frac{1}{(i\omega)^2 + \gamma\,(i\omega) + v^2}\right]$$

$$= b^{-1}\,\exp(-a\,t)\,\sin(b\,t)\,u(t), \tag{2.47}$$

where $a = \gamma/2$, $b = \sqrt{v^2 - \gamma^2/4}$, and $u(t)$ is the Heaviside step function, which is zero for $t < 0$ and one for $t \geq 0$. Then the full solution can be expressed as

$$x(t) = \int_{-\infty}^{\infty} h(t - \tau)\, w(\tau)\, d\tau, \qquad (2.48)$$

which can be interpreted such that we construct $x(t)$ by feeding the signal $w(t)$ through a linear system (filter) with impulse response $h(t)$.

We can also use the Fourier transform to solve general linear time-invariant nonhomogeneous equations

$$\frac{d\mathbf{x}(t)}{dt} = \mathbf{F}\,\mathbf{x}(t) + \mathbf{L}\,\mathbf{w}(t). \qquad (2.49)$$

Taking Fourier transforms componentwise gives

$$(i\,\omega)\,\mathbf{X}(i\,\omega) = \mathbf{F}\,\mathbf{X}(i\,\omega) + \mathbf{L}\,\mathbf{W}(i\,\omega). \qquad (2.50)$$

Solving for $\mathbf{X}(i\,\omega)$ then gives

$$\mathbf{X}(i\,\omega) = ((i\,\omega)\,\mathbf{I} - \mathbf{F})^{-1}\,\mathbf{L}\,\mathbf{W}(i\,\omega). \qquad (2.51)$$

Comparing to Equation (2.31) and recalling $\mathbf{x}(0) = \mathbf{0}$ now reveals that actually we have

$$\mathfrak{F}^{-1}\left[((i\,\omega)\,\mathbf{I} - \mathbf{F})^{-1}\right] = \exp(\mathbf{F}\,t)\,u(t), \qquad (2.52)$$

where $u(t)$ is the Heaviside step function. This identity also provides one way to compute matrix exponentials.

Example 2.3 (Matrix exponential via Fourier transform). *The matrix exponential considered in Example 2.1 can also be computed as*

$$\exp\left(\begin{pmatrix} 0 & 1 \\ 0 & 0 \end{pmatrix} t\right) = \mathfrak{F}^{-1}\left[\left(\begin{pmatrix} (i\,\omega) & 0 \\ 0 & (i\,\omega) \end{pmatrix} - \begin{pmatrix} 0 & 1 \\ 0 & 0 \end{pmatrix}\right)^{-1}\right] = \begin{pmatrix} 1 & t \\ 0 & 1 \end{pmatrix}.$$
$$(2.53)$$

2.5 Laplace Transforms

Another often encountered method for solving linear time-invariant differential equations is the *Laplace transform* (see, e.g., Kreyszig, 1993). The Laplace transform of a function $f(t)$, defined for all $t \geq 0$, is

$$F(s) = \mathcal{L}[f(t)] = \int_0^{\infty} f(t)\exp(-st)\, dt \qquad (2.54)$$

and the inverse transform is $f(t) = \mathcal{L}^{-1}[F(s)]$.

Just as for the Fourier transform, the usefulness of the Laplace transform comes from its property of reducing several often encountered "hard" differential equations into a "simple" subsidiary form that can be solved by algebraic manipulations. By inverse-transforming the solution of the subsidiary equation, the solution to the original problem can be retrieved.

The Laplace transform also has a similar derivative property as the Fourier transform (cf. Eq. 2.39), except that we can also take the initial conditions into account. If $x(0) =$ given, then

$$\mathcal{L}\left[\frac{dx(t)}{dt}\right] = s\,X(s) - x(0),\tag{2.55}$$

where $X(s) = \mathcal{L}[x(t)]$. If we apply this n times, we get

$$\mathcal{L}\left[\frac{d^n x(t)}{dt^n}\right] = s^n\,X(s) - s^{n-1}\,x(0) - \cdots - \frac{dx^{n-1}}{dt^{n-1}}(0).\tag{2.56}$$

The Laplace transform is especially useful with zero initial conditions, and it leads to a very similar solution method as the Fourier transform.

Example 2.4 (Laplace transform solution to spring model). *We use the Laplace transform to solve the following differential equation that we already considered in Example 2.2:*

$$\frac{d^2 x(t)}{dt^2} + \gamma\,\frac{dx(t)}{dt} + v^2\,x(t) = w(t)\tag{2.57}$$

with the initial conditions $x(0) = 0$, $(dx/dt)(0) = 0$. *Taking Laplace transforms of both sides gives*

$$s^2\,X(s) + \gamma\,s\,X(s) + v^2\,X(s) = W(s),\tag{2.58}$$

where $W(s) = \mathcal{L}[w(t)]$. *The solution to the differential equation is given as*

$$X(s) = \left[\frac{1}{s^2 + \gamma\,s + v^2}\right] W(s).\tag{2.59}$$

We can also identify the Laplace domain transfer function as

$$H(s) = \frac{1}{s^2 + \gamma\,s + v^2},\tag{2.60}$$

and by using the convolution property

$$\mathcal{L}[h(t) * x(t)] = \mathcal{L}[h(t)]\,\mathcal{L}[x(t)]\tag{2.61}$$

of the Laplace transform, we can write down the solution to the differential equation as

$$x(t) = h(t) * w(t) = \int_{-\infty}^{\infty} h(t - \tau)\, w(\tau)\, d\tau, \qquad (2.62)$$

where

$$h(t) = \mathcal{L}^{-1}[H(s)] = b^{-1} \exp(-a\, t)\, \sin(b\, t)\, u(t) \qquad (2.63)$$

is exactly the impulse response that we already obtained in Equation (2.47).

Let us now consider the vector differential equation

$$\frac{d\mathbf{x}(t)}{dt} = \mathbf{F}\,\mathbf{x}(t) + \mathbf{L}\,\mathbf{w}(t) \qquad (2.64)$$

with a nonzero initial condition $\mathbf{x}(0) \neq \mathbf{0}$. Taking the vector-valued Laplace transform of the equation and using the derivative property leads to

$$s\,\mathbf{X}(s) - \mathbf{x}(0) = \mathbf{F}\,\mathbf{X}(s) + \mathbf{L}\,\mathbf{W}(s), \qquad (2.65)$$

where $\mathbf{W}(s) = \mathcal{L}[\mathbf{w}(t)]$. Solving for $\mathbf{X}(s)$ gives

$$\mathbf{X}(s) = (s\,\mathbf{I} - \mathbf{F})^{-1}\,\mathbf{x}(0) + (s\,\mathbf{I} - \mathbf{F})^{-1}\,\mathbf{L}\,\mathbf{W}(s), \qquad (2.66)$$

which can be seen to correspond to the solution in Equation (2.31) with $t_0 = 0$. We can now see that the matrix exponential has the representation

$$\exp(\mathbf{F}\,t) = \mathcal{L}^{-1}\left[(s\,\mathbf{I} - \mathbf{F})^{-1}\right] \qquad (2.67)$$

for $t \geq 0$. As with the Fourier transform, this expression is sometimes useful for computing matrix exponentials, as can be seen by the following example.

Example 2.5 (Matrix exponential via Laplace transform). *Consider again solving the matrix exponential in Example 2.1. We can now compute*

$$(s\,\mathbf{I} - \mathbf{F})^{-1} = \begin{pmatrix} 1/s & 1/s^2 \\ 0 & 1/s \end{pmatrix}, \qquad (2.68)$$

which should be the Laplace transform of the matrix exponential. Recalling that $\mathcal{L}[1] = \int_0^{\infty} \exp(-s\,t)\, dt = 1/s$ and $\mathcal{L}[t] = \int_0^{\infty} t\, \exp(-s\,t)\, dt = 1/s^2$, gives

$$\exp(\mathbf{F}\,t) = \begin{pmatrix} 1 & t \\ 0 & 1 \end{pmatrix}. \qquad (2.69)$$

2.6 Numerical Solutions of Differential Equations

For a generic nonlinear differential equation of the form

$$\frac{d\mathbf{x}(t)}{dt} = \mathbf{f}(\mathbf{x}(t), t), \quad \mathbf{x}(t_0) = \text{given}, \tag{2.70}$$

there is no general way to find an analytic solution. However, it is possible to approximate the solution numerically.

If we integrate the equation from t to $t + \Delta t$, we get

$$\mathbf{x}(t + \Delta t) = \mathbf{x}(t) + \int_t^{t+\Delta t} \mathbf{f}(\mathbf{x}(\tau), \tau) \, d\tau. \tag{2.71}$$

If we knew how to compute the integral on the right-hand side, we could generate the solution at time steps $t_0, t_1 = t_0 + \Delta t, t_2 = t_0 + 2\Delta t$ iterating the preceding equation, which would give a series of solutions:

$$\mathbf{x}(t_0 + \Delta t) = \mathbf{x}(t_0) + \int_{t_0}^{t_0+\Delta t} \mathbf{f}(\mathbf{x}(\tau), \tau) \, d\tau,$$

$$\mathbf{x}(t_0 + 2\Delta t) = \mathbf{x}(t_0 + \Delta t) + \int_{t_0+\Delta t}^{t+2\Delta t} \mathbf{f}(\mathbf{x}(\tau), \tau) \, d\tau, \tag{2.72}$$

$$\mathbf{x}(t_0 + 3\Delta t) = \mathbf{x}(t_0 + 2\Delta t) + \int_{t_0+2\Delta t}^{t+3\Delta t} \mathbf{f}(\mathbf{x}(\tau), \tau) \, d\tau.$$

$$\vdots$$

It is now possible to derive various numerical methods by constructing approximations to the integrals on the right-hand side. In the Euler method, we use the approximation

$$\int_t^{t+\Delta t} \mathbf{f}(\mathbf{x}(\tau), \tau) \, d\tau \approx \mathbf{f}(\mathbf{x}(t), t) \, \Delta t, \tag{2.73}$$

which leads to the following algorithm.

Algorithm 2.6 (Euler method). *Start from $\hat{\mathbf{x}}(t_0) = \mathbf{x}(t_0)$ and divide the integration interval $[t_0, t]$ into M steps $t_0 < t_1 < t_2 < \ldots < t_M = t$ such that $\Delta t = t_{k+1} - t_k$. At each step k, approximate the solution as follows:*

$$\hat{\mathbf{x}}(t_{k+1}) = \hat{\mathbf{x}}(t_k) + \mathbf{f}(\hat{\mathbf{x}}(t_k), t_k) \, \Delta t. \tag{2.74}$$

The (global) order of a numerical integration methods can be defined to be the largest exponent ρ such that if we numerically solve an ODE using $M = 1/\Delta t$ steps of length Δt, then there exists a constant K such that

$$|\hat{\mathbf{x}}(t_M) - \mathbf{x}(t_M)| \leq K \, \Delta t^\rho, \tag{2.75}$$

where $\hat{\mathbf{x}}(t_M)$ is the approximation and $\mathbf{x}(t_M)$ is the true solution. Because in the Euler method the first discarded term of the Taylor series is of order Δt^2, the error of integrating over $1/\Delta t$ steps is proportional to Δt. Thus the Euler method has order $\rho = 1$.

We can also improve the approximation by using a trapezoidal approximation

$$\int_t^{t+\Delta t} \mathbf{f}(\mathbf{x}(\tau), \tau) \, d\tau \approx \frac{\Delta t}{2} \left[\mathbf{f}(\mathbf{x}(t), t) + \mathbf{f}(\mathbf{x}(t + \Delta t), t + \Delta t) \right], \quad (2.76)$$

which leads to the approximate integration rule

$$\mathbf{x}(t_{k+1}) \approx \mathbf{x}(t_k) + \frac{\Delta t}{2} \left[\mathbf{f}(\mathbf{x}(t_k), t_k) + \mathbf{f}(\mathbf{x}(t_{k+1}), t_{k+1}) \right], \quad (2.77)$$

which is an *implicit* rule in the sense that $\mathbf{x}(t_{k+1})$ appears also on the right-hand side. To actually use such an implicit rule, we would need to solve a nonlinear equation at each integration step, which tends to be computationally too expensive when the dimensionality of $\mathbf{x}$ is high. Thus here we consider *explicit* rules only, where the next value $\mathbf{x}(t_{k+1})$ does not appear on the right-hand side. If we now replace the term $\mathbf{x}(t_{k+1})$ on the right-hand side with its Euler approximation, we get the *Heun method*.

Algorithm 2.7 (Heun method). *Start from $\hat{\mathbf{x}}(t_0) = \mathbf{x}(t_0)$ and divide the integration interval $[t_0, t]$ into M steps $t_0 < t_1 < t_2 < \ldots < t_M = t$ such that $\Delta t = t_{k+1} - t_k$. At each step k, approximate the solution as follows:*

$$\begin{aligned} \tilde{\mathbf{x}}(t_{k+1}) &= \hat{\mathbf{x}}(t_k) + \mathbf{f}(\hat{\mathbf{x}}(t_k), t_k) \, \Delta t, \\ \hat{\mathbf{x}}(t_{k+1}) &= \hat{\mathbf{x}}(t_k) + \frac{\Delta t}{2} \left[\mathbf{f}(\hat{\mathbf{x}}(t_k), t_k) + \mathbf{f}(\tilde{\mathbf{x}}(t_{k+1}), t_{k+1}) \right]. \end{aligned} \quad (2.78)$$

It can be shown that the Heun method has global order $\rho = 2$.

Another useful class of methods are the *Runge–Kutta* methods. The classical *fourth-order Runge–Kutta method* is the following.

Algorithm 2.8 (Fourth-order Runge–Kutta method). *Start from $\hat{\mathbf{x}}(t_0) = \mathbf{x}(t_0)$ and divide the integration interval $[t_0, t]$ into M steps $t_0 < t_1 < t_2 < \ldots < t_M = t$ such that $\Delta t = t_{k+1} - t_k$. At each step k, approximate the* *solution as follows:*

$$\Delta \mathbf{x}_k^1 = \mathbf{f}(\hat{\mathbf{x}}(t_k), t_k) \, \Delta t,$$
$$\Delta \mathbf{x}_k^2 = \mathbf{f}(\hat{\mathbf{x}}(t_k) + \Delta \mathbf{x}_k^1/2, t_k + \Delta t/2) \, \Delta t,$$
$$\Delta \mathbf{x}_k^3 = \mathbf{f}(\hat{\mathbf{x}}(t_k) + \Delta \mathbf{x}_k^2/2, t_k + \Delta t/2) \, \Delta t, \tag{2.79}$$
$$\Delta \mathbf{x}_k^4 = \mathbf{f}(\hat{\mathbf{x}}(t_k) + \Delta \mathbf{x}_k^3, t_k + \Delta t) \, \Delta t,$$

$$\hat{\mathbf{x}}(t_{k+1}) = \hat{\mathbf{x}}(t_k) + \frac{1}{6}(\Delta \mathbf{x}_k^1 + 2\Delta \mathbf{x}_k^2 + 2\Delta \mathbf{x}_k^3 + \Delta \mathbf{x}_k^4).$$

The preceding Runge–Kutta method can be derived by writing down the Taylor series expansion for the solution and by selecting coefficients such that many of the lower-order terms cancel out. The order of this method is $\rho = 4$.

In fact, all the preceding integration methods are based on the Taylor series expansions of the solution. This is slightly problematic, because in the case of SDEs the Taylor series expansion does not exist and all of the methods need to be modified at least to some extent. However, it is possible to replace the Taylor series with a so-called Itô–Taylor series and then work out the analogous algorithms. The resulting algorithms are more complicated than the deterministic counterparts, because the Itô–Taylor series is considerably more complicated than the Taylor series. We will come back to this issue in Chapter 8.

There exists a wide class of other numerical ODE solvers as well. For example, all the aforementioned methods have a fixed step length, but there exist variable step size methods that automatically adapt the step size. However, constructing variable step size methods for stochastic differential equations is much more involved than for deterministic equations, and thus we shall not consider them here.

Example 2.9 (Numerical solution of ODEs). *Consider the spring–mass model given in Equation (2.1). Let the parameters be $v = 2$ and $\gamma = 1$, and the forcing term be zero. When we write this equation in state-space form (2.4), we get a differential equation $\mathrm{d}\mathbf{x}/\mathrm{d}t = \mathbf{f}(\mathbf{x}, t)$, where*

$$\mathbf{f}(\mathbf{x}, t) = \begin{pmatrix} 0 & 1 \\ -v^2 & -\gamma \end{pmatrix} \mathbf{x}. \tag{2.80}$$

In the following, the solution is approximated numerically on the interval $t \in [0, 10]$. Figure 2.1a shows results for the Euler method, the Heun method, and the fourth-order Runge–Kutta (RK4) method with step size $\Delta t = 0.1$, where the effect of a too long step size for the Euler method is

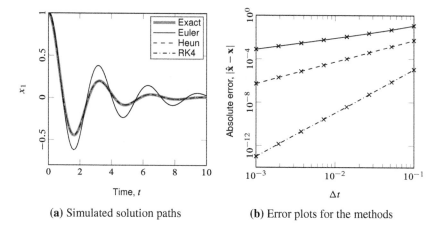

(a) Simulated solution paths **(b)** Error plots for the methods

Figure 2.1 Comparison of the Euler, Heun, and fourth-order Runge–Kutta method from Example 2.9. In (a), the step size of $\Delta t = 0.1$ is clearly too big for the Euler method. In (b), the effect of the step size on the absolute error is visualized.

clearly visible. Figure 2.1b demonstrates in more detail the effect of step size versus error in the simulated path.

2.7 Picard–Lindelöf Theorem

One important question in differential equations is whether the solution exists and is unique. To analyze these questions, consider a generic equation of the form

$$\frac{\mathrm{d}\mathbf{x}(t)}{\mathrm{d}t} = \mathbf{f}(\mathbf{x}(t), t), \quad \mathbf{x}(t_0) = \mathbf{x}_0, \tag{2.81}$$

where $\mathbf{f}(\mathbf{x}, t)$ is some given function. Integrating both sides from t_0 to t gives

$$\mathbf{x}(t) = \mathbf{x}_0 + \int_{t_0}^{t} \mathbf{f}(\mathbf{x}(\tau), \tau)\,\mathrm{d}\tau. \tag{2.82}$$

We can now use this identity to find an approximate solution to the differential equation by the following *Picard iteration* (see, e.g., Tenenbaum and Pollard, 1985).

Algorithm 2.10 (Picard iteration). *Start from the initial guess* $\boldsymbol{\varphi}_0(t) = \mathbf{x}_0$.

Then compute approximations $\varphi_1(t), \varphi_2(t), \varphi_3(t), \ldots$ *via the recursion*

$$\varphi_{n+1}(t) = \mathbf{x}_0 + \int_{t_0}^{t} \mathbf{f}(\varphi_n(\tau), \tau) \, d\tau. \tag{2.83}$$

The preceding recursion, which we already used for finding the solution to linear differential equations in Section 2.2, can be shown to converge to the unique solution

$$\lim_{n \to \infty} \varphi_n(t) = \mathbf{x}(t), \tag{2.84}$$

provided that $\mathbf{f}(\mathbf{x}, t)$ is continuous in both arguments and Lipschitz continuous in the first argument.

The implication of this recursion is the *Picard–Lindelöf theorem*, which says that under the preceding continuity conditions the differential equation has a solution and it is unique in a certain interval around $t = t_0$. We emphasize the innocent-looking but important issue in the theorem: the function $\mathbf{f}(\mathbf{x}, t)$ needs to be *continuous*. This is important, because in the case of stochastic differential equations the corresponding function will be *discontinuous everywhere* and thus we need a completely new existence theory for them.

2.8 Exercises

2.1   Consider the initial value problem

$$\frac{dx(t)}{dt} = \theta_1(\theta_2 - x(t)), \quad x(0) = x_0,$$

where θ_1 and θ_2 are constants.

(a) Derive the solution using the integrating factor method.
(b) Derive the solution using the Laplace transform.

2.2   One way of solving ODEs is to use an ansatz (an educated guess). Show that $x(t) = c_1 \sin \omega t + c_2 \cos \omega t$ is the solution to the second-order ODE

$$\ddot{x} + \omega^2 x = 0, \quad \omega > 0. \tag{2.85}$$

2.3   Solve the differential equation

$$\frac{dy}{dt} = -y^2 - 1$$

with initial condition $y(0) = 0$ using the method of separation of variables.

2.4   In classical ODE literature, ODE systems are typically characterized by the nature of their fixed points (roots of the differential equation). Find out what is meant by this, sketch the behavior of the following ODEs, and classify all their fixed points:

(a) $\dot{x} = 1 + 2\cos x$.

(b) $\dot{x} = x - y$ and $\dot{y} = 1 - e^x$.

(c) $\dot{x} = x - y$ and $\dot{y} = x^2 - 4$.

2.5   Study the behavior of the following differential equation

$$\ddot{x} + \dot{x} - (\alpha - x^2)x = 0, \quad \alpha \geq 0,$$

as follows:

(a) Rewrite the problem in terms of a first-order ODE.

(b) Find the fixed points of the ODE.

(c) Characterize the nature of the fixed points.

(d) Sketch the behavior of trajectories in the $(x, \dot{x})$ plane.

2.6   We wish to find the Laplace domain solution to the equation

$$a \frac{d^2 y}{dt^2} + b \frac{dy}{dt} + c\,y = u,$$

where $y(t)$ is an unknown time-varying function and $u(t)$ is a given function. Solve the equation by following these steps:

(a) Calculate the Laplace transform of the equation.

(b) Solve the Laplace domain equation for $Y(s)$.

(c) Take the inverse Laplace transform and provide a solution for $y(t)$.

2.7   We wish to find the matrix exponential $\exp(\mathbf{F}\,t)$, where $t \geq 0$ with

$$\mathbf{F} = \begin{pmatrix} 0 & 1 & 0 \\ 0 & 0 & 1 \\ 0 & 0 & 0 \end{pmatrix}.$$

(a) Solve it using the series expansion definition of the matrix exponential (note that the matrix is nilpotent).

(b) Solve it using the Laplace transform as in Example 2.5.

2.8   Use a computer algebra system (e.g., MATLAB®) to compute the matrix exponential

$$\exp\left\{ \begin{pmatrix} 0 & 1 \\ -\nu^2 & -\gamma \end{pmatrix} t \right\}.$$

How can you extract the impulse response (2.47) from the result?

2.9   Consider the initial value problem $\dot{x} = -x$, where $x(0) = 1$.

(a) Solve the problem analytically. What is the exact value of $x(1)$?

(b) Implement the Euler method for this ODE. Using a step size of 1, estimate $x(1)$ numerically. Repeat this for step sizes 10^{-n}, where $n = 1, 2, 3, 4$.

2.10   Implement the fourth-order Runge–Kutta (RK4) and Heun methods for solving the following second-order ODE:

$$\ddot{x} + \dot{x} - (\alpha - x^2) x = 0, \quad \alpha \geq 0,$$

with $\alpha = 1$. How does the choice of Δt affect the results obtained from the methods?

2.11   Show that the Picard iteration converges for the linear system

$$\frac{dx(t)}{dt} = F x(t)$$

by showing that it is a contraction mapping. Assume that $|F| < 1$.

3

Pragmatic Introduction to Stochastic Differential Equations

In this chapter, our aim is to look at SDEs as noise-driven ODEs. This is the point of view usually encountered in applications. We start from a physical construction of Brownian motion by Einstein (1905), which was also the topic of his PhD thesis. We then proceed to give additional examples of real-world phenomena leading to SDEs – or noise-driven differential equations. Based on the intuition in solution methods for inhomogeneous ODEs in the previous chapter, we then derive "leap of faith" solutions to SDEs – they are leaps of faith in the sense that the classical ODE theory does not guarantee the existence of their solutions. And in fact, the leap-of-faith solutions start to go awfully wrong at some point, which then motivates us to look into proper stochastic calculus in the next chapter.

3.1 Stochastic Processes in Physics, Engineering, and Other Fields

The history of SDEs can be seen to have started from the classical paper of Einstein (1905), where he presented a mathematical connection between microscopic random motion of particles and the macroscopic diffusion equation. This is one of the results that proved the existence of the atom. Einstein's reasoning was roughly the following.

Example 3.1 (Microscopic motion of Brownian particles). *Let τ be a small time interval and consider n particles suspended in liquid. During the time interval τ, the x-coordinates of the particles will change by displacement Δ. The number of particles with a displacement between Δ and $\Delta + d\Delta$ is then*

$$\mathrm{d}n = n\,\phi(\Delta)\,\mathrm{d}\Delta, \qquad (3.1)$$

where $\phi(\Delta)$ is the probability density of Δ, which can be assumed to be symmetric, $\phi(\Delta) = \phi(-\Delta)$, and differ from zero only for very small values of Δ (see Figure 3.1).

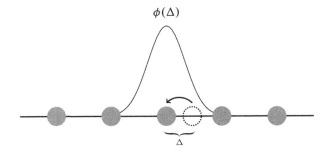

Figure 3.1 Illustration of Einstein's model of Brownian motion from Example 3.1, where particles move by displacement Δ within small time intervals τ. The probability density $\phi(\Delta)$ is shown by the curve.

Let $u(x, t)$ be the number of particles per unit volume. Then the number of particles at time $t + \tau$ located between x and $x + \mathrm{d}x$ is given as

$$u(x, t + \tau)\, \mathrm{d}x = \left[\int_{-\infty}^{\infty} u(x + \Delta, t)\, \phi(\Delta)\, \mathrm{d}\Delta \right] \mathrm{d}x. \qquad (3.2)$$

Because τ is small, we can put

$$u(x, t + \tau) = u(x, t) + \tau\, \frac{\partial u(x, t)}{\partial t}. \qquad (3.3)$$

We can expand $u(x + \Delta, t)$ in powers of Δ as

$$u(x + \Delta, t) = u(x, t) + \Delta\, \frac{\partial u(x, t)}{\partial x} + \frac{\Delta^2}{2}\, \frac{\partial^2 u(x, t)}{\partial x^2} + \cdots \qquad (3.4)$$

Substituting (3.3) and (3.4) into (3.2) gives

$$\begin{aligned}
u(x, t) + \tau\, \frac{\partial u(x, t)}{\partial t} &= u(x, t) \int_{-\infty}^{\infty} \phi(\Delta)\, \mathrm{d}\Delta \\
&+ \frac{\partial u(x, t)}{\partial x} \int_{-\infty}^{\infty} \Delta\, \phi(\Delta)\, \mathrm{d}\Delta \\
&+ \frac{\partial^2 u(x, t)}{\partial x^2} \int_{-\infty}^{\infty} \frac{\Delta^2}{2}\, \phi(\Delta)\, \mathrm{d}\Delta + \cdots, \quad (3.5)
\end{aligned}$$

where all the odd order terms vanish. Let us assume that the rest of the terms with order $n \geq 3$ vanish as well. If we recall that $\int_{-\infty}^{\infty} \phi(\Delta)\, \mathrm{d}\Delta = 1$ and we put

$$\frac{1}{\tau} \int_{-\infty}^{\infty} \frac{\Delta^2}{2}\, \phi(\Delta)\, \mathrm{d}\Delta = D, \qquad (3.6)$$

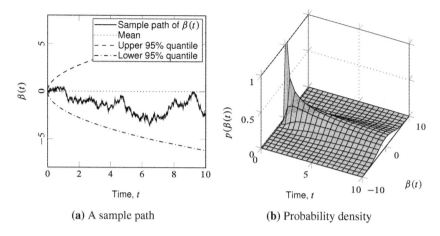

(a) A sample path   (b) Probability density

Figure 3.2 Two views of Brownian motion: (a) a sample path and 95% quantiles, and (b) evolution of the probability density.

we get the diffusion equation

$$\frac{\partial u(x,t)}{\partial t} = D\, \frac{\partial^2 u(x,t)}{\partial x^2}. \tag{3.7}$$

This connection was significant during the time, because the diffusion equation was only known as a macroscopic equation. Einstein was also able to derive a formula for D in terms of microscopic quantities. From this, Einstein was able to compute the prediction for mean squared displacement of the particles as function of time:

$$z(t) = \frac{R\,T}{N}\, \frac{1}{3\,\pi\,\eta\,r}\, t, \tag{3.8}$$

where η is the viscosity of the liquid, r is the diameter of the particles, T is the temperature, R is the gas constant, and N is the Avogadro constant.

In modern terms, Brownian motion[1] (see Figure 3.2) is an abstraction of a random walk process that has the property that each increment is independent. That is, direction and magnitude of each change of the process are completely random and independent of the previous changes. One way to think about Brownian motion is that it is the solution to the following stochastic differential equation:

$$\frac{d\beta(t)}{dt} = w(t), \tag{3.9}$$

---

[1] In mathematics, Brownian motion is also often called the Wiener process.

where $w(t)$ is a white random process. The term *white* here means that each of the values $w(t)$ and $w(t')$ are independent whenever $t \neq t'$. We will later see that the probability density of the solution of this equation will solve the diffusion equation. However, in Einstein's time the theory of stochastic differential equations did not exist and therefore the reasoning was completely different.

A couple of years after Einstein's contribution, Langevin (1908) presented an alternative construction of Brownian motion that leads to the same macroscopic properties (see Figure 3.3). The reasoning in the article, which is outlined in the following, was more mechanical than in Einstein's derivation.

Example 3.2 (Langevin's model of Brownian motion). *Consider a small particle suspended in liquid. Assume that there are two kinds of forces acting on the particle:*

1. *Friction force F_f, which by the Stoke's law has the following form:*

$$F_f = -6 \pi \eta r v, \tag{3.10}$$

   where η is the viscosity, r is the diameter of the particle, and v is its velocity.
2. *Random force F_r caused by random collisions of the particles.*

Newton's law then gives

$$m \frac{d^2 x}{dt^2} = -6 \pi \eta r \frac{dx}{dt} + F_r, \tag{3.11}$$

where m is the mass of the particle. Recall that

$$\begin{aligned}
\frac{1}{2} \frac{d(x^2)}{dt} &= \frac{dx}{dt} x, \\
\frac{1}{2} \frac{d^2(x^2)}{dt^2} &= \frac{d^2 x}{dt^2} x + \left(\frac{dx}{dt} \right)^2.
\end{aligned} \tag{3.12}$$

Thus if we multiply Equation (3.11) with x, substitute the aforementioned identities, and take expectations, we get

$$\frac{m}{2} E\left[\frac{d^2(x^2)}{dt^2} \right] - m E\left[\left(\frac{dx}{dt} \right)^2 \right] = -3 \pi \eta r E\left[\frac{d(x^2)}{dt} \right] + E[F_r x]. \tag{3.13}$$

From statistical physics, we know the relationship between the average

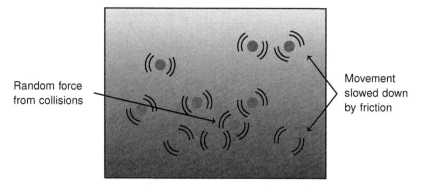

Figure 3.3 Illustration of Langevin's model of Brownian motion from Example 3.2.

kinetic energy and temperature:

$$m \, \mathrm{E}\left[\left(\frac{\mathrm{d}x}{\mathrm{d}t}\right)^2\right] = \frac{RT}{N}. \tag{3.14}$$

If we then assume that the random force has zero mean $\mathrm{E}[F_\mathrm{r}] = 0$, *the random force and the position are uncorrelated,* $\mathrm{E}[F_\mathrm{r} \, x] = 0$, *and we define a new variable* $\dot{z} = \mathrm{d}\,\mathrm{E}[x^2]/\mathrm{d}t$, *we get the differential equation*

$$\frac{m}{2} \frac{\mathrm{d}\dot{z}}{\mathrm{d}t} - \frac{RT}{N} = -3\,\pi\,\eta\,r\,\dot{z}, \tag{3.15}$$

which has the general solution

$$\dot{z}(t) = \frac{RT}{N} \frac{1}{3\,\pi\,\eta\,r}\left[1 - \exp\left(\frac{6\,\pi\,\eta\,r}{m} t\right)\right]. \tag{3.16}$$

The preceding exponential goes to zero very quickly, and thus the resulting mean squared displacement is nominally just the resulting constant multiplied with time:

$$z(t) = \frac{RT}{N} \frac{1}{3\,\pi\,\eta\,r} t, \tag{3.17}$$

which is exactly the same as Einstein obtained.

In the preceding model, Brownian motion is not actually seen as a solution to the white noise driven differential equation

$$\frac{\mathrm{d}\beta(t)}{\mathrm{d}t} = w(t), \tag{3.18}$$

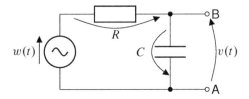

Figure 3.4 Example resistor-capacitor (RC) circuit from Example 3.3.

but instead as the solution to an equation of the form

$$\frac{d^2\tilde{\beta}(t)}{dt^2} = -c\,\frac{d\tilde{\beta}(t)}{dt} + w(t) \tag{3.19}$$

in the limit of the vanishing time constant. The latter (Langevin's version) is sometimes called the physical Brownian motion and the former (Einstein's version) the mathematical Brownian motion. In this book the term Brownian motion always means the mathematical Brownian motion.

Stochastic differential equations also arise in other contexts. For example, the effect of thermal noise in electrical circuits and various kinds of disturbances in telecommunications systems can be modeled as SDEs. This is illustrated in the following example.

Example 3.3 (RC circuit). *Consider the simple RC circuit shown in Figure 3.4. In Laplace domain, the output voltage $V(s)$ can be expressed in terms of the input voltage $W(s)$ as follows:*

$$V(s) = \frac{1}{1 + RCs}\,W(s). \tag{3.20}$$

An inverse Laplace transform gives the differential equation

$$\frac{dv(t)}{dt} = -\frac{1}{RC}\,v(t) + \frac{1}{RC}\,w(t). \tag{3.21}$$

For the purposes of studying the response of the circuit to noise, we can now replace the input voltage with a white noise process $w(t)$ and analyze the properties of the resulting equation.

Stochastic differential equations can also be used for modeling dynamic phenomena, where the exact dynamics of the system are uncertain. For example, the motion model of a car cannot be exactly written down if we do not know all the external forces affecting the car and the input from the

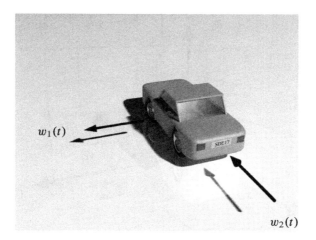

Figure 3.5 Illustration of a car's dynamic model from Example 3.4.

driver. However, the unknown subphenomena can be modeled as stochastic processes, which leads to stochastic differential equations. This kind of modeling principle of representing uncertainties as random variables is sometimes called *Bayesian modeling*. Stochastic differential equation models of this kind are commonly used in navigation and control systems (see, e.g., Jazwinski, 1970; Bar-Shalom et al., 2001; Grewal and Andrews, 2001). Stock prices can also be modeled using stochastic differential equations, and these kinds of models are indeed commonly used in analysis and pricing of stocks and related quantities (Øksendal, 2003).

Example 3.4 (Dynamic model of a car). *The dynamics of a car in two dimensions* (x_1, x_2) *are governed by* Newton's law *(see Figure 3.5):*

$$\mathbf{f}(t) = m\,\mathbf{a}(t), \tag{3.22}$$

where $\mathbf{a}(t)$ *is the acceleration, m is the mass of the car, and* $\mathbf{f}(t)$ *is a vector of (unknown) forces acting on the car. Let us now model* $\mathbf{f}(t)/m$ *as a two-dimensional white random process* $\mathbf{w}(t) = (w_1(t), w_2(t))$:

$$\frac{\mathrm{d}^2 x_1}{\mathrm{d}t^2} = w_1(t) \quad and \quad \frac{\mathrm{d}^2 x_2}{\mathrm{d}t^2} = w_2(t). \tag{3.23}$$

If we define $x_3 = \mathrm{d}x_1/\mathrm{d}t$, $x_4 = \mathrm{d}x_2/\mathrm{d}t$, *the model can be written as a*

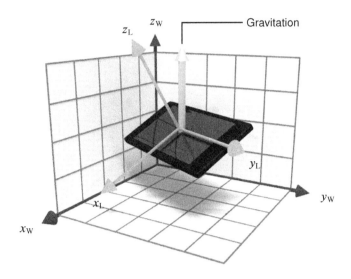

Figure 3.6 Tracking of the movement of a smartphone device in local (L) and world coordinates (W) from Example 3.5.

first-order system of differential equations:

$$\frac{d}{dt}\begin{pmatrix} x_1 \\ x_2 \\ x_3 \\ x_4 \end{pmatrix} = \underbrace{\begin{pmatrix} 0 & 0 & 1 & 0 \\ 0 & 0 & 0 & 1 \\ 0 & 0 & 0 & 0 \\ 0 & 0 & 0 & 0 \end{pmatrix}}_{\mathbf{F}}\begin{pmatrix} x_1 \\ x_2 \\ x_3 \\ x_4 \end{pmatrix} + \underbrace{\begin{pmatrix} 0 & 0 \\ 0 & 0 \\ 1 & 0 \\ 0 & 1 \end{pmatrix}}_{\mathbf{L}}\begin{pmatrix} w_1 \\ w_2 \end{pmatrix}. \tag{3.24}$$

In shorter matrix form, *this can be written as a linear differential equation model:*

$$\frac{d\mathbf{x}}{dt} = \mathbf{F}\,\mathbf{x} + \mathbf{L}\,\mathbf{w}.$$

Example 3.5 (Smartphone tracking). *Many modern smartphones contain three-dimensional acceleration sensors (i.e., accelerometers) as well as three-dimensional angular velocity sensors (i.e., gyroscopes). It turns out that the nominal component seen by the accelerometer is the gravitation, which from the sensor's point of view points up from the ground. However, as the sensors are attached to the smartphone, "up" will not usually correspond to the z-direction (or any other fixed direction) of the sensor, but points to a direction that depends on the orientation of the smartphone*

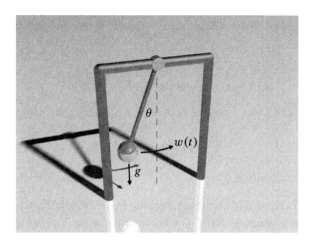

Figure 3.7 An illustration of the noisy pendulum model from Example 3.6, where the forces acting upon the pendulum are gravity and a random noise process $w(t)$.

(see Figure 3.6). These observations allow us to track the orientation of a smartphone (Särkkä et al., 2015a).

In ideal conditions, we can write down the following differential equation for the relation between the locally seen gravitation $\mathbf{g}_L$ and the angular velocity $\boldsymbol{\omega}_L$ measured by the gyroscope given by the following cross-product:

$$\frac{d\mathbf{g}_L}{dt} = -\boldsymbol{\omega}_L \times \mathbf{g}_L. \tag{3.25}$$

We can now take into account the inaccuracies in the sensors by adding a white noise term $\mathbf{w}(t)$ to the model, which leads to the SDE

$$\frac{d\mathbf{g}_L}{dt} = -\boldsymbol{\omega}_L \times \mathbf{g}_L + \mathbf{w}(t). \tag{3.26}$$

The acceleration sensors can now be modeled to produce observations of the form $\mathbf{a}_L = \mathbf{g}_L + noise$, where the noise term models the local accelerations. A Kalman filter can then be used to track the orientation of the smartphone (Särkkä et al., 2015a).

Example 3.6 (Noisy pendulum). *The differential equation for a simple pendulum (see Figure 3.7) with unit length and mass can be written as*

$$\ddot{\theta} = -g \sin(\theta) + w(t), \tag{3.27}$$

where θ is the angle, g is the gravitational acceleration, and $w(t)$ is a random noise process. This model can be converted into the following state-space model:

$$\frac{\mathrm{d}}{\mathrm{d}t}\begin{pmatrix}\theta_1 \\ \theta_2\end{pmatrix} = \begin{pmatrix} \theta_2 \\ -g\,\sin(\theta_1)\end{pmatrix} + \begin{pmatrix}0 \\ 1\end{pmatrix} w(t). \tag{3.28}$$

This can be seen to be a special case of an equation of the form

$$\frac{\mathrm{d}\mathbf{x}}{\mathrm{d}t} = \mathbf{f}(\mathbf{x}) + \mathbf{L}\,\mathbf{w}(t), \tag{3.29}$$

where $\mathbf{f}(\mathbf{x})$ is a nonlinear function.

Example 3.7 (Heart and breathing tracking in the brain). *Functional magnetic resonance imaging (fMRI) is a noninvasive neuroimaging method that is capable of capturing functional brain activity by measuring the variation of blood flow in the brain through excitation of nuclear spins and measuring their relaxation times. Physiological signals related to heartbeats and respiratory cycles cause structured temporal artifacts in the data. These structured noise components can be modeled – and removed – by using stochastic differential equations and Kalman filters (Särkkä et al., 2012).*
Treating each three-dimensional pixel (voxel) as independent, we can model the signal as a sum of a slowly moving brain activity (blood-oxygen-level dependent, BOLD) signal, quasiperiodic (almost periodic) physiological noise signals, and white noise. The quasiperiodic signals can be modeled as a sum of stochastic oscillators (which are SDEs):

$$\frac{\mathrm{d}\mathbf{x}_n}{\mathrm{d}t} = \begin{pmatrix} 0 & 2\pi n\,f(t) \\ -2\pi n\,f(t) & 0\end{pmatrix}\mathbf{x}_n + \begin{pmatrix}0 \\ 1\end{pmatrix} w(t), \tag{3.30}$$

where $f(t)$ is the time-dependent frequency and $w(t)$ a white random process. Figure 3.8 shows example realizations of quasiperiodic oscillations.

Example 3.8 (Black–Scholes model). *In the Black–Scholes model, the asset (e.g., a stock price) x is assumed to follow geometric Brownian motion*

$$\mathrm{d}x = \mu\,x\,\mathrm{d}t + \sigma\,x\,\mathrm{d}\beta, \tag{3.31}$$

where $\mathrm{d}\beta$ is a Brownian motion increment, μ is a drift constant, and σ is a volatility constant. If we formally divide by $\mathrm{d}t$, this equation can be heuristically interpreted as a differential equation

$$\frac{\mathrm{d}x}{\mathrm{d}t} = \mu\,x + \sigma\,x\,w, \tag{3.32}$$

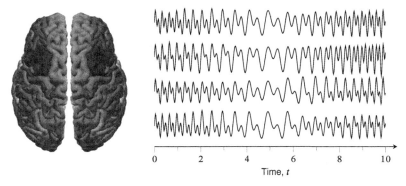

(a) Segmented brain surface   **(b)** Superpositions of stochastic oscillators

Figure 3.8 Quasiperiodic realizations of superpositions of stochastic oscillators with time-varying frequencies from Example 3.7. The trajectories share a common frequency trajectory, but the amplitudes and phases are allowed to vary – a good model for periodic noise confounds in the brain.

where $w(t)$ is a white random process. This equation is an example of more general multiplicative noise models of the form

$$\frac{\mathrm{d}\mathbf{x}}{\mathrm{d}t} = \mathbf{f}(\mathbf{x}) + \mathbf{L}(\mathbf{x})\,\mathbf{w}. \tag{3.33}$$

3.2 Differential Equations with Driving White Noise

As discussed in the previous section, many time-varying phenomena in various fields in science and engineering can be modeled as differential equations of the form

$$\frac{\mathrm{d}\mathbf{x}}{\mathrm{d}t} = \mathbf{f}(\mathbf{x}, t) + \mathbf{L}(\mathbf{x}, t)\,\mathbf{w}(t), \tag{3.34}$$

where $\mathbf{w}(t)$ is a vector-valued forcing function that we often can model as white noise.

We can indeed think of an SDE as an equation of the preceding form where the forcing function is a stochastic process. A particularly useful special case of a stochastic forcing function is the white noise process. As we saw in the previous section, one motivation for studying such equations is that various physical phenomena can be modeled as random processes (e.g., thermal motion), and when such a phenomenon enters a physical system, we get a model of the preceding SDE form. Another motivation is that

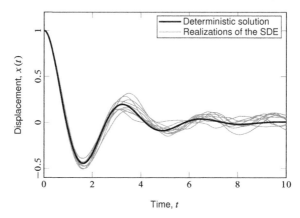

Figure 3.9 Solutions of the spring model in Equation (2.1) when the input is white noise. The solution of the SDE is different for each realization of the noise process and hence the solution can be summarized in terms of its statistics such as the mean and covariance. For example, the mean in the case of a linear SDE corresponds to the deterministic solution with zero noise.

in Bayesian statistical modeling *unknown* forces are naturally modeled as *random* forces, which again leads to SDE types of models. Because the forcing function is random, the solution to the SDE is a random process as well. With a different realization of the noise process, we get a different solution. For this reason, the particular solutions of the equations are not often of interest, but instead, we aim to determine the statistics of the solutions over all realizations. An example of an SDE solution is given in Figure 3.9.

In the context of SDEs, the term $\mathbf{f}(\mathbf{x}, t)$ in Equation (3.34) is called the *drift function*, which determines the nominal dynamics of the system, and $\mathbf{L}(\mathbf{x}, t)$ is the *dispersion matrix*, which determines how the noise $\mathbf{w}(t)$ enters the system. This is the most general form of SDEs that we discuss in this book. Although it would be tempting to generalize these equations to $d\mathbf{x}/dt = \mathbf{f}(\mathbf{x}, \mathbf{w}, t)$, it is not possible in the present theory. We shall discuss the reason for this later.

The stochastic noise process $\mathbf{w}(t)$ is usually modeled as Gaussian and "white" in the sense that $\mathbf{w}(t)$ and $\mathbf{w}(t')$ are uncorrelated (and independent) for all $t \neq t'$. The term *white* arises from the property that the power spectrum (or actually, the spectral density) of white noise is constant (flat)

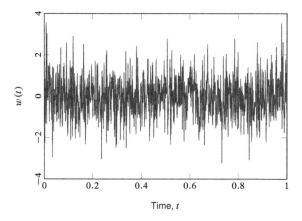

Figure 3.10 A sketch of a realization of white noise.

over all frequencies. White light is another phenomenon that has this same property and hence the name.

In mathematical sense, a white noise process can be defined as follows.

Definition 3.9 (White noise). *A white noise process* $\mathbf{w}(t) \in \mathbb{R}^S$ *is a random function with the following properties:*

1. *The two values* $\mathbf{w}(t)$ *and* $\mathbf{w}(t')$ *are independent if* $t \neq t'$.
2. *The mapping* $t \mapsto \mathbf{w}(t)$ *is a Gaussian process with zero mean and Dirac delta correlation:*

$$
\begin{aligned}
\mathbf{m_w}(t) &= \mathrm{E}[\mathbf{w}(t)] = \mathbf{0}, \\
\mathbf{C_w}(t, s) &= \mathrm{E}[\mathbf{w}(t)\,\mathbf{w}^\mathsf{T}(s)] = \delta(t - s)\,\mathbf{Q},
\end{aligned}
\tag{3.35}
$$

where $\mathbf{Q}$ *is the spectral density of the process.*

From the preceding properties we can also deduce the following somewhat peculiar properties of white noise:

• The sample path $t \mapsto \mathbf{w}(t)$ is discontinuous almost everywhere.
• White noise is unbounded and it takes arbitrarily large positive and negative values at any finite interval.

An illustration of a scalar white noise process realization is shown in Figure 3.10.

It is also possible to use non-Gaussian driving functions in SDEs such as Poisson processes or more general Lévy processes (see, e.g., Applebaum,

2009), but here we will always assume that the driving function is Gaussian. However, we briefly discuss Lévy process-driven SDEs in Chapter 13.

3.3 Heuristic Solutions of Linear SDEs

Let us first consider linear time-invariant stochastic differential equations (LTI SDEs) of the form

$$\frac{d\mathbf{x}(t)}{dt} = \mathbf{F}\,\mathbf{x}(t) + \mathbf{L}\,\mathbf{w}(t), \qquad \mathbf{x}(t_0) \sim \mathrm{N}(\mathbf{m}_0, \mathbf{P}_0), \qquad (3.36)$$

where $\mathbf{x}(t) \in \mathbb{R}^D$ is the state, $\mathbf{F} \in \mathbb{R}^{D \times D}$ and $\mathbf{L} \in \mathbb{R}^{D \times S}$ are some constant matrices, and the white noise process $\mathbf{w}(t) \in \mathbb{R}^S$ has zero mean and a given spectral density matrix $\mathbf{Q} \in \mathbb{R}^{S \times S}$. In Equation (3.36), we have specified a random initial condition for the equation such that at initial time t_0 the state $\mathbf{x}(t_0)$ should be Gaussian with a given mean $\mathbf{m}_0$ and covariance $\mathbf{P}_0$. The white noise process and the initial condition are assumed to be independent.

If we pretend for a while that the driving process $\mathbf{w}(t)$ is deterministic and continuous, we can form the general solution to the differential equation as follows:

$$\mathbf{x}(t) = \exp\left(\mathbf{F}\,(t - t_0)\right)\mathbf{x}(t_0) + \int_{t_0}^{t} \exp\left(\mathbf{F}\,(t - \tau)\right)\mathbf{L}\,\mathbf{w}(\tau)\,d\tau, \quad (3.37)$$

where $\exp(\bullet)$ is the matrix exponential function.

We can now take a "leap of faith" and hope that this solutions is valid also when $\mathbf{w}(t)$ is a white noise process. It turns out that it indeed is, but just because the differential equation happens to be linear (we will come back to this issue in the next chapter). However, it is enough for our purposes for now. The solution also turns out to be Gaussian, because the noise process is Gaussian and a linear differential equation can be considered as a linear operator acting on the noise process (and the initial condition).

Because the white noise process has zero mean, taking expectations of the both sides of Equation (3.37) gives

$$\mathrm{E}[\mathbf{x}(t)] = \exp\left(\mathbf{F}\,(t - t_0)\right)\mathbf{m}_0, \qquad (3.38)$$

which is thus the expected value of the SDE solutions over all realizations of noise. The mean function is here denoted as $\mathbf{m}(t) = \mathrm{E}[\mathbf{x}(t)]$.

The covariance of the solution can be derived by substituting the solution into the definition of covariance and by using the delta-correlation property of white noise, which results in

$$E\left[(\mathbf{x}(t) - \mathbf{m}(t))(\mathbf{x}(t) - \mathbf{m}(t))^\mathsf{T}\right]$$
$$= \exp(\mathbf{F}(t - t_0))\,\mathbf{P}_0\,\exp(\mathbf{F}(t - t_0))^\mathsf{T}$$
$$+ \int_{t_0}^t \exp(\mathbf{F}(t - \tau))\,\mathbf{L}\,\mathbf{Q}\,\mathbf{L}^\mathsf{T}\,\exp(\mathbf{F}(t - \tau))^\mathsf{T}\,d\tau. \quad (3.39)$$

In this book, we will hereafter denote the covariance as $\mathbf{P}(t) = E\left[(\mathbf{x}(t) - \mathbf{m}(t))(\mathbf{x}(t) - \mathbf{m}(t))^\mathsf{T}\right]$.

By differentiating the mean and covariance solutions and collecting the terms, we can also derive the following differential equations for the mean and covariance:

$$\frac{d\mathbf{m}(t)}{dt} = \mathbf{F}\,\mathbf{m}(t),$$
$$\frac{d\mathbf{P}(t)}{dt} = \mathbf{F}\,\mathbf{P}(t) + \mathbf{P}(t)\,\mathbf{F}^\mathsf{T} + \mathbf{L}\,\mathbf{Q}\,\mathbf{L}^\mathsf{T}. \quad (3.40)$$

Example 3.10 (Stochastic spring model). *Consider the spring model in Equation (2.4). If we replace the unknown input force with a white noise with spectral density q, we get the following LTI SDE:*

$$\frac{d}{dt}\underbrace{\begin{pmatrix} x_1(t) \\ x_2(t) \end{pmatrix}}_{d\mathbf{x}(t)/dt} = \underbrace{\begin{pmatrix} 0 & 1 \\ -\nu^2 & -\gamma \end{pmatrix}}_{\mathbf{F}}\underbrace{\begin{pmatrix} x_1(t) \\ x_2(t) \end{pmatrix}}_{\mathbf{x}(t)} + \underbrace{\begin{pmatrix} 0 \\ 1 \end{pmatrix}}_{\mathbf{L}}w(t). \quad (3.41)$$

The equations for the mean and covariance are then given as

$$\begin{pmatrix} \frac{dm_1}{dt} \\ \frac{dm_2}{dt} \end{pmatrix} = \begin{pmatrix} 0 & 1 \\ -\nu^2 & -\gamma \end{pmatrix}\begin{pmatrix} m_1 \\ m_2 \end{pmatrix},$$

$$\begin{pmatrix} \frac{dP_{11}}{dt} & \frac{dP_{12}}{dt} \\ \frac{dP_{21}}{dt} & \frac{dP_{22}}{dt} \end{pmatrix} = \begin{pmatrix} 0 & 1 \\ -\nu^2 & -\gamma \end{pmatrix}\begin{pmatrix} P_{11} & P_{12} \\ P_{21} & P_{22} \end{pmatrix}$$
$$+ \begin{pmatrix} P_{11} & P_{12} \\ P_{21} & P_{22} \end{pmatrix}\begin{pmatrix} 0 & 1 \\ -\nu^2 & -\gamma \end{pmatrix}^\mathsf{T} + \begin{pmatrix} 0 & 0 \\ 0 & q \end{pmatrix}. \quad (3.42)$$

Figure 3.11 shows the theoretical mean and the 95% quantiles computed from the variances $P_{11}(t)$ along with trajectories from the stochastic spring model.

Despite the heuristic derivation, Equations (3.40) are indeed the correct differential equations for the mean and covariance. However, as we see in the following, it is easy to demonstrate that one has to be extremely careful

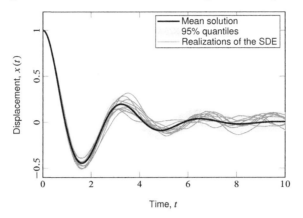

Figure 3.11 Solutions, exact mean, and the 95% quantiles for the spring model in Equation (2.1), when the input is white noise.

in extrapolating deterministic differential equation results to the stochastic setting.

Note that we can indeed derive the first of the preceding equations simply by taking the expectations of both sides of Equation (3.36):

$$E\left[\frac{dx(t)}{dt}\right] = E[F\,x(t)] + E[L\,w(t)]. \tag{3.43}$$

Exchanging the order of expectation and differentiation, using the linearity of expectation, and recalling that white noise has zero mean then results in a correct mean differential equation. We can attempt to do the same for the covariance. By the chain rule of ordinary calculus, we get

$$\frac{d}{dt}\left[(x-m)\,(x-m)^{\mathsf{T}}\right]$$

$$= \left(\frac{dx}{dt} - \frac{dm}{dt}\right)(x-m)^{\mathsf{T}} + (x-m)\left(\frac{dx}{dt} - \frac{dm}{dt}\right)^{\mathsf{T}}. \tag{3.44}$$

Substituting the time derivatives on the right-hand side and taking expectation results in

$$\frac{d}{dt}E\left[(x-m)\,(x-m)^{\mathsf{T}}\right] = F\,E\left[(x-m)\,(x-m)^{\mathsf{T}}\right]$$

$$+ E\left[(x-m)\,(x-m)^{\mathsf{T}}\right]F^{\mathsf{T}}, \tag{3.45}$$

which implies the covariance differential equation

$$\frac{d\mathbf{P}(t)}{dt} = \mathbf{F}\,\mathbf{P}(t) + \mathbf{P}(t)\,\mathbf{F}^\mathsf{T}. \tag{3.46}$$

But this equation is *wrong*, because the term $\mathbf{L}\,\mathbf{Q}\,\mathbf{L}^\mathsf{T}$ is missing on the right-hand side. Our mistake was to assume that we can use the product rule in Equation (3.44), but in fact we cannot. This is one of the peculiar features of stochastic calculus and it is also a warning sign that we should not take our "leaps of faith" too far when analyzing solutions of SDEs via formal extensions of deterministic ODE solutions.

3.4 Heuristic Solutions of Nonlinear SDEs

We can also attempt to analyze differential equations of the form

$$\frac{d\mathbf{x}}{dt} = \mathbf{f}(\mathbf{x}, t) + \mathbf{L}(\mathbf{x}, t)\,\mathbf{w}(t), \tag{3.47}$$

where $\mathbf{x}(t) \in \mathbb{R}^D$, $\mathbf{f}(\mathbf{x}, t)$, and $\mathbf{L}(\mathbf{x}, t)$ are nonlinear functions, and $\mathbf{w}(t) \in \mathbb{R}^S$ is a white noise process with a spectral density $\mathbf{Q}$. Unfortunately, in the nonlinear case we cannot take the same kind of "leap of faith" from deterministic solutions as in the case of linear differential equations, because in general, we cannot solve even the corresponding deterministic differential equation.

An attempt to generalize the numerical methods for deterministic differential equations discussed in the previous chapter will fail as well, because the basic requirement in almost all of those methods is continuity of the right-hand side, and in fact, even differentiability of several orders. Because white noise is discontinuous everywhere, the right-hand side is discontinuous everywhere, and is certainly not differentiable anywhere either.

We can, however, take a shortcut and generalize the Euler method (leading to the so-called Euler–Maruyama method) to the present stochastic setting, because it does not explicitly require continuity. From that, we get an iteration of the form

$$\hat{\mathbf{x}}(t_{k+1}) = \hat{\mathbf{x}}(t_k) + \mathbf{f}(\hat{\mathbf{x}}(t_k), t_k)\,\Delta t + \mathbf{L}(\hat{\mathbf{x}}(t_k), t_k)\,\Delta\boldsymbol{\beta}_k, \tag{3.48}$$

where $\Delta\boldsymbol{\beta}_k$ is a Gaussian random variable with distribution $\mathrm{N}(\mathbf{0}, \mathbf{Q}\,\Delta t)$. Note that it is the variance that is proportional to Δt, not the standard deviation as one might expect. This results from the peculiar properties of stochastic differential equations. We can use the preceding method for simulating trajectories from stochastic differential equations and the result converges to the true solution in the limit $\Delta t \to 0$. However, the convergence is quite slow, as the (strong) order of convergence is only $\gamma = 1/2$.

In the case of SDEs, the convergence order definition is a bit more complicated than in the deterministic case, because we have a couple of different types of convergence orders. We can talk about pathwise approximations, which correspond to approximating the solution with fixed $\mathbf{w}(t)$. These are also called strong solutions and give rise to the strong order of convergence γ. However, we can also think of approximating the probability density or the moments of the solutions. These give rise to weak solutions and weak order of convergence α. We will come back to these and numerical solution methods in general later in Chapter 8.

3.5 The Problem of Solution Existence and Uniqueness

One can attempt to analyze the uniqueness and existence of the equation

$$\frac{d\mathbf{x}}{dt} = \mathbf{f}(\mathbf{x}, t) + \mathbf{L}(\mathbf{x}, t) \mathbf{w}(t) \tag{3.49}$$

using the Picard–Lindelöf theorem presented in the previous chapter. The basic assumption in the theorem for the right-hand side of the differential equation consisted of the following:

- Continuity in both arguments
- Lipschitz continuity in the first argument

Unfortunately, the first of these fails miserably, because white noise is discontinuous everywhere. However, a small glimpse of hope is implied by the fact that $\mathbf{f}(\mathbf{x}, t)$ might indeed be Lipschitz continuous in the first argument, as well as $\mathbf{L}(\mathbf{x}, t)$. However, without extending the Picard–Lindelöf theorem, we cannot determine the existence or uniqueness of stochastic differential equations.

It turns out that a stochastic analogue of the Picard iteration will indeed lead to the solution to the existence and uniqueness question also in the stochastic case. However, before going into that, we need to make the theory of stochastic differential equations mathematically meaningful.

3.6 Exercises

3.1   Mean and covariance equations of linear SDEs:

   (a) Complete the missing steps in the derivation of the covariance (3.39).

(b) Derive the mean and covariance differential equations (3.40) by differentiating the equations (3.38) and (3.39).

3.2 Solution of an Ornstein–Uhlenbeck process:

(a) Find the complete solution $x(t)$ as well as the mean $m(t)$ and variance $P(t)$ of the following scalar stochastic differential equation:

$$\frac{dx(t)}{dt} = -\lambda x(t) + w(t), \quad x(0) = x_0, \qquad (3.50)$$

where x_0 and $\lambda > 0$ are given constants and the white noise $w(t)$ has spectral density q.

(b) Compute the limit of the mean and variance when $t \to \infty$ (i) directly via $\lim_{t\to\infty} m(t)$, $\lim_{t\to\infty} P(t)$, and (ii) by solving the stationary state of the differential equations $dm/dt = 0$ and $dP/dt = 0$.

3.3 Simulate 1,000 trajectories on the time interval $t \in [0, 1]$ from the Ornstein–Uhlenbeck process in the previous exercise using the Euler–Maruyama method with $\lambda = 1/2$, $q = 1$, $\Delta t = 1/100$, $x_0 = 1$, and check that the mean and covariance trajectories approximately agree with the theoretical values.

3.4 Simulate 1,000 trajectories from the Black–Scholes model (see Example 3.8) on the time interval $[0, 1]$ using the Euler–Maruyama method with $\mu = 1/10$ and $\sigma = 1$. By comparing to the exact solution given in Example 4.7 (approximate $\beta(t)$ as sum of $\Delta\beta_k$), study the scaling of the error as a function of Δt.

3.5 The covariance function of the Ornstein–Uhlenbeck process is

$$C(t, t') = \frac{q}{2\lambda} \exp(-\lambda |t - t'|).$$

(a) Show that in the limit of $\lambda \to \infty$, the Ornstein–Uhlenbeck process reverts to white noise.

(b) Show that the power spectral density of the Ornstein–Uhlenbeck process becomes flat in the limit.

3.6 Let us consider the rotation formula in (3.25) and assume that the angular velocity ω_L is constant. Then it is useful to rewrite (3.25) as

$$\frac{d\mathbf{g}_L}{dt} = [-\omega_L]_\times \mathbf{g}_L,$$

where $[\bullet]_\times$ denotes the cross-product matrix. Show that the matrix exponential of the cross-product matrix is given by the Rodrigues formula

$$\exp([-\omega_L]_\times) = \mathbf{I} + \sin(\|\omega_L\|)\frac{[-\omega_L]_\times}{\|\omega_L\|} + (1 - \cos(\|\omega_L\|))\frac{[-\omega_L]_\times^2}{\|\omega_L\|^2}.$$

Write down the solution of the differential equation (3.25) in terms of this. What is the solution to the differential equation (3.26)?

4

Itô Calculus and Stochastic Differential Equations

Itô calculus or stochastic calculus (see, e.g., Karatzas and Shreve, 1991; Øksendal, 2003, for full theory) is the theory of integration and differentiation for stochastic processes. In fact, it is more a theory of integration than differentiation and the theory of differentiation of stochastic processes belongs to the domain of Malliavin calculus (Nualart, 2006), which is beyond the scope of this book. In this chapter, we first see how SDEs can be expressed as equivalent integral equations, which then leads to the requirement of defining a stochastic integral with respect to a Brownian motion appearing in the equation. We proceed to demonstrate why this integral cannot be defined as a Riemann integral, Stieltjes integral, or even a Lebesgue integral. Instead, we must study a completely new integral, which becomes the *Itô integral*. Given the definition of this integral, we can derive the computation rules for the resulting stochastic calculus. The cornerstone is the Itô formula, the chain rule for stochastic processes, from which we can derive almost all of the other results. With the formula, we can properly derive the solutions for linear SDEs that we heuristically obtained in the previous chapter. We also discuss the difficulty of solving nonlinear SDEs even when we know how to define them and finally discuss other stochastic calculi, of which the most famous one is the Stratonovich stochastic calculus.

4.1 The Stochastic Integral of Itô

As discussed in the previous chapter, a stochastic differential equation can be heuristically considered as a vector differential equation of the form

$$\frac{d\mathbf{x}}{dt} = \mathbf{f}(\mathbf{x}, t) + \mathbf{L}(\mathbf{x}, t)\,\mathbf{w}(t), \qquad (4.1)$$

where $\mathbf{w}(t)$ is a zero mean white Gaussian process. Even though this is somewhat true, it is not the whole truth. The aim in this section is to clarify what really is going on behind stochastic differential equations and how they should be treated.

The problem in the preceding equation is that it cannot be a differential equation in the traditional sense, because the ordinary theory of differential equations does not permit discontinuous functions such as $\mathbf{w}(t)$ in differential equations (recall the problem with the Picard–Lindelöf theorem). This problem is not purely theoretical, because the solution actually turns out to depend on infinitesimally small differences in mathematical definitions of the noise, and thus without further restrictions the solution would not be unique even with a given realization of white noise $\mathbf{w}(t)$.

Fortunately, there is a solution to this problem, but in order to find it we need to reduce the problem to definition of a new kind of integral called the *Itô integral*, which is an integral with respect to a stochastic process. In order to do that, let us first formally integrate the differential equation from some initial time t_0 to final time t:

$$\mathbf{x}(t) - \mathbf{x}(t_0) = \int_{t_0}^{t} \mathbf{f}(\mathbf{x}(t), t) \, dt + \int_{t_0}^{t} \mathbf{L}(\mathbf{x}(t), t) \, \mathbf{w}(t) \, dt. \tag{4.2}$$

The first integral on the right-hand side is just a normal integral with respect to time and can be defined as a Riemann integral of $t \mapsto \mathbf{f}(\mathbf{x}(t), t)$, or as a Lebesgue integral with respect to the Lebesgue measure, if more generality is desired.

The second integral is the problematic one. First of all, it cannot be defined as a Riemann integral due to the unboundedness and discontinuity of the white noise process. Recall that in the Riemannian sense the integral would be defined as the following kind of limit:

$$\int_{t_0}^{t} \mathbf{L}(\mathbf{x}(t), t) \, \mathbf{w}(t) \, dt = \lim_{n \to \infty} \sum_{k} \mathbf{L}(\mathbf{x}(t_k^*), t_k^*) \, \mathbf{w}(t_k^*) \, (t_{k+1} - t_k), \tag{4.3}$$

where $t_0 < t_1 < \ldots < t_n = t$ and $t_k^* \in [t_k, t_{k+1}]$. In the context of Riemann integrals, so-called upper and lower sums are defined as the selections of t_k^* such that the integrand $\mathbf{L}(\mathbf{x}(t_k^*), t_k^*) \, \mathbf{w}(t_k^*)$ has its maximum and minimum values, respectively. The Riemann integral is defined if the upper and lower sums converge to the same value, which is then defined to be the value of the integral. In the case of white noise, it happens that $\mathbf{w}(t_k^*)$ is not bounded and takes arbitrarily small and large values at every finite interval, and thus the Riemann integral does not converge.

We could also attempt to define the second integral as a Stieltjes integral, which is more general than the Riemann integral. For that definition, we need to interpret the increment $\mathbf{w}(t) \, dt$ as an increment of another process

$\boldsymbol{\beta}(t)$ such that the integral becomes

$$\int_{t_0}^t \mathbf{L}(\mathbf{x}(t), t) \, \mathbf{w}(t) \, dt = \int_{t_0}^t \mathbf{L}(\mathbf{x}(t), t) \, d\boldsymbol{\beta}(t). \qquad (4.4)$$

It turns out that a suitable process for this purpose is Brownian motion, which we already discussed in the previous chapter:

Definition 4.1 (Brownian motion). *Brownian motion* $\boldsymbol{\beta}(t) \in \mathbb{R}^S$ *is a continuous stochastic process with the following properties:*

1. *Any increment* $\Delta\boldsymbol{\beta}_k = \boldsymbol{\beta}(t_{k+1}) - \boldsymbol{\beta}(t_k)$ *is a zero mean Gaussian random variable with covariance* $\mathbf{Q}\,\Delta t_k$, *where* $\mathbf{Q}$ *is the diffusion matrix of the Brownian motion and* $\Delta t_k = t_{k+1} - t_k$.
2. *When the time spans of increments do not overlap, the increments are independent.*
3. *The process starts at origin:* $\boldsymbol{\beta}(0) = \mathbf{0}$.

When the Brownian motion is one-dimensional, we denote it as $\beta(t)$. In the scalar case, we often refer to the corresponding diffusion matrix as the *diffusion constant* or *diffusion coefficient*, and denote it as q. An example of a scalar Brownian motion realization is shown in Figure 4.1.

Some further properties of Brownian motion that result from the preceding definition are the following:

1. Brownian motion $t \mapsto \boldsymbol{\beta}(t)$ is nowhere differentiable.
2. White noise can be considered as the formal (or weak) derivative of Brownian motion, $\mathbf{w}(t) = d\boldsymbol{\beta}(t)/dt$.

Unfortunately, the definition of the latter integral in Equation (4.2) in terms of increments of Brownian motion as in Equation (4.4) does not solve our existence problem. The problem is the everywhere discontinuous derivative of $\boldsymbol{\beta}(t)$, which makes it too irregular for the defining sum of the Stieltjes integral to converge. Unfortunately, the same happens with the Lebesgue integral. Recall that both Stieltjes and Lebesgue integrals are essentially defined as limits of the form

$$\int_{t_0}^t \mathbf{L}(\mathbf{x}(t), t) \, d\boldsymbol{\beta} = \lim_{n \to \infty} \sum_k \mathbf{L}(\mathbf{x}(t_k^*), t_k^*) \, [\boldsymbol{\beta}(t_{k+1}) - \boldsymbol{\beta}(t_k)], \qquad (4.5)$$

where $t_0 < t_1 < \ldots < t_n$ and $t_k^* \in [t_k, t_{k+1}]$. Both of these definitions would require the limit to be independent of the position on the interval $t_k^* \in [t_k, t_{k+1}]$. However, for integration with respect to Brownian motion,

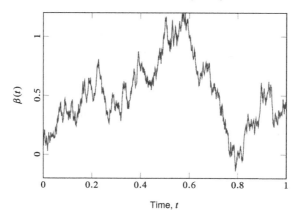

Figure 4.1 A realization of trajectory of Brownian motion, where the derivative is discontinuous everywhere. White noise can be considered the formal derivative of Brownian motion.

this is not the case. Thus, the Stieltjes or Lebesgue integral definitions do not work either.

A solution to the problem is the Itô stochastic integral, which is based on the observation that if we fix the choice to $t_k^* = t_k$, then the limit becomes unique. The Itô integral can thus be defined as the limit (in the L_2 sense)

$$\int_{t_0}^{t} \mathbf{L}(\mathbf{x}(t), t) \, d\boldsymbol{\beta}(t) = \lim_{n \to \infty} \sum_{k} \mathbf{L}(\mathbf{x}(t_k), t_k) \left[\boldsymbol{\beta}(t_{k+1}) - \boldsymbol{\beta}(t_k) \right], \quad (4.6)$$

where $t_0 < t_1 < \ldots < t_n = t$, which is a sensible definition of the stochastic integral required for the SDE.

The stochastic differential equation (3.34) can now be defined to actually refer to the corresponding (Itô) integral equation

$$\mathbf{x}(t) - \mathbf{x}(t_0) = \int_{t_0}^{t} \mathbf{f}(\mathbf{x}(t), t) \, dt + \int_{t_0}^{t} \mathbf{L}(\mathbf{x}(t), t) \, d\boldsymbol{\beta}(t), \quad (4.7)$$

which should be true for arbitrary t_0 and t.

We can now take a step backward and return from this stochastic integral equation to the differential equation as follows. If we choose the integration limits in Equation (4.7) to be t and $t + dt$, where dt is "small", we can write the equation in the differential form

$$d\mathbf{x} = \mathbf{f}(\mathbf{x}, t) \, dt + \mathbf{L}(\mathbf{x}, t) \, d\boldsymbol{\beta}, \quad (4.8)$$

which should be interpreted as shorthand for the integral equation. The preceding is the form that is most often used in literature on SDEs (e.g., Øksendal, 2003; Karatzas and Shreve, 1991). We can now formally divide by dt to obtain a differential equation

$$\frac{d\mathbf{x}}{dt} = \mathbf{f}(\mathbf{x}, t) + \mathbf{L}(\mathbf{x}, t) \frac{d\boldsymbol{\beta}}{dt}, \tag{4.9}$$

which shows that also here white noise can be interpreted as the formal derivative of Brownian motion. However, due to non-classical transformation properties of the Itô differentials, one has to be very careful in working with such formal manipulations.

Now we also see why we are not permitted to consider more general differential equations of the form

$$\frac{d\mathbf{x}(t)}{dt} = \mathbf{f}(\mathbf{x}(t), \mathbf{w}(t), t), \tag{4.10}$$

where the white noise $\mathbf{w}(t)$ enters the system through a nonlinear transformation. There is no way to rewrite this equation as a stochastic integral with respect to a Brownian motion, and thus we cannot define the mathematical meaning of this equation. More generally, white noise should not be thought of as an entity as such, but it only exists as the formal derivative of Brownian motion. Therefore, only linear functions of white noise have a meaning, whereas nonlinear functions do not.

4.2 Itô Formula

Consider the stochastic integral

$$\int_0^t \beta(t) \, d\beta(t), \tag{4.11}$$

where $\beta(t)$ is a standard Brownian motion, that is, a scalar Brownian motion with diffusion constant $q = 1$. Based on ordinary calculus, we would expect the value of this integral to be $\beta^2(t)/2$, but this is a wrong answer. If we select a partition $0 = t_0 < t_1 < \ldots < t_n = t$, we can rearrange the terms as follows:

$$\int_0^t \beta(t) \, d\beta(t) = \lim_{n \to \infty} \sum_k \beta(t_k)[\beta(t_{k+1}) - \beta(t_k)]$$

$$= \lim_{n \to \infty} \sum_k \left[-\frac{1}{2}(\beta(t_{k+1}) - \beta(t_k))^2 + \frac{1}{2}(\beta^2(t_{k+1}) - \beta^2(t_k)) \right]$$

$$= -\frac{1}{2}t + \frac{1}{2}\beta^2(t), \tag{4.12}$$

where we have used the result that the limit of the first term is $\lim_{n\to\infty} \sum_k (\beta(t_{k+1}) - \beta(t_k))^2 = t$. The Itô differential of $\beta^2(t)/2$ is analogously

$$d\left[\frac{1}{2}\beta^2(t)\right] = \beta(t)\,d\beta(t) + \frac{1}{2}\,dt, \tag{4.13}$$

not $\beta(t)\,d\beta(t)$ as we might expect. This is a consequence and also a drawback of the selection of the fixed $t_k^* = t_k$.

The general rule for calculating the Itô differentials and thus Itô integrals can be summarized as the following Itô formula, which corresponds to chain rule in ordinary calculus.

Theorem 4.2 (Itô formula). *Assume that $\mathbf{x}(t)$ is an Itô process, and consider an arbitrary (scalar) function $\phi(\mathbf{x}(t), t)$ of the process. Then the Itô differential of ϕ, that is, the Itô SDE for ϕ is given as*

$$\begin{aligned}
d\phi &= \frac{\partial\phi}{\partial t}\,dt + \sum_i \frac{\partial\phi}{\partial x_i}\,dx_i + \frac{1}{2}\sum_{i,j}\left(\frac{\partial^2\phi}{\partial x_i\,\partial x_j}\right) dx_i\,dx_j \\
&= \frac{\partial\phi}{\partial t}\,dt + (\nabla\phi)^\mathsf{T}\,d\mathbf{x} + \frac{1}{2}\,\mathrm{tr}\left\{\left(\nabla\nabla^\mathsf{T}\phi\right) d\mathbf{x}\,d\mathbf{x}^\mathsf{T}\right\},
\end{aligned} \tag{4.14}$$

provided that the required partial derivatives exist, where the mixed differentials are combined according to the rules

$$\begin{aligned}
d\boldsymbol{\beta}\,dt &= \mathbf{0}, \\
dt\,d\boldsymbol{\beta} &= \mathbf{0}, \\
d\boldsymbol{\beta}\,d\boldsymbol{\beta}^\mathsf{T} &= \mathbf{Q}\,dt.
\end{aligned} \tag{4.15}$$

Proof   See, for example, Øksendal (2003) or Karatzas and Shreve (1991). □

Although the preceding Itô formula is defined only for scalar ϕ, it works for each of the components of a vector-valued function separately and thus also includes the vector case. Also note that every Itô process has a representation as the solution of an SDE of the form

$$d\mathbf{x} = \mathbf{f}(\mathbf{x}, t)\,dt + \mathbf{L}(\mathbf{x}, t)\,d\boldsymbol{\beta} \tag{4.16}$$

and an explicit expression for the differential in terms of the functions $\mathbf{f}(\mathbf{x}, t)$ and $\mathbf{L}(\mathbf{x}, t)$ can be derived by substituting the preceding equation for $d\mathbf{x}$ in the Itô formula.

The Itô formula can be conceptually derived by a Taylor series expansion:

$$\phi(\mathbf{x} + d\mathbf{x}, t + dt) = \phi(\mathbf{x}, t) + \frac{\partial \phi(\mathbf{x}, t)}{\partial t} dt$$

$$+ \sum_i \frac{\partial \phi(\mathbf{x}, t)}{\partial x_i} dx_i + \frac{1}{2} \sum_{i,j} \left(\frac{\partial^2 \phi}{\partial x_i \partial x_j} \right) dx_i \, dx_j + \dots. \quad (4.17)$$

For the first-order in dt and second-order in $d\mathbf{x}$, we have

$$d\phi = \phi(\mathbf{x} + d\mathbf{x}, t + dt) - \phi(\mathbf{x}, t)$$

$$\approx \frac{\partial \phi(\mathbf{x}, t)}{\partial t} dt + \sum_i \frac{\partial \phi(x, t)}{\partial x_i} dx_i + \frac{1}{2} \sum_{i,j} \left(\frac{\partial^2 \phi}{\partial x_i \partial x_j} \right) dx_i \, dx_j. \quad (4.18)$$

In the deterministic case, we could ignore the second-order and higher-order terms, because $d\mathbf{x} \, d\mathbf{x}^\mathsf{T}$ would already be of the order dt^2. Thus the deterministic counterpart is

$$d\phi = \frac{\partial \phi}{\partial t} dt + \frac{\partial \phi}{\partial x} dx. \quad (4.19)$$

In the stochastic case, we know that $d\mathbf{x} \, d\mathbf{x}^\mathsf{T}$ is potentially of order dt, because $d\boldsymbol{\beta} \, d\boldsymbol{\beta}^\mathsf{T}$ is of the same order. This implies that we need to retain the second-order term also.

Example 4.3 (Itô differential of $\beta^2(t)/2$). *If we apply the Itô formula to $\phi(x) = \frac{1}{2}x^2$, with $x(t) = \beta(t)$, where $\beta(t)$ is a standard Brownian motion, we get*

$$d\phi = \beta \, d\beta + \frac{1}{2} d\beta^2 = \beta \, d\beta + \frac{1}{2} dt, \quad (4.20)$$

as expected.

Example 4.4 (Itô differential of $\sin(\omega \, x)$). *Assume that $x(t)$ is the solution to the scalar SDE*

$$dx = f(x) \, dt + d\beta, \quad (4.21)$$

where $\beta(t)$ is a Brownian motion with diffusion constant q. The Itô differential of $\sin(\omega\,x(t))$, where $\omega > 0$ is a constant, is then given as

$$d[\sin(\omega\,x)] = \omega\,\cos(\omega\,x)\,dx - \frac{1}{2}\omega^2\,\sin(\omega\,x)\,dx^2$$

$$= \omega\,\cos(\omega\,x)\,[f(x)\,dt + d\beta] - \frac{1}{2}\omega^2\,\sin(\omega\,x)\,[f(x)\,dt + d\beta]^2$$

$$= \omega\,\cos(\omega\,x)\,[f(x)\,dt + d\beta] - \frac{1}{2}\omega^2\,\sin(\omega\,x)\,q\,dt. \qquad (4.22)$$

4.3 Explicit Solutions to Linear SDEs

In this section, we derive the full solution to a general time-varying linear SDE. The SDE is assumed to have the form

$$d\mathbf{x} = \mathbf{F}(t)\,\mathbf{x}\,dt + \mathbf{u}(t)\,dt + \mathbf{L}(t)\,d\boldsymbol{\beta}, \qquad (4.23)$$

where $\mathbf{x}(t) \in \mathbb{R}^D$ is the state, $\mathbf{u}(t) \in \mathbb{R}^D$ is an input, and $\boldsymbol{\beta}(t) \in \mathbb{R}^S$ is a Brownian motion with diffusion matrix $\mathbf{Q}$. The initial condition is assumed to be given as a distribution or fixed value for $\mathbf{x}(t_0)$, and it is assumed to be independent of $\boldsymbol{\beta}$.

We can now proceed by defining a transition matrix $\boldsymbol{\Psi}(\tau, t)$ in the same way as we did in Equation (2.34). Multiplying the preceding SDE with the integrating factor $\boldsymbol{\Psi}(t_0, t)$ and rearranging gives

$$\boldsymbol{\Psi}(t_0, t)\,d\mathbf{x} - \boldsymbol{\Psi}(t_0, t)\,\mathbf{F}(t)\,\mathbf{x}\,dt = \boldsymbol{\Psi}(t_0, t)\,\mathbf{u}(t)\,dt + \boldsymbol{\Psi}(t_0, t)\,\mathbf{L}(t)\,d\boldsymbol{\beta}. \qquad (4.24)$$

Applying the Itô formula and the properties of the integrating factor gives

$$d[\boldsymbol{\Psi}(t_0, t)\,\mathbf{x}] = -\boldsymbol{\Psi}(t_0, t)\,\mathbf{F}(t)\,\mathbf{x}\,dt + \boldsymbol{\Psi}(t_0, t)\,d\mathbf{x}. \qquad (4.25)$$

Thus the SDE in Equation (4.24) can be rewritten as

$$d[\boldsymbol{\Psi}(t_0, t)\,\mathbf{x}] = \boldsymbol{\Psi}(t_0, t)\,\mathbf{u}(t)\,dt + \boldsymbol{\Psi}(t_0, t)\,\mathbf{L}(t)\,d\boldsymbol{\beta}, \qquad (4.26)$$

where the differential is an Itô differential. Integration (in the Itô sense) from t_0 to t gives

$$\boldsymbol{\Psi}(t_0, t)\,\mathbf{x}(t) - \boldsymbol{\Psi}(t_0, t_0)\,\mathbf{x}(t_0)$$

$$= \int_{t_0}^{t} \boldsymbol{\Psi}(t_0, \tau)\,\mathbf{u}(\tau)\,d\tau + \int_{t_0}^{t} \boldsymbol{\Psi}(t_0, \tau)\,\mathbf{L}(\tau)\,d\boldsymbol{\beta}(\tau), \quad (4.27)$$

which can be further written in the form

$$\mathbf{x}(t) = \boldsymbol{\Psi}(t, t_0)\,\mathbf{x}(t_0) + \int_{t_0}^t \boldsymbol{\Psi}(t, \tau)\,\mathbf{u}(\tau)\,d\tau + \int_{t_0}^t \boldsymbol{\Psi}(t, \tau)\,\mathbf{L}(\tau)\,d\boldsymbol{\beta}(\tau).$$

$$(4.28)$$

This is the desired full solution of the linear SDE.

In the case of a linear time-invariant SDE

$$d\mathbf{x} = \mathbf{F}\,\mathbf{x}\,dt + \mathbf{L}\,d\boldsymbol{\beta}, \tag{4.29}$$

where $\mathbf{F}$ and $\mathbf{L}$ are constants, and $\boldsymbol{\beta}$ has a constant diffusion matrix $\mathbf{Q}$, the solution simplifies to

$$\mathbf{x}(t) = \exp\left(\mathbf{F}\,(t - t_0)\right)\mathbf{x}(t_0) + \int_{t_0}^t \exp\left(\mathbf{F}\,(t - \tau)\right)\mathbf{L}\,d\boldsymbol{\beta}(\tau). \tag{4.30}$$

By comparing this to Equation (3.37) in Section 3.3, this solution is exactly what we would have expected – it is what we would obtain if we formally replaced $\mathbf{w}(\tau)\,d\tau$ with $d\boldsymbol{\beta}(\tau)$ in the deterministic solution. However, it is just because the use of the Itô formula in Equation (4.25) happened to result in the same result as a deterministic differentiation would. In the nonlinear case, we cannot expect to get the right result with this kind of formal replacement.

Example 4.5 (Solution of the Ornstein–Uhlenbeck process). *The complete solution to the scalar SDE*

$$dx = -\lambda\,x\,dt + d\beta, \qquad x(0) = x_0, \tag{4.31}$$

where $\lambda > 0$ is a given constant and $\beta(t)$ is a Brownian motion, is

$$x(t) = \exp(-\lambda\,t)\,x_0 + \int_0^t \exp(-\lambda\,(t - \tau))\,d\beta(\tau). \tag{4.32}$$

Let us choose the parameter values $\lambda = 1/2$ and $x_0 = 4$. The solution, called the Ornstein–Uhlenbeck process, is illustrated in Figure 4.2.

It is also possible to extend the preceding results to linear SDEs with multiplicative noise, which in general can be written as

$$d\mathbf{x} = \mathbf{F}(t)\,\mathbf{x}\,dt + \mathbf{u}(t)\,dt + \mathbf{L}(t)\,d\boldsymbol{\beta} + \sum_i \mathbf{B}_i(t)\,x_i\,d\boldsymbol{\beta}. \tag{4.33}$$

However, the solutions to these equations are no longer Gaussian processes, and constructing the solutions is not as straightforward as with additive noise. For that reason, in this book, we only briefly discuss the multiplicative noise case and guide the reader to the books of Karatzas and

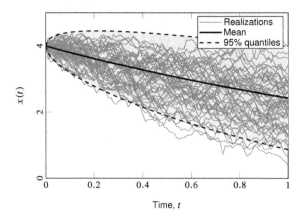

Figure 4.2 Simulated realizations, the mean, and 95% quantiles of an Ornstein–Uhlenbeck process with $\lambda = 1/2$ and $x_0 = 4$ from Example 4.5.

Shreve (1991), Kloeden and Platen (1999), and Gardiner (2004) for more information.

In the linear time-invariant scalar case the solution to a multiplicative noise SDE is quite simple, as is illustrated in the following example.

Example 4.6 (Linear time-invariant model with multiplicative noise). *The solution to the SDE*

$$\mathrm{d}x = (a\,x + c)\,\mathrm{d}t + (b\,x + d)\,\mathrm{d}\beta, \quad x(0) = x_0, \tag{4.34}$$

where a, b, c, and d are constants and β is a standard Brownian motion, is given as (see Kloeden and Platen, 1999)

$$x(t) = \psi(t)\left(x_0 + (c - b\,d)\int_0^t \frac{1}{\psi(s)}\,\mathrm{d}s + d\int_0^t \frac{1}{\psi(s)}\,\mathrm{d}\beta(s)\right), \tag{4.35}$$

where

$$\psi(t) = \exp\left(\left(a - \frac{1}{2}b^2\right)t + b\,\beta(t)\right). \tag{4.36}$$

In financial applications, scalar linear SDEs with multiplicative noise are quite typical, because they can be used to model processes that are strictly positive. An example of such an SDE is the Black–Scholes model.

Example 4.7 (Black–Scholes solution). *Following the results from the previous example (Example 4.6), the Black–Scholes model considered in Example 3.8 has the solution*

$$x(t) = \exp\left(\left(\mu - \frac{1}{2}\sigma^2\right) t + \sigma \beta(t)\right) x_0. \qquad (4.37)$$

4.4 Finding Solutions to Nonlinear SDEs

Let us now turn our attention to nonlinear SDEs of the form

$$d\mathbf{x} = \mathbf{f}(\mathbf{x}, t) \, dt + \mathbf{L}(\mathbf{x}, t) \, d\boldsymbol{\beta}, \qquad (4.38)$$

where $\mathbf{x}(t) \in \mathbb{R}^D$ is the state, $\mathbf{f}(\mathbf{x}, t)$ is a given vector-valued nonlinear function, $\mathbf{L}(\mathbf{x}, t)$ a given matrix-valued function, and $\boldsymbol{\beta}(t) \in \mathbb{R}^S$ is a Brownian motion with diffusion matrix $\mathbf{Q}$. The initial condition is assumed to be given for $\mathbf{x}(t_0)$ as a distribution or fixed value, and it is assumed to be independent of the Brownian motion $\boldsymbol{\beta}$.

For this kind of nonlinear SDEs, we do not have any general method for finding the exact solution. However, one way to proceed is to try to determine the solution to the corresponding deterministic ODE first. Then one can attempt to apply the Itô formula to the deterministic solution to check if the solution also applies to the stochastic case. This approach indeed works for linear SDEs, and this is why we were able to correctly determine the solutions to linear SDEs in Chapter 3 even before we had the proper theory of SDEs at hand. However, more generally, it is hard to even solve nonhomogeneous deterministic ODEs in the nonlinear case, and hence this approach is hard to apply in practice.

The preceding approach is also a special case of an *ansatz* or educated guess–based approaches. That is, one way to solve a nonlinear SDE is to guess the solution and then use the Itô formula to check it. It is easy to see that this approach is quite limited. However, this approach can be used to determine solvable classes of the SDEs by applying the Itô formula to arbitrary functions of Brownian motions, but using the approach to solve a given SDE is hard. For a list of solution–SDE pairs, see, for example, Kloeden and Platen (1999).

Example 4.8 (Matching a nonlinear SDE to a solution). *Let us say that the solution to our SDE is*

$$x(t) = \sin(\omega \beta(t)), \qquad (4.39)$$

where $\omega > 0$ is a constant and $\beta(t)$ has the diffusion constant q. The Itô *formula then gives*

$$dx = \omega \cos(\omega \beta) \, d\beta - \frac{1}{2} q \, \omega^2 \sin(\omega \beta)$$

$$= \omega \sqrt{1 - x^2(t)} \, d\beta - \frac{1}{2} q \, \omega^2 x(t) \, dt. \qquad (4.40)$$

That is, if we had been given an SDE of the form

$$dx = -\frac{1}{2} q \, \omega^2 x(t) \, dt + \omega \sqrt{1 - x^2(t)} \, d\beta, \qquad x(0) = 0, \qquad (4.41)$$

we could start our solution attempt by using an ansatz $x(t) = \sin(\omega \, \beta(t))$. Further examples of solvable SDEs constructed this way can be found in the book of Kloeden and Platen (1999).

One quite generic approach for solving nonlinear SDEs is by change of variables. In that approach, we do a change of variables such that $y = h(x)$ for some nonlinear function h. The goal is that the SDE in terms of y should be easier to solve than the SDE in x. In practice, this usually means that we aim to transform the SDE into a linear SDE, which we then are able to solve. Furthermore, this approach mainly works for scalar SDEs.

As a generic class of scalar SDEs where the preceding approach can be used, Kloeden and Platen (1999) give

$$dx = \left(\alpha \, b(x) \, h(x) + \frac{1}{2} q \, b(x) \, \frac{db}{dx} \right) dt + b(x) \, d\beta, \quad x(0) = x_0, \quad (4.42)$$

where α is a constant, $b(x)$ is a given function, $\beta(t)$ is a Brownian motion with the diffusion coefficient q, and

$$h(x) = \int_\xi^x \frac{1}{b(u)} \, du, \qquad (4.43)$$

where ξ is an arbitrary point. The equation can now be solved by the change of variable $y = h(x)$. By using the Itô formula, we get that y should solve

$$dy = \alpha \, y \, dt + d\beta, \qquad (4.44)$$

and hence the final solution is

$$x(t) = h^{-1} \left(\exp(\alpha \, t) \, h(x_0) + \int_0^t \exp(\alpha \, (t - s)) \, d\beta(s) \right). \qquad (4.45)$$

Example 4.9 (Nonlinear scalar SDE). *The SDE*

$$dx = \left(\alpha \, x \, \log x + \frac{1}{2} x \right) dt + x \, d\beta, \quad x(0) = x_0, \qquad (4.46)$$

where $\beta(t)$ is a standard Brownian motion, is a special case of the previous equation with $b(x) = x$, and hence we get $h(x) = \log x$ and thus

$$x(t) = \exp\left(\exp(\alpha\, t)\, \log x_0 + \int_0^t \exp(\alpha\,(t - s))\, \mathrm{d}\beta(s)\right). \qquad (4.47)$$

It is also possible to use the integrating factor method to solve certain classes of nonlinear equations with multiplicative linear drift. For details of this approach, see Henderson and Plaschko (2006). Furthermore, some additional special cases of solvable SDEs can also be found, for example, in the books of Kloeden and Platen (1999) and Iacus (2008).

4.5 Existence and Uniqueness of Solutions

A solution to a stochastic differential equation is called *strong* if for a given Brownian motion $\boldsymbol{\beta}(t)$, it is possible to construct a solution $\mathbf{x}(t)$, which is driven by the given Brownian motion. A solution is called *weak* if it is possible to construct some Brownian motion $\tilde{\boldsymbol{\beta}}(t)$ and a stochastic process $\tilde{\mathbf{x}}(t)$ such that the pair is a solution to the stochastic differential equation. The existence of a strong solution always implies the existence of a weak solution (every strong solution is also a weak solution), but the converse is not true.

The solution is *strongly unique* if the whole path of the process is unique for a given Brownian motion. Hence strong uniqueness is also called pathwise uniqueness. *Weak uniqueness* means that the probability law of the solution is unique, that is, there cannot be two solutions with different finite-dimensional distributions. Determining if an equation has a unique weak solution when it does not have a unique strong solution is considerably harder than the criterion for the strong solution (see, e.g., Ikeda and Watanabe, 1981).

The strong uniqueness of a solution to an SDE of the general form

$$\mathrm{d}\mathbf{x} = \mathbf{f}(\mathbf{x}, t)\,\mathrm{d}t + \mathbf{L}(\mathbf{x}, t)\,\mathrm{d}\boldsymbol{\beta}, \quad \mathbf{x}(t_0) = \mathbf{x}_0, \qquad (4.48)$$

can be determined using the *stochastic Picard iteration*, which is a direct extension of the deterministic Picard iteration. Thus we first rewrite the equation in integral form

$$\mathbf{x}(t) = \mathbf{x}_0 + \int_{t_0}^t \mathbf{f}(\mathbf{x}(\tau), \tau)\,\mathrm{d}\tau + \int_{t_0}^t \mathbf{L}(\mathbf{x}(\tau), \tau)\,\mathrm{d}\boldsymbol{\beta}(\tau). \qquad (4.49)$$

Then the solution can be approximated with the following iteration.

Algorithm 4.10 (Stochastic Picard iteration). *Start from the initial guess* $\varphi_0(t) = \mathbf{x}_0$. *With a given $\boldsymbol{\beta}$, compute approximations $\varphi_1(t), \varphi_2(t), \ldots$ via the following recursion:*

$$\varphi_{n+1}(t) = \mathbf{x}_0 + \int_{t_0}^{t} \mathbf{f}(\varphi_n(\tau), \tau)\, d\tau + \int_{t_0}^{t} \mathbf{L}(\varphi_n(\tau), \tau)\, d\boldsymbol{\beta}(\tau). \quad (4.50)$$

It can be shown that this iteration converges to the exact solution in a mean squared sense if both of the functions $\mathbf{f}$ and $\mathbf{L}$ grow at most linearly in $\mathbf{x}$ and they are Lipschitz continuous in the same variable (see, e.g., Øksendal, 2003). If these conditions are met, there exists a unique strong solution to the SDE.

4.6 Stratonovich Calculus

It is also possible to define a stochastic integral in such a way that the chain rule from ordinary calculus is valid. The symmetrized stochastic integral or the *Stratonovich integral* (Stratonovich, 1968) can be defined as follows (Protter, 2013):

$$\int_{t_0}^{t} \mathbf{L}(\mathbf{x}(t), t) \circ d\boldsymbol{\beta}(t) = \lim_{n \to \infty} \sum_{k} \mathbf{L}(\mathbf{x}(t_k^*), t_k^*)\, [\boldsymbol{\beta}(t_{k+1}) - \boldsymbol{\beta}(t_k)], \quad (4.51)$$

where $t_0 < t_1 < \ldots < t_n = t$ and $t_k^* = (t_k + t_{k+1})/2$. The difference to the Itô integral is that we do not select the starting point of the interval as the evaluation point but the middle point. This ensures that the calculation rules of ordinary calculus apply. The disadvantage of the Stratonovich integral over the Itô integral is that the Stratonovich integral is not a martingale (we briefly discuss them later), which makes its theoretical analysis harder.

Stratonovich stochastic differential equations (Stratonovich, 1968; Øksendal, 2003; Protter, 2013) are similar to Itô differential equations, but instead of Itô integrals they involve stochastic integrals in the Stratonovich sense. To distinguish between Itô and Stratonovich stochastic differential equations, the Stratonovich integral is denoted by a small circle before the Brownian differential as follows:

$$d\mathbf{x} = \mathbf{f}(\mathbf{x}, t)\, dt + \mathbf{L}(\mathbf{x}, t) \circ d\boldsymbol{\beta}. \quad (4.52)$$

The white noise interpretation of SDEs naturally leads to stochastic differential equations in the Stratonovich sense. This is because, broadly speaking, discrete-time and smooth approximations of white noise–driven differential equations converge to stochastic differential equations in the Stratonovich sense, not in the Itô sense. However, this result of Wong and

Zakai (1965) is strictly true only for scalar SDEs, and thus this result should not be extrapolated too far.

A Stratonovich SDE can be converted into an equivalent Itô equation by using simple transformation formulas (Stratonovich, 1968; Øksendal, 2003). If the dispersion term is independent of the state $\mathbf{L}(\mathbf{x}, t) = \mathbf{L}(t)$, then the Itô and Stratonovich interpretations of the SDE are the same.

Algorithm 4.11 (Conversion of Stratonovich SDEs to Itô SDEs). *The following SDE in the Stratonovich sense*

$$d\mathbf{x} = \mathbf{f}(\mathbf{x}, t)\, dt + \mathbf{L}(\mathbf{x}, t) \circ d\boldsymbol{\beta}, \tag{4.53}$$

is equivalent to the following SDE in the Itô sense

$$d\mathbf{x} = \tilde{\mathbf{f}}(\mathbf{x}, t)\, dt + \mathbf{L}(\mathbf{x}, t)\, d\boldsymbol{\beta}, \tag{4.54}$$

where

$$\tilde{f}_i(\mathbf{x}, t) = f_i(\mathbf{x}, t) + \frac{1}{2} \sum_{j,l} \frac{\partial L_{ij}(\mathbf{x}, t)}{\partial x_l} L_{lj}(\mathbf{x}, t). \tag{4.55}$$

It is also possible to define other forms of stochastic integrals and calculi by evaluating the integrand at a point $t_k^* = (1 - \alpha)\, t_k + \alpha\, t_{k+1}$, where $\alpha \in [0, 1]$. With the selection $\alpha = 0$, we get the Itô calculus, and with $\alpha = 1/2$, we get the Stratonovich calculus as special cases. More details can be found, for example, in Protter (2013).

4.7 Exercises

4.1   The Itô formula:

    (a) Compute the Itô differential of $\phi(\beta, t) = t + \exp(\beta)$, where $\beta(t)$ is a Brownian motion with diffusion constant q.

    (b) Compute the Itô differential of $\phi(x) = x^2$, where x solves the scalar SDE

$$dx = f(x)\, dt + L\, d\beta,$$

    the parameter L is a constant, and $\beta(t)$ is a standard Brownian motion ($q = 1$).

    (c) Compute the Itô differential of $\phi(\mathbf{x}) = \mathbf{x}^\mathsf{T} \mathbf{x}$, where

$$d\mathbf{x} = \mathbf{F}\, \mathbf{x}\, dt + d\boldsymbol{\beta},$$

    where $\mathbf{F}$ is a constant matrix and the diffusion matrix of $\boldsymbol{\beta}$ is $\mathbf{Q}$.

4.2   Check that

$$x(t) = \exp(\beta(t))$$

solves the SDE

$$dx = \frac{1}{2} x \, dt + x \, d\beta,$$

where $\beta(t)$ is a standard Brownian motion ($q = 1$).

4.3   Consider the Black–Scholes model

$$dx = \mu \, x \, dt + \sigma \, x \, d\beta,$$

where μ and $\sigma > 0$ are constants, and $\beta(t)$ is a standard Brownian motion.

(a) Solve $x(t)$ by changing the variable to $y = \log x$ and transforming back.

(b) Notice that the solution is log-normal and compute its mean $m(t)$ and variance $P(t)$.

(c) Compare (and check) your expressions for the mean and variance against the numerical results obtained by the Euler–Maruyama method (cf. Exercise 3.4).

4.4   Find the solution to the mean-reverting Ornstein–Uhlenbeck process given in a parameterization favoured in financial applications (in finance, the model is known as the Vasicek model):

$$dx = \theta \, (\mu - x) \, dt + \sigma \, d\beta(t), \quad x(0) = x_0, \qquad (4.56)$$

where θ gives the speed of reversion, μ is the long-run equlibrium, σ stands for the volatility, and $\beta(t)$ is a standard Brownian motion with diffusion coefficient $q = 1$.

4.5   Derive the Black–Scholes solution given in Example 4.7 from the solution given for linear time-invariant models with multiplicative noise in Example 4.6.

4.6   Consider the nonlinear SDE:

$$dx = -\tanh(x) \left(a + \frac{1}{2} b^2 \, \mathrm{sech}^2(x) \right) dt + b \, \mathrm{sech}(x) \, d\beta,$$

where $\beta(t)$ is a standard Brownian motion. Solve it using the methods given for one-dimensional nonlinear SDEs in Section 4.4.

Hint: Try $h(x) = \sinh(x)$.

4.7   It is also possible to derive an integration by parts formula for Itô processes. Let $\beta(t)$ be a standard Brownian motion.

(a) By computing the Itô differential of $\phi(\beta) \zeta(\beta)$ and integrating the result, show that the following holds:

$$\int_0^t \left[\frac{\partial \phi(x)}{\partial x} \zeta(x) \right]_{x=\beta(\tau)} d\beta(\tau)$$

$$= \phi(\beta(t)) \zeta(\beta(t)) - \phi(0) \zeta(0) - \int_0^t \left[\phi(x) \frac{\partial \zeta(x)}{\partial x} \right]_{x=\beta(\tau)} d\beta(\tau)$$

$$- \frac{1}{2} \int_0^t \left[\frac{\partial^2 \phi(x)}{\partial x^2} \zeta(x) + 2 \frac{\partial \phi(x)}{\partial x} \frac{\partial \zeta(x)}{\partial x} + \phi(x) \frac{\partial^2 \zeta(x)}{\partial x^2} \right]_{x=\beta(\tau)} dt.$$

(b) Use this result to express the following integral in terms of ordinary integrals (*hint*: put $\zeta(x) = 1$):

$$\int_0^t \tanh(\beta) \, d\beta.$$

4.8   Convert the following Stratonovich SDE into the equivalent Itô SDE:

$$dx = \tanh(x) \, dt + x \circ d\beta,$$

where $\beta(t)$ is a scalar Brownian motion.

4.9   Convert the following Stratonovich SDE into the equivalent Itô SDE:

$$dx_1 = -x_2 \circ d\beta,$$
$$dx_2 = \quad x_1 \circ d\beta,$$

where $\beta(t)$ is a scalar Brownian motion.

5

Probability Distributions and Statistics of SDEs

The solutions of SDEs are stochastic processes, and therefore their solutions have certain probability distributions and statistics. In this chapter, we start by briefly discussing martingale properties of Itô processes (see, e.g., Rogers and Williams, 2000a,b, for the full theory) and discuss generators of SDEs that are related to statistics of SDEs. We then proceed to deriving the Fokker–Planck–Kolmogorov equation, which is a partial differential equation for the evolution of the probability density of the SDE solution. We also discuss the Markov properties and the transition densities of SDEs as well as derive the general equations for the means, covariances, and other moments of SDEs.

5.1 Martingale Properties and Generators of SDEs

In this section, we discuss martingale properties and generators of Itô processes, which are important concepts in theoretical analysis of SDEs. We denote the *history* of the Itô process $\mathbf{x}(t)$ up to the time t as

$$\mathcal{X}_t = \{\mathbf{x}(\tau) \mid 0 \leq \tau \leq t\}. \tag{5.1}$$

More formally, the history of an Itô process should not be defined through its explicit path, but via the *sigma-algebra* generated by it (see, e.g., Øksendal, 2003). The history as a function of increasing t is then an object called *filtration*, which means an increasing family of sigma-algebras. However, for pedagogical reasons we simply talk about the history of an Itô process.

The definition of a martingale is the following.

Definition 5.1 (Martingale). *A stochastic process* $\mathbf{x}(t)$ *with bounded expectation* $E[|\mathbf{x}(t)|] < \infty$ *is called a* martingale *if*

$$E[\mathbf{x}(t) \mid \mathcal{X}_s] = \mathbf{x}(s), \qquad \text{for all } t \geq s. \tag{5.2}$$

It turns out that all Itô integrals are martingales, and this follows from the fact that Brownian motion is a martingale as well. However, solutions to Itô SDEs are martingales only if the drift $\mathbf{f}(\mathbf{x}, t) = \mathbf{0}$. For more information on martingales and their role in stochastic calculus, the reader is referred to the books of Øksendal (2003), Karatzas and Shreve (1991), and Rogers and Williams (2000a,b).

Another useful concept in the theory of Itô processes and more general stochastic processes is the (infinitesimal) generator, which in the case of Itô processes is the following.

Definition 5.2 (Generator). *The (infinitesimal) generator of a stochastic process* $\mathbf{x}(t)$ *for a function* $\phi(\mathbf{x})$ *can be defined as*

$$\mathcal{A}\,\phi(\mathbf{x}) = \lim_{s \downarrow 0} \frac{\mathrm{E}[\phi(\mathbf{x}(t + s))] - \phi(\mathbf{x}(t))}{s}, \tag{5.3}$$

where $\phi(\bullet)$ *is a suitably regular function. For an Itô process defined as the solution to the SDE*

$$\mathrm{d}\mathbf{x} = \mathbf{f}(\mathbf{x}, t)\,\mathrm{d}t + \mathbf{L}(\mathbf{x}, t)\,\mathrm{d}\boldsymbol{\beta}, \tag{5.4}$$

the generator (operating on some function $\bullet$*) is given as*

$$\mathcal{A}(\bullet) = \sum_i \frac{\partial(\bullet)}{\partial x_i} f_i(\mathbf{x}, t)$$

$$+ \frac{1}{2} \sum_{i,j} \left(\frac{\partial^2(\bullet)}{\partial x_i \partial x_j} \right) [\mathbf{L}(\mathbf{x}, t)\,\mathbf{Q}\,\mathbf{L}^\mathsf{T}(\mathbf{x}, t)]_{ij}. \tag{5.5}$$

For a time-invariant stochastic process (such as a time-invariant SDE), we can also equivalently define

$$\mathcal{A}\,\phi(\mathbf{x}) = \lim_{t \downarrow 0} \frac{\mathrm{E}[\phi(\mathbf{x}(t))] - \phi(\mathbf{x}(0))}{t}, \tag{5.6}$$

which is the classical definition and equivalent when the SDE does not explicitly depend on time (i.e., when $\mathbf{f}(\mathbf{x})$ *and* $\mathbf{L}(\mathbf{x})$ *only depend on* $\mathbf{x}$*).*

We can also define a similar operator for time-dependent functions $\phi(\mathbf{x}, t)$ as follows. Because generators are not classically defined this way, we call the operator the *generalized generator*.

Definition 5.3 (Generalized generator). *The* generalized *(infinitesimal)* generator *for a time-dependent process for a time-dependent function*

$\phi(\mathbf{x}, t)$ *can be defined as*

$$\mathscr{A}_t\, \phi(\mathbf{x}, t) = \lim_{s\downarrow 0} \frac{E[\phi(\mathbf{x}(t+s), t+s)] - \phi(\mathbf{x}(t), t)}{s}, \qquad (5.7)$$

which for the solution of the time-dependent SDE

$$\mathrm{d}\mathbf{x} = \mathbf{f}(\mathbf{x}, t)\, \mathrm{d}t + \mathbf{L}(\mathbf{x}, t)\, \mathrm{d}\boldsymbol{\beta} \qquad (5.8)$$

is given as (note the appearance of the time derivative in the beginning)

$$\mathscr{A}_t(\bullet) = \frac{\partial(\bullet)}{\partial t} + \sum_i \frac{\partial(\bullet)}{\partial x_i}\, f_i(\mathbf{x}, t)$$

$$+ \frac{1}{2} \sum_{i,j} \left(\frac{\partial^2(\bullet)}{\partial x_i\, \partial x_j} \right) [\mathbf{L}(\mathbf{x}, t)\, \mathbf{Q}\, \mathbf{L}^{\mathsf{T}}(\mathbf{x}, t)]_{ij}. \quad (5.9)$$

5.2 Fokker–Planck–Kolmogorov Equation

In this section, we derive the equation for the probability density of an Itô process $\mathbf{x}(t)$, when the process is defined as the solution to the SDE

$$\mathrm{d}\mathbf{x} = \mathbf{f}(\mathbf{x}, t)\, \mathrm{d}t + \mathbf{L}(\mathbf{x}, t)\, \mathrm{d}\boldsymbol{\beta}, \qquad (5.10)$$

where $\mathbf{x}(t) \in \mathbb{R}^D$ is the state, $\mathbf{f}(\mathbf{x}, t)$ is a given vector-valued function, $\mathbf{L}(\mathbf{x}, t)$ is a given matrix-valued function, and $\boldsymbol{\beta}(t) \in \mathbb{R}^S$ is a Brownian motion with diffusion matrix $\mathbf{Q}$. The initial condition is $\mathbf{x}(t_0) \sim p(\mathbf{x}(t_0))$, and it is assumed to be independent of the Brownian motion.

The corresponding probability density is usually denoted as $p(\mathbf{x}(t))$, but in this section, to emphasize that the density is actually a function of both $\mathbf{x}$ and t, we will occasionally write it as $p(\mathbf{x}, t)$.

Theorem 5.4 (Fokker–Planck–Kolmogorov equation). *The probability density $p(\mathbf{x}, t)$ of the solution of the SDE in Equation (5.10) solves the partial differential equation (PDE):*

$$\frac{\partial p(\mathbf{x}, t)}{\partial t} = -\sum_i \frac{\partial}{\partial x_i}[f_i(\mathbf{x}, t)\, p(\mathbf{x}, t)]$$

$$+ \frac{1}{2} \sum_{i,j} \frac{\partial^2}{\partial x_i\, \partial x_j} \left\{ [\mathbf{L}(\mathbf{x}, t)\, \mathbf{Q}\, \mathbf{L}^{\mathsf{T}}(\mathbf{x}, t)]_{ij}\, p(\mathbf{x}, t) \right\}. \quad (5.11)$$

This PDE is here called the Fokker–Planck–Kolmogorov (FPK) equation. In physics literature, it is often called the Fokker–Planck equation, and in stochastics it is the forward Kolmogorov equation, hence the name. With the probability density $p(\mathbf{x}(t_0)) \triangleq p(\mathbf{x}, t_0)$ on the state initial state $\mathbf{x}(t_0)$, the PDE is an initial value problem with the given initial condition $p(\mathbf{x}, t_0)$ at time $t = t_0$.

Proof   Let $\phi(\mathbf{x})$ be an arbitrary twice differentiable function. The Itô differential of $\phi(\mathbf{x}(t))$ is, by the Itô formula, given as follows:

$$
\begin{aligned}
d\phi &= \sum_i \frac{\partial \phi}{\partial x_i} \, dx_i + \frac{1}{2} \sum_{i,j} \left(\frac{\partial^2 \phi}{\partial x_i \partial x_j} \right) dx_i \, dx_j \\
&= \sum_i \frac{\partial \phi}{\partial x_i} \, f_i(\mathbf{x}, t) \, dt + \sum_i \frac{\partial \phi}{\partial x_i} \, [\mathbf{L}(\mathbf{x}, t) \, d\boldsymbol{\beta}]_i \\
&\quad + \frac{1}{2} \sum_{i,j} \left(\frac{\partial^2 \phi}{\partial x_i \partial x_j} \right) [\mathbf{L}(\mathbf{x}, t) \, \mathbf{Q} \, \mathbf{L}^{\mathsf{T}}(\mathbf{x}, t)]_{ij} \, dt.
\end{aligned} \tag{5.12}
$$

Taking the expectation of both sides with respect to $\mathbf{x}$ and formally dividing by dt gives the following:

$$
\begin{aligned}
\frac{d \, \mathrm{E}[\phi]}{dt} &= \sum_i \mathrm{E} \left[\frac{\partial \phi}{\partial x_i} \, f_i(\mathbf{x}, t) \right] \\
&\quad + \frac{1}{2} \sum_{i,j} \mathrm{E} \left[\left(\frac{\partial^2 \phi}{\partial x_i \partial x_j} \right) [\mathbf{L}(\mathbf{x}, t) \, \mathbf{Q} \, \mathbf{L}^{\mathsf{T}}(\mathbf{x}, t)]_{ij} \right].
\end{aligned} \tag{5.13}
$$

The left-hand side can now be written as

$$
\begin{aligned}
\frac{d \, \mathrm{E}[\phi]}{dt} &= \frac{d}{dt} \int \phi(\mathbf{x}) \, p(\mathbf{x}, t) \, d\mathbf{x} \\
&= \int \phi(\mathbf{x}) \, \frac{\partial p(\mathbf{x}, t)}{\partial t} \, d\mathbf{x}.
\end{aligned} \tag{5.14}
$$

Recall the multidimensional integration by parts formula

$$
\int_\Omega \frac{\partial u(\mathbf{x})}{\partial x_i} \, v(\mathbf{x}) \, d\mathbf{x} = \int_{\partial \Omega} u(\mathbf{x}) \, v(\mathbf{x}) \, n_i \, dS - \int_\Omega u(\mathbf{x}) \, \frac{\partial v(\mathbf{x})}{\partial x_i} \, d\mathbf{x}, \tag{5.15}
$$

where $\mathbf{n}$ is the outer normal of the boundary $\partial \Omega$ of integration domain Ω and dS is its area element. As the integration area is whole $\mathbb{R}^D$, the boundary term on the right-hand side should vanish and the formula becomes

$$
\int \frac{\partial u(\mathbf{x})}{\partial x_i} \, v(\mathbf{x}) \, d\mathbf{x} = -\int u(\mathbf{x}) \, \frac{\partial v(\mathbf{x})}{\partial x_i} \, d\mathbf{x}. \tag{5.16}
$$

For the first term on the right-hand side of Equation (5.13), we now get

$$\mathrm{E}\left[\frac{\partial \phi}{\partial x_i} f_i(\mathbf{x}, t)\right] = \int \frac{\partial \phi}{\partial x_i} f_i(\mathbf{x}, t)\, p(\mathbf{x}, t)\, d\mathbf{x}$$

$$= -\int \phi(\mathbf{x}) \frac{\partial}{\partial x_i}[f_i(\mathbf{x}, t)\, p(\mathbf{x}, t)]\, d\mathbf{x}, \qquad (5.17)$$

where we have used the integration by parts formula with $u(\mathbf{x}) = \phi(\mathbf{x})$ and $v(\mathbf{x}) = f_i(\mathbf{x}, t)\, p(\mathbf{x}, t)$. For the second term, we get the following:

$$\mathrm{E}\left[\left(\frac{\partial^2 \phi}{\partial x_i \partial x_j}\right)[\mathbf{L}(\mathbf{x}, t)\, \mathbf{Q}\, \mathbf{L}^{\mathsf{T}}(\mathbf{x}, t)]_{ij}\right]$$

$$= \int \left(\frac{\partial^2 \phi}{\partial x_i \partial x_j}\right)[\mathbf{L}(\mathbf{x}, t)\, \mathbf{Q}\, \mathbf{L}^{\mathsf{T}}(\mathbf{x}, t)]_{ij}\, p(\mathbf{x}, t)\, d\mathbf{x}$$

$$= -\int \left(\frac{\partial \phi}{\partial x_j}\right)\frac{\partial}{\partial x_i}\left\{[\mathbf{L}(\mathbf{x}, t)\, \mathbf{Q}\, \mathbf{L}^{\mathsf{T}}(\mathbf{x}, t)]_{ij}\, p(\mathbf{x}, t)\right\} d\mathbf{x}$$

$$= \int \phi(\mathbf{x}) \frac{\partial^2}{\partial x_i\, \partial x_j}\left\{[\mathbf{L}(\mathbf{x}, t)\, \mathbf{Q}\, \mathbf{L}^{\mathsf{T}}(\mathbf{x}, t)]_{ij}\, p(\mathbf{x}, t)\right\} d\mathbf{x}, \qquad (5.18)$$

where we have first used the integration by parts formula with $u(\mathbf{x}) = \partial \phi(\mathbf{x})/\partial x_j$ and $v(\mathbf{x}) = [\mathbf{L}(\mathbf{x}, t)\, \mathbf{Q}\, \mathbf{L}^{\mathsf{T}}(\mathbf{x}, t)]_{ij}\, p(\mathbf{x}, t)$, and then again with $u(\mathbf{x}) = \phi(\mathbf{x})$ and $v(\mathbf{x}) = \frac{\partial}{\partial x_i}\{[\mathbf{L}(\mathbf{x}, t)\, \mathbf{Q}\, \mathbf{L}^{\mathsf{T}}(\mathbf{x}, t)]_{ij}\, p(\mathbf{x}, t)\}$.

If we substitute Equations (5.14), (5.17), and (5.18) into (5.13), we get

$$\int \phi(\mathbf{x}) \frac{\partial p(\mathbf{x}, t)}{\partial t}\, d\mathbf{x} = -\sum_i \int \phi(\mathbf{x}) \frac{\partial}{\partial x_i}[f_i(\mathbf{x}, t)\, p(\mathbf{x}, t)]\, d\mathbf{x}$$

$$+ \frac{1}{2}\sum_{i, j} \int \phi(\mathbf{x}) \frac{\partial^2}{\partial x_i\, \partial x_j}\left\{[\mathbf{L}(\mathbf{x}, t)\, \mathbf{Q}\, \mathbf{L}^{\mathsf{T}}(\mathbf{x}, t)]_{ij}\, p(\mathbf{x}, t)\right\} d\mathbf{x}, \qquad (5.19)$$

which can also be written as

$$\int \phi(\mathbf{x})\left[\frac{\partial p(\mathbf{x}, t)}{\partial t} + \sum_i \frac{\partial}{\partial x_i}[f_i(\mathbf{x}, t)\, p(\mathbf{x}, t)]\right.$$

$$\left. - \frac{1}{2}\sum_{i, j} \frac{\partial^2}{\partial x_i\, \partial x_j}\left\{[\mathbf{L}(\mathbf{x}, t)\, \mathbf{Q}\, \mathbf{L}^{\mathsf{T}}(\mathbf{x}, t)]_{ij}\, p(\mathbf{x}, t)\right\}\right] d\mathbf{x} = 0. \qquad (5.20)$$

The only way this equation can be true for an arbitrary $\phi(\mathbf{x})$ is if the term in the brackets vanishes, which gives the FPK equation. □

Example 5.5 (Diffusion equation). *In Example 3.1, we derived the diffusion equation by considering random Brownian movement occurring during small time intervals. Note that Brownian motion can be defined as a* *solution to the SDE*

$$\mathrm{d}x = \mathrm{d}\beta. \tag{5.21}$$

If we set the diffusion constant of the Brownian motion to be $q = 2\,D$, then the FPK reduces to

$$\frac{\partial p}{\partial t} = D\,\frac{\partial^2 p}{\partial x^2}, \tag{5.22}$$

which is the same result as in Equation (3.7).

When the SDE is time independent, that is, has the form

$$\mathrm{d}\mathbf{x} = \mathbf{f}(\mathbf{x})\,\mathrm{d}t + \mathbf{L}(\mathbf{x})\,\mathrm{d}\boldsymbol{\beta}, \tag{5.23}$$

the solution to the FPK often converges to a stationary solution that satisfies $\partial p(\mathbf{x}, t)/\partial t = 0$. If we denote this stationary solution as $p(\mathbf{x})$, then it should be the solution to the stationary FPK equation

$$-\sum_i \frac{\partial}{\partial x_i}[f_i(\mathbf{x})\,p(\mathbf{x})]$$

$$+ \frac{1}{2}\sum_{i,j} \frac{\partial^2}{\partial x_i\,\partial x_j}\{[\mathbf{L}(\mathbf{x})\,\mathbf{Q}\,\mathbf{L}^{\mathsf{T}}(\mathbf{x})]_{ij}\,p(\mathbf{x})\} = 0. \tag{5.24}$$

Even though in general solving this equation is hard, solving the stationary probability density $p(\mathbf{x})$ is tractable if the SDE happens to have the special form

$$\mathrm{d}\mathbf{x} = -\frac{1}{2}\nabla v(\mathbf{x})\,\mathrm{d}t + \mathrm{d}\boldsymbol{\beta}, \tag{5.25}$$

where $v(\mathbf{x})$ is a potential function such that the drift function $\mathbf{f}(\mathbf{x}) = -\nabla v(\mathbf{x})$ and $\boldsymbol{\beta}(t)$ is a Brownian motion with diffusion matrix $\mathbf{Q} = q\,\mathbf{I}$ for some $q > 0$. We have the following theorem.

Theorem 5.6 (Stationary solution to FPK). *The stationary solution to the FPK equation corresponding to (5.25) is given as*

$$p(\mathbf{x}) = \frac{1}{Z}\,\exp(-v(\mathbf{x})/q), \tag{5.26}$$

where the normalization constant is

$$Z = \int \exp(-v(\mathbf{x})/q)\,\mathrm{d}\mathbf{x}. \tag{5.27}$$

For more general stationary FPK solutions the reader is referred to the book by Gardiner (2004).

Example 5.7 (Stationary distribution of the Ornstein–Uhlenbeck process).
*The Ornstein–Uhlenbeck process (see Example 4.5) has the form (5.25)
with $v(x) = \lambda\,x^2$. Hence the stationary solution should have the form*

$$p(x) \propto \exp\left(-\frac{\lambda\,x^2}{q}\right), \tag{5.28}$$

*which is the probability density of a normal distribution with zero mean
and variance $q/2\lambda$.*

5.3 Operator Formulation of the FPK Equation

In this section, we discuss the derivation and interpretation of the FPK
equation using operator theory. This operator point of view allows us to
gain insight on what actually happened in the FPK derivation in the previous section. Let us define the L_2 inner product between two functions ϕ
and φ as follows:

$$\langle \phi, \varphi \rangle = \int \phi(\mathbf{x})\,\varphi(\mathbf{x})\,\mathrm{d}\mathbf{x}. \tag{5.29}$$

The expectation of a function $\phi(\mathbf{x}(t))$ can now be written in terms of the
inner product as

$$\mathrm{E}[\phi(\mathbf{x}(t))] = \langle \phi, p \rangle, \tag{5.30}$$

where $p = p(\mathbf{x}, t)$. This also means that Equation (5.13), which was derived from the Itô formula, can be compactly written as

$$\frac{\mathrm{d}}{\mathrm{d}t}\langle \phi, p \rangle = \langle \mathcal{A}\,\phi, p \rangle, \tag{5.31}$$

where $\mathcal{A}$ is the generator defined in Equation (5.5).

Recall that the (formal) adjoint of an operator $\mathcal{A}$ – with respect to the
given inner product – is defined to be an operator $\mathcal{A}^*$ such that for all ϕ
and φ we have

$$\langle \mathcal{A}\,\phi, \varphi \rangle = \langle \phi, \mathcal{A}^*\,\varphi \rangle. \tag{5.32}$$

Using the adjoint operator, we can now write Equation (5.31) as

$$\frac{\mathrm{d}}{\mathrm{d}t}\langle \phi, p \rangle = \langle \phi, \mathcal{A}^*\,p \rangle. \tag{5.33}$$

As ϕ is independent of time, this can also be written as

$$\langle \phi, \frac{\partial p}{\partial t} \rangle = \langle \phi, \mathcal{A}^*\,p \rangle. \tag{5.34}$$

Because ϕ can be arbitrary, the preceding can only be true if in fact

$$\frac{\partial p}{\partial t} = \mathcal{A}^* \, p. \qquad (5.35)$$

It now turns out that the adjoint operator $\mathcal{A}^*$ is

$$\mathcal{A}^*(\bullet) = -\sum_i \frac{\partial}{\partial x_i} [f_i(\mathbf{x}, t) (\bullet)]$$

$$+ \frac{1}{2} \sum_{i,j} \frac{\partial^2}{\partial x_i \, \partial x_j} \{[\mathbf{L}(\mathbf{x}, t) \, \mathbf{Q} \, \mathbf{L}^\mathsf{T}(\mathbf{x}, t)]_{ij} (\bullet)\}, \quad (5.36)$$

which is exactly the operator appearing in the Fokker–Planck–Kolmogorov equation in Theorem 5.4 and hence Equation (5.35) is in fact the Fokker–Planck–Kolmogorov equation. In the previous section, we used brute-force integration by parts to derive the adjoint of the operator $\mathcal{A}$. We could also have used the properties of the adjoints directly as is illustrated in the following example.

Example 5.8 (Operator adjoint derivation of FPK). *Let us consider a one-dimensional SDE*

$$\mathrm{d}x = f(x) \, \mathrm{d}t + L(x) \, \mathrm{d}\beta, \qquad (5.37)$$

in which case the operator $\mathcal{A}$ takes the form

$$\mathcal{A} = f(x) \frac{\partial}{\partial x} + \frac{1}{2} L^2(x) \, q \, \frac{\partial^2}{\partial x^2}. \qquad (5.38)$$

Now recall the following L_2 adjoint computation rules:

- *The operation of multiplication with a function $f(x)$ is its own adjoint (i.e., the operator is self-adjoint).*
- *The operation of differentiation obeys $\left(\frac{\partial}{\partial x}\right)^* = -\frac{\partial}{\partial x}$ and hence the second derivative operator is self-adjoint $\left(\frac{\partial^2}{\partial x^2}\right)^* = \frac{\partial^2}{\partial x^2}$.*
- *The adjoint of a sum is $(\mathcal{A}_1 + \mathcal{A}_2)^* = \mathcal{A}_1^* + \mathcal{A}_2^*$ and the product of two operators is $(\mathcal{A}_1 \, \mathcal{A}_2)^* = \mathcal{A}_2^* \, \mathcal{A}_1^*$.*

Thus we get

$$
\begin{aligned}
\langle \mathcal{A}\phi, p \rangle &= \langle \left[f(x)\frac{\partial}{\partial x} + \frac{1}{2}L^2(x)\,q\,\frac{\partial^2}{\partial x^2} \right]\phi, p \rangle \\
&= \langle f(x)\frac{\partial}{\partial x}\phi, p \rangle + \frac{1}{2}q\,\langle L^2(x)\frac{\partial^2}{\partial x^2}\phi, p \rangle \\
&= \langle \frac{\partial}{\partial x}\phi, f(x)\,p \rangle + \frac{1}{2}q\,\langle \frac{\partial^2}{\partial x^2}\phi, L^2(x)\,p \rangle \\
&= \langle \phi, -\frac{\partial}{\partial x}f(x)\,p \rangle + \frac{1}{2}q\,\langle \phi, \frac{\partial^2}{\partial x^2}L^2(x)\,p \rangle \\
&= \langle \phi, -\frac{\partial}{\partial x}f(x)\,p + \frac{1}{2}q\,\frac{\partial^2}{\partial x^2}L^2(x)\,p \rangle, \tag{5.39}
\end{aligned}
$$

where we can thus recover the adjoint operator

$$
\mathcal{A}^*(\bullet) = -\frac{\partial}{\partial x}\left[f(x)\,(\bullet) \right] + \frac{1}{2}q\,\frac{\partial^2}{\partial x^2}\left[L^2(x)\,(\bullet) \right]. \tag{5.40}
$$

Note that Equation (5.35) resembles a linear differential equation of the form $d\mathbf{x}(t)/dt = \mathbf{A}\,\mathbf{x}(t)$, which we can easily solve as $\mathbf{x}(t) = \exp(\mathbf{A}\,(t - t_0))\,\mathbf{x}(t_0)$, where $\exp(\bullet)$ is the matrix exponential. Analogously, when $\mathcal{A}^*$ does not explicitly depend on time we can (at least formally) express the solution to the FPK as

$$
p(\bullet, t) = \exp(\mathcal{A}^*\,(t - t_0))\,p(\bullet, t_0), \tag{5.41}
$$

where $\exp(\bullet)$ is now the exponential of an operator. The operator $\exp(\mathcal{A}^*\,t)$ is actually the semigroup generated by the operator $\mathcal{A}^*$, which can be rigorously defined. The kernel of the semigroup operator is sometimes called the propagator in physics.

5.4 Markov Properties and Transition Densities of SDEs

In this section, we cover the Markov property of Itô processes and the corresponding transition kernels. The definition of a *Markov process* is the following.

Definition 5.9 (Markov process). *A stochastic process* $\mathbf{x}(t)$ *is a* Markov *process if its future is independent of its past given the present:*

$$
p(\mathbf{x}(t) \mid \mathcal{X}_s) = p(\mathbf{x}(t) \mid \mathbf{x}(s)), \quad \text{for all } t \geq s. \tag{5.42}
$$

It turns out that all Itô processes, that is, solutions to Itô stochastic differential equations, are Markov processes. The proof of this can be found, for example, in Øksendal (2003, theorem 7.1.2). This means that all Itô processes are, in a probabilistic sense, completely characterized by the transition densities $p(\mathbf{x}(t) \mid \mathbf{x}(s))$. The transition density is also a solution to the Fokker–Planck–Kolmogorov equation with a degenerate (Dirac delta) initial density concentrated on $\mathbf{x}(s)$ at time s.

Theorem 5.10 (Transition density of an SDE). *The transition density* $p(\mathbf{x}(t) \mid \mathbf{x}(s))$ *of the SDE* (5.10), *where* $t \geq s$, *is the solution to the Fokker–Planck–Kolmogorov equation* (5.11) *with the initial condition* $p(\mathbf{x}(t) \mid \mathbf{x}(s)) = \delta(\mathbf{x}(t) - \mathbf{x}(s))$ *at* $t = s$. *More explicitly, if we denote the transition density from* $\mathbf{y}(s)$ *to* $\mathbf{x}(t)$ *as* $p(\mathbf{x}, t \mid \mathbf{y}, s)$, *then it solves*

$$\frac{\partial p(\mathbf{x}, t \mid \mathbf{y}, s)}{\partial t} = \mathcal{A}^* p(\mathbf{x}, t \mid \mathbf{y}, s), \quad p(\mathbf{x}, s \mid \mathbf{y}, s) = \delta(\mathbf{x} - \mathbf{y}), \quad (5.43)$$

where $\mathcal{A}^*$ *was defined in Equation* (5.36).

The transition density also satisfies another equation, the backward Kolmogorov equation (recall that FPK is also called the forward Kolmogorov equation).

Theorem 5.11 (Backward Kolmogorov equation). *If we denote the transition density from* $\mathbf{x}(s)$ *to* $\mathbf{y}(t)$ *as* $p(\mathbf{y}, t \mid \mathbf{x}, s)$, *then it solves the backward Kolmogorov equation*

$$-\frac{\partial p(\mathbf{y}, t \mid \mathbf{x}, s)}{\partial s} = \mathcal{A} p(\mathbf{y}, t \mid \mathbf{x}, s), \quad p(\mathbf{y}, s \mid \mathbf{x}, s) = \delta(\mathbf{y} - \mathbf{x}), \quad (5.44)$$

where $\mathcal{A}$ *is the generator defined in Equation* (5.5).

Once we know the transition densities of an SDE, we can also use Markov properties to form an explicit formula for the *finite-dimensional distributions* of the SDE.

Remark 5.12 (Finite-dimensional distributions of SDEs). *For an arbitrary finite set of time indices* $t_0 < t_1 < \ldots < t_T$, *the joint distribution of the values of the process (i.e., the finite-dimensional distribution) is characterized by the probability density*

$$p(\mathbf{x}(t_0), \mathbf{x}(t_1), \ldots, \mathbf{x}(t_T)) = p(\mathbf{x}(t_0)) \prod_{k=1}^{T} p(\mathbf{x}(t_k) \mid \mathbf{x}(t_{k-1})). \quad (5.45)$$

The preceding result is important in Bayesian filtering theory (Särkkä, 2013), because it states that a Bayesian filtering problem on an SDE model with discrete time measurements can always be converted into an equivalent discrete-time Bayesian filtering problem. We will return to this in the later chapters.

Remark 5.13 (Chapman–Kolmogorov equation). *The Markov property also implies that the transition densities have the following group property. For any three time instants* $t_1 < t_2 < t_3$, *we have the Chapman–Kolmogorov equation*

$$p(\mathbf{x}(t_3) \mid \mathbf{x}(t_1)) = \int p(\mathbf{x}(t_3) \mid \mathbf{x}(t_2))\, p(\mathbf{x}(t_2) \mid \mathbf{x}(t_1))\, d\mathbf{x}(t_2). \quad (5.46)$$

Although the Chapman–Kolmogorov equation follows from the FPK equation, it is also possible to derive the FPK equation from it as was done, for example, in Jazwinski (1970). Thus, in that sense, they are equivalent.

5.5 Means and Covariances of SDEs

In Section 5.2, we derived the FPK equation, which, in principle, is the complete probabilistic description of the state. The mean, covariance, and other moments of the state distribution can be derived from its solution. However, we are often primarily interested in the mean and covariance of the distribution and would like to avoid solving the FPK equation as an intermediate step.

If we take a look at Equation (5.13) in Section 5.2, we can see that it can be interpreted as an equation for the general moments of the state distribution. This equation can be generalized to time-dependent $\phi(\mathbf{x}, t)$ by including the time derivative (i.e., by replacing the generator on the right-hand side with generalized generator in Definition 5.3):

$$\frac{d\, \mathrm{E}[\phi]}{dt} = \mathrm{E}\left[\frac{\partial \phi}{\partial t}\right] + \sum_i \mathrm{E}\left[\frac{\partial \phi}{\partial x_i} f_i(\mathbf{x}, t)\right]$$
$$+ \frac{1}{2} \sum_{i,j} \mathrm{E}\left[\left(\frac{\partial^2 \phi}{\partial x_i \partial x_j}\right) [\mathbf{L}(\mathbf{x}, t)\, \mathbf{Q}\, \mathbf{L}^{\mathsf{T}}(\mathbf{x}, t)]_{ij}\right]. \quad (5.47)$$

If we select the function as $\phi(\mathbf{x}, t) = x_u$ (the uth component of $\mathbf{x}$), then the Equation (5.47) reduces to

$$\frac{d\, \mathrm{E}[x_u]}{dt} = \mathrm{E}\left[f_u(\mathbf{x}, t)\right], \quad (5.48)$$

which can be seen as the differential equation for the components of the mean of the state. If we denote the mean function as $\mathbf{m}(t) = \mathrm{E}[\mathbf{x}(t)]$ and select the function as $\phi(\mathbf{x}, t) = x_u x_v - m_u(t) m_v(t)$, then Equation (5.47) gives

$$
\frac{\mathrm{d}\,\mathrm{E}[x_u x_v - m_u(t) m_v(t)]}{\mathrm{d}t}
$$
$$
= \mathrm{E}\left[(x_v - m_v(t))\, f_u(\mathbf{x}, t)\right] + \mathrm{E}\left[(x_u - m_u(v))\, f_v(\mathbf{x}, t)\right]
$$
$$
+ [\mathbf{L}(\mathbf{x}, t)\,\mathbf{Q}\,\mathbf{L}^{\mathsf{T}}(\mathbf{x}, t)]_{uv}. \quad (5.49)
$$

If we denote the covariance as $\mathbf{P}(t) = \mathrm{E}[(\mathbf{x}(t) - \mathbf{m}(t))\,(\mathbf{x}(t) - \mathbf{m}(t))^{\mathsf{T}}]$, then Equations (5.48) and (5.49) can be written in the following matrix form:

$$
\frac{\mathrm{d}\mathbf{m}}{\mathrm{d}t} = \mathrm{E}\left[\mathbf{f}(\mathbf{x}, t)\right], \quad (5.50)
$$
$$
\frac{\mathrm{d}\mathbf{P}}{\mathrm{d}t} = \mathrm{E}\left[\mathbf{f}(\mathbf{x}, t)\,(\mathbf{x} - \mathbf{m})^{\mathsf{T}}\right] + \mathrm{E}\left[(\mathbf{x} - \mathbf{m})\,\mathbf{f}^{\mathsf{T}}(\mathbf{x}, t)\right]
$$
$$
+ \mathrm{E}\left[\mathbf{L}(\mathbf{x}, t)\,\mathbf{Q}\,\mathbf{L}^{\mathsf{T}}(\mathbf{x}, t)\right], \quad (5.51)
$$

which are the differential equations for the mean and covariance of the state. However, these equations cannot be used in practice as such, because the expectations should be taken with respect to the actual distribution of the state – which is given by the solution to the FPK equation. Only in the Gaussian case do the first two moments actually characterize the solution. Even though we cannot use these equations as such in the nonlinear case, they provide a useful starting point for forming Gaussian approximations to SDEs.

It is worth noting that because $\mathrm{E}[\mathrm{E}[\mathbf{f}(\mathbf{x}, t)]\,(\mathbf{x} - \mathbf{m})^{\mathsf{T}}] = \mathbf{0}$, it follows that

$$
\mathrm{E}\left[\mathbf{f}(\mathbf{x}, t)\,(\mathbf{x} - \mathbf{m})^{\mathsf{T}}\right] = \mathrm{E}\left[(\mathbf{f}(\mathbf{x}, t) - \mathrm{E}[\mathbf{f}(\mathbf{x}, t)])\,(\mathbf{x} - \mathbf{m})^{\mathsf{T}}\right], \quad (5.52)
$$

and hence the covariance equation can be equivalently written as

$$
\frac{\mathrm{d}\mathbf{P}}{\mathrm{d}t} = \mathrm{E}\left[(\mathbf{f}(\mathbf{x}, t) - \mathrm{E}[\mathbf{f}(\mathbf{x}, t)])\,(\mathbf{x} - \mathbf{m})^{\mathsf{T}}\right]
$$
$$
+ \mathrm{E}\left[(\mathbf{x} - \mathbf{m})\,(\mathbf{f}(\mathbf{x}, t) - \mathrm{E}[\mathbf{f}(\mathbf{x}, t)])^{\mathsf{T}}\right]
$$
$$
+ \mathrm{E}\left[\mathbf{L}(\mathbf{x}, t)\,\mathbf{Q}\,\mathbf{L}^{\mathsf{T}}(\mathbf{x}, t)\right], \quad (5.53)
$$

which is useful, for example, when deriving the equations for linear SDEs.

Example 5.14 (Moments of the Ornstein–Uhlenbeck process). *Let us* again consider the Ornstein–Uhlenbeck process, which we solved in Example 4.5:

$$dx = -\lambda x \, dt + d\beta, \qquad x(0) = x_0, \tag{5.54}$$

where $\lambda > 0$ and $\beta(t)$ is a Brownian motion with diffusion constant q. We have $f(x) = -\lambda x$ and thus

$$\mathrm{E}[f(x)] = -\lambda \, \mathrm{E}[x] = -\lambda \, m,$$
$$\mathrm{E}[f(x)(x - m)] = \mathrm{E}[-\lambda x (x - m)] = -\lambda \, \mathrm{E}[(x - m)^2] = -\lambda \, P. \tag{5.55}$$

The differential equations for the mean and variance are thus given as

$$\frac{dm}{dt} = -\lambda \, m,$$
$$\frac{dP}{dt} = -2\lambda \, P + q, \tag{5.56}$$

with the initial conditions $m(0) = x_0$, $P(0) = 0$. Because the solution of the Ornstein–Uhlenbeck process is a Gaussian process, these first two moments characterize the whole state distribution, which has the density

$$p(x,t) \triangleq p(x(t)) = \mathrm{N}(x(t) \mid m(t), P(t)). \tag{5.57}$$

As discussed previously, in the case of a nonlinear SDE it often happens that the mean and covariance differential equations are not closed in the sense that only with knowledge of the mean and covariance it is not possible to compute the expectations on the right-hand side of the equations. An example of that kind of SDE is given in the following example.

Example 5.15 (Moments of the sine diffusion model). *An example of a nonlinear SDE, which is often used for demonstrating numerical methods for SDEs, is*

$$dx = \sin(x) \, dt + d\beta, \tag{5.58}$$

where $\beta(t)$ is a Brownian motion with diffusion constant q. The mean and variance equations are now given as

$$\frac{dm}{dt} = \mathrm{E}[\sin(x)],$$
$$\frac{dP}{dt} = 2 \, \mathrm{E}[\sin(x)(x - m)] + q. \tag{5.59}$$

With knowledge of only the mean and variance of x, it is impossible to *compute the expectations on the right-hand side. This is because the expectation of* $\sin(x)$ *depends on the higher-order moments of* x, *which we cannot know without knowing the full distribution of* x:

$$\sin(x) = x - \frac{x^3}{3!} + \frac{x^5}{5!} + \cdots . \tag{5.60}$$

It also sometimes happens that even though the process is non-Gaussian, it is still possible to solve the mean and covariance from a closed set of differential equations. This happens, for example, in the case of linear SDEs with multiplicative noise.

Example 5.16 (Statistics of the Black–Scholes model). *Consider the following linear Black–Scholes SDE* (3.31), *which has multiplicative noise:*

$$dx = \mu \, x \, dt + \sigma \, x \, d\beta, \tag{5.61}$$

where β *is a standard Brownian motion. We can now write down the mean and variance equations as*

$$\begin{aligned} \frac{dm}{dt} &= \mu \, m, \\ \frac{dP}{dt} &= 2\mu \, P + \sigma^2 \, P + \sigma^2 \, m^2, \end{aligned} \tag{5.62}$$

which can be solved to give the mean and variance of the solution at each time step. However, unlike additive noise linear equations, the solution to this SDE is not Gaussian.

5.6 Higher-Order Moments of SDEs

It is also possible to derive differential equations for the higher-order moments of SDEs. However, the required number of equations quickly becomes large, because if the state dimension is D, the number of independent third moments is cubic, D^3, in the number of state dimensions. The number of fourth-order moments is quartic, D^4, and so on. The general moment equations can be found, for example, in the book of Socha (2008).

To illustrate the idea, let us consider the scalar SDE

$$dx = f(x) \, dt + L(x) \, d\beta. \tag{5.63}$$

Recall that the expectation of an arbitrary twice differentiable function $\phi(x)$ satisfies

$$\frac{d \, E[\phi(x)]}{dt} = E\left[\frac{\partial \phi(x)}{\partial x} \, f(x)\right] + \frac{q}{2} \, E\left[\frac{\partial^2 \phi(x)}{\partial x^2} \, L^2(x)\right]. \tag{5.64}$$

If we apply this to $\phi(x) = x^n$, where $n \geq 2$, we get

$$\frac{d E[x^n]}{dt} = n\, E[x^{n-1}\, f(x,t)] + \frac{q}{2} n\,(n-1)\, E[x^{n-2}\, L^2(x)], \qquad (5.65)$$

which, in principle, gives the equations for the third-order moments, fourth-order moments, and so on. It is also possible to derive similar differential equations for the central moments, cumulants, or quasimoments.

However, unless $f(x)$ and $L(x)$ are linear (or affine) functions, the equation for the nth-order moment depends on the moments of higher-order (greater than n). Thus in order to actually compute the required expectations, we would need to integrate an infinite number of moment equations, which is impossible in practice. This problem can be approximately solved by using moment closure methods, which typically are based on replacing the higher-order moments (or cumulants or quasimoments) with suitable approximations. For example, it is possible to set the cumulants above a certain order to zero, or to approximate the moments, cumulants, and quasimoments with their steady-state values (Socha, 2008).

In the scalar case, given a set of moments, cumulants, or quasimoments, it is possible to form a distribution that has these moments, cumulants, and quasimoments, for example, as the maximum entropy distribution (Cover and Thomas, 2006). Unfortunately, in the multidimensional case the situation is much more complicated.

5.7 Exercises

5.1 Write the generator $\mathcal{A}$ for the Beneš SDE

$$dx = \tanh(x)\, dt + d\beta, \qquad x(0) = 0, \qquad (5.66)$$

where $\beta(t)$ is a standard Brownian motion.

5.2 FPK equation:

(a) Write down the FPK for the Beneš Equation (5.66), and check that the following probability density solves it:

$$p(x,t) = \frac{1}{\sqrt{2\pi\, t}}\, \cosh(x)\, \exp\left(-\frac{1}{2}t\right) \exp\left(-\frac{1}{2t}x^2\right).$$

(b) Plot the evolution of the probability density when $t \in [0, 5]$.

(c) Simulate 1,000 trajectories from the SDE using the Euler–Maruyama method and check visually that the histogram matches the correct density at time $t = 5$.

5.3 Numerical solution of the FPK equation: Use a finite-differences method to solve the FPK for the Beneš SDE in Equation (5.66). For simplicity, you can select a finite range $x \in [-L, L]$ and use the Dirichlet boundary conditions $p(-L, t) = p(L, t) = 0$.

(a) Divide the range to n grid points and let $h = 2L/(n + 1)$. On the grid, approximate the partial derivatives of $p(x, t)$ via

$$\frac{\partial p(x,t)}{\partial x} \approx \frac{p(x+h,t) - p(x-h,t)}{2h},$$

$$\frac{\partial^2 p(x,t)}{\partial x^2} \approx \frac{p(x+h,t) - 2\,p(x,t) + p(x-h,t)}{h^2}.$$

(b) Let $\mathbf{p}(t) = (p(h - L, t)\ p(2h - L, t) \cdots p(nh - L, t))^\mathsf{T}$ and from the preceding, form an equation of the form

$$\frac{d\mathbf{p}}{dt} = \mathbf{F}\,\mathbf{p}.$$

(c) Solve the preceding equation using (i) the implicit backward Euler method, (ii) by numerical computation of $\exp(\mathbf{F}\,t)$, and (iii) by the forward Euler method. Check that the results match the solution in the previous exercise.

5.4   Consider the Langevin model of Brownian motion (with $m = 1$), which we already considered in Example 3.2:

$$\frac{d^2 x}{dt^2} = -c\,\frac{dx}{dt} + w, \qquad x(0) = (dx/dt)(0) = 0, \qquad (5.67)$$

where $c = 6\,\pi\,\eta\,r$ and the white noise $w(t)$ has some spectral density q.

(a) Interpret the preceding model as an Itô SDE and write it as a two-dimensional state-space form SDE.

(b) Write down the differential equations for the elements of the mean $\mathbf{m}(t)$ and covariance $\mathbf{P}(t)$. Conclude that the mean is zero and find the closed-form solutions for the elements $P_{11}(t)$, $P_{12}(t)$, $P_{21}(t)$, and $P_{22}(t)$ of the covariance matrix $\mathbf{P}(t)$.

   Hint: Start by solving $P_{22}(t)$, then use it to find the solutions for $P_{12}(t) = P_{21}(t)$, and finally solve $P_{11}(t)$.

(c) Find the limiting solution $P_{22}(t)$ when $t \to \infty$, and use the following to determine the diffusion coefficient (spectral density) q:

$$\mathrm{E}\left[\left(\frac{dx}{dt}\right)^2\right] = \frac{R\,T}{N}.$$

(d) Plot the solution $P_{11}(t)$ and conclude that it asymptotically approaches a straight line. Compute the asymptotic solution $P_{11}(t)$ when $t \to \infty$, and conclude that it gives Langevin's result.

5.5   Stationary FPK equation: Show that Equation (5.26) solves the corresponding stationary FPK (Eq. 5.24).

5.6　In so-called Metropolis-adjusted Langevin Monte Carlo methods (see, e.g., Girolami and Calderhead, 2011), the idea is to construct an SDE whose stationary solution is a given probability distribution $\pi(x)$.

(a) Construct an SDE of the form (5.25) such that the stationary solution is the Gamma distribution with the probability density

$$\pi(x) = \frac{b^a}{\Gamma(a)} x^{a-1} \exp(-b\,x),$$

where $a, b > 0$ are constants and the gamma function is defined as

$$\Gamma(a) = \int_0^\infty x^{a-1} \exp(-x)\,dx.$$

(b) Simulate the SDE using the Euler–Maruyama method and check that the solution indeed stabilizes to this distribution.

5.7　Show that the operator $\mathcal{A}$ in Exercise 5.1 is the adjoint of the operator $\mathcal{A}^*$ in the FPK equation in Exercise 5.2.

5.8　Recall the mean-reverting Ornstein–Uhlenbeck process is

$$dx = \theta\,(\mu - x)\,dt + \sigma\,d\beta(t), \quad x(0) = x_0,$$

where θ gives the speed of reversion, μ is the long-run equlibrium, and σ stands for the volatility.

(a) Recall (from Exercise 4.4) or find the complete solution $x(t)$.
(b) Derive the expressions for the mean $m(t)$ and variance $P(t)$ of the solution.

5.9　The Cox–Ingersoll–Ross (CIR) model is used in finance for modeling interest rates. The model is given by

$$dx = (\theta_1 - \theta_2\,x)\,dt + \theta_3\,\sqrt{x}\,d\beta(t), \quad x(0) = x_0 > 0,$$

where $\theta_1, \theta_2, \theta_3 > 0$. The transition density for the Cox–Ingersoll–Ross model can be written down in closed form: Consider the transformed process $y(t) = 2\,c\,x(t)$ with $c = 2\theta_2/(\theta_3^2(1 - e^{-\theta_2 t}))$. Its transition density is a noncentral χ^2 probability density

$$\frac{1}{2}\exp\left(-\frac{y+\lambda}{2}\right)\left(\frac{y}{\lambda}\right)^{\nu/4-1/2} I_{\nu/2-1}\left(\sqrt{\lambda y}\right) \tag{5.68}$$

with $\nu = 4\theta_1/\theta_3^2$ degrees of freedom, and a noncentrality of $\lambda = y_0 \exp(-\theta_2 t)$. I_α denotes the modified Bessel function of the first kind. Derive the expression for the transition density of the original process $x(t)$.

5.10　Show that the modified CIR model

$$dx = -\theta_1\,x\,dt + \theta_2\,\sqrt{1 + x^2}\,d\beta, \tag{5.69}$$

where $2\theta_1 > \theta_2^2$, has a stationary distribution with a density proportional to

$$\frac{1}{(1+x^2)^{1+\theta_1/\theta_2^2}}.$$

Reparametrize this density to show that it is in fact the probability density of a Student's t distribution.

5.11   Statistics of the Black–Scholes model:

(a) Solve the differential equation (5.62) for the mean $m(t)$ and variance $P(t)$ with given initial conditions $m(0) = m_0$ and $P(0) = P_0$.

(b) Verify using the Euler–Maruyama method that your solutions are correct.

5.12   Derive the moments up to the order four for the Ornstein–Uhlenbeck process

(a) by solving the moment differential equations, and

(b) by expressing the moments of the Gaussian process in terms of the mean and variance.

6

Statistics of Linear Stochastic Differential Equations

Linear SDEs are useful in applications as surprisingly many physical and electrical systems can be modeled as linear SDEs. We already derived the full solutions to linear SDEs in Section 4.3. However, we did not fully characterize their statistics, transition densities, and computational methods. In this chapter, we aim to first compute the means and covariances of linear SDEs – which completely characterize the solutions of linear SDEs as they are Gaussian processes. We also discuss equivalent discretizations, which are related to the transition densities of linear SDEs. We also encounter the matrix fraction decomposition, which is a useful numerical method for finding solutions and equivalent discretizations of LTI SDEs. Finally, we discuss covariance functions and stationary solutions of LTI SDEs in terms of Lyapunov equations and Fourier analysis. For more details, the reader is referred to various books on Kalman filtering (e.g., Grewal and Andrews, 2001) – linear SDEs tend to be of more interest in target tracking applications than in mathematical books on SDEs, and therefore they are more thoroughly treated in Kalman filtering books.

6.1 Means, Covariances, and Transition Densities of Linear SDEs

Consider a linear stochastic differential equation of the general form

$$\mathrm{d}\mathbf{x} = \mathbf{F}(t)\,\mathbf{x}\,\mathrm{d}t + \mathbf{u}(t)\,\mathrm{d}t + \mathbf{L}(t)\,\mathrm{d}\boldsymbol{\beta}, \tag{6.1}$$

where $\mathbf{x}(t) \in \mathbb{R}^D$ is the state, the initial conditions are $\mathbf{x}(t_0) \sim \mathrm{N}(\mathbf{m}_0, \mathbf{P}_0)$, $\mathbf{F}(t)$ and $\mathbf{L}(t)$ are matrix valued functions of time, $\mathbf{u}(t) \in \mathbb{R}^D$ is a vector valued function of time, and $\boldsymbol{\beta}(t) \in \mathbb{R}^S$ is a Brownian motion with diffusion matrix $\mathbf{Q}$. The initial conditions are assumed to be independent of the Brownian motion.

The mean and covariance can be solved from Equations (5.50) and

(5.51), which in this case reduce to

$$\frac{d\mathbf{m}}{dt} = \mathbf{F}(t)\,\mathbf{m} + \mathbf{u}(t),$$
$$\frac{d\mathbf{P}}{dt} = \mathbf{F}(t)\,\mathbf{P} + \mathbf{P}\,\mathbf{F}^{\mathsf{T}}(t) + \mathbf{L}(t)\,\mathbf{Q}\,\mathbf{L}^{\mathsf{T}}(t) \tag{6.2}$$

with the initial conditions $\mathbf{m}(t_0) = \mathbf{m}_0$ and $\mathbf{P}(t_0) = \mathbf{P}_0$. The general solutions to these differential equations are (recall the definition of the transition matrix $\boldsymbol{\Psi}(\tau, t)$ in Eq. 2.34):

$$\mathbf{m}(t) = \boldsymbol{\Psi}(t, t_0)\,\mathbf{m}(t_0) + \int_{t_0}^{t} \boldsymbol{\Psi}(t, \tau)\,\mathbf{u}(\tau)\,d\tau, \tag{6.3}$$

$$\mathbf{P}(t) = \boldsymbol{\Psi}(t, t_0)\,\mathbf{P}(t_0)\,\boldsymbol{\Psi}^{\mathsf{T}}(t, t_0)$$
$$+ \int_{t_0}^{t} \boldsymbol{\Psi}(t, \tau)\,\mathbf{L}(\tau)\,\mathbf{Q}\,\mathbf{L}^{\mathsf{T}}(\tau)\,\boldsymbol{\Psi}^{\mathsf{T}}(t, \tau)\,d\tau, \tag{6.4}$$

which could also be obtained by computing the mean and covariance of the explicit solution in Equation (4.28).

Because the solution is a linear transformation of Brownian motion, which is a Gaussian process, the solution is Gaussian

$$p(\mathbf{x}, t) \triangleq p(\mathbf{x}(t)) = \mathrm{N}(\mathbf{x}(t) \mid \mathbf{m}(t), \mathbf{P}(t)), \tag{6.5}$$

which can be verified by checking that this density indeed solves the corresponding FPK equation (5.11). Furthermore, the transition density can be recovered by formally using the initial conditions $\mathbf{m}(s) = \mathbf{x}(s)$ and $\mathbf{P}(s) = \mathbf{0}$, which gives

$$p(\mathbf{x}(t) \mid \mathbf{x}(s)) = \mathrm{N}(\mathbf{x}(t) \mid \mathbf{m}(t \mid s), \mathbf{P}(t \mid s)), \tag{6.6}$$

where

$$\mathbf{m}(t \mid s) = \boldsymbol{\Psi}(t, s)\,\mathbf{x}(s) + \int_{s}^{t} \boldsymbol{\Psi}(t, \tau)\,\mathbf{u}(\tau)\,d\tau,$$
$$\mathbf{P}(t \mid s) = \int_{s}^{t} \boldsymbol{\Psi}(t, \tau)\,\mathbf{L}(\tau)\,\mathbf{Q}\,\mathbf{L}^{\mathsf{T}}(\tau)\,\boldsymbol{\Psi}^{\mathsf{T}}(t, \tau)\,d\tau. \tag{6.7}$$

It is now useful to note that the preceding implies that the original linear SDE is (weakly, in distribution) equivalent to the following discrete-time system:

$$\mathbf{x}(t_{k+1}) = \mathbf{A}_k\,\mathbf{x}(t_k) + \mathbf{u}_k + \mathbf{q}_k, \qquad \mathbf{q}_k \sim \mathrm{N}\left(\mathbf{0}, \boldsymbol{\Sigma}_k\right), \tag{6.8}$$

where

$$\mathbf{A}_k \triangleq \boldsymbol{\Psi}(t_{k+1}, t_k), \tag{6.9}$$

$$\mathbf{u}_k \triangleq \int_{t_k}^{t_{k+1}} \boldsymbol{\Psi}(t_{k+1}, \tau) \mathbf{u}(\tau) \, d\tau, \tag{6.10}$$

$$\boldsymbol{\Sigma}_k \triangleq \boldsymbol{\Sigma}(t_{k+1}, t_k) = \int_{t_k}^{t_{k+1}} \boldsymbol{\Psi}(t_{k+1}, \tau) \mathbf{L}(\tau) \mathbf{Q} \mathbf{L}^\mathsf{T}(\tau) \boldsymbol{\Psi}^\mathsf{T}(t_{k+1}, \tau) \, d\tau, \tag{6.11}$$

which is sometimes called the *equivalent discretization* of SDEs in Kalman filtering context (cf. Grewal and Andrews, 2001; Särkkä, 2006, 2013). This system is equivalent in the sense that the distributions of the two systems coincide at the points $\{t_k\}$.

Example 6.1 (Time-varying Ornstein–Uhlenbeck process). *Let us consider the following time-varying Ornstein–Uhlenbeck process, where $\lambda(t)$ is a given function of time and the diffusion constant of the Brownian motion is q:*

$$dx = -\lambda(t) \, x \, dt + d\beta. \tag{6.12}$$

In this one-dimensional case, we can solve the transition matrix (which is scalar in this case) explicitly using the properties in Equation (2.34):

$$\frac{\partial \psi(\tau, t)}{\partial \tau} = -\lambda(\tau) \, \psi(\tau, t),$$

$$\frac{\partial \psi(\tau, t)}{\partial t} = \lambda(t) \, \psi(\tau, t), \tag{6.13}$$

$$\psi(t, t) = 1.$$

The solution is

$$\psi(s, t) = \exp\left(-\int_t^s \lambda(\tau) \, d\tau\right). \tag{6.14}$$

It is worth noting that this expression does not generalize to the multivariate case simply by replacing the parameter $\lambda(t)$ with a matrix, because the matrix product does not commute as the scalar product does. The equivalent discretization matrices (scalars in this case) are now given as

$$a_k = \exp\left(-\int_{t_k}^{t_{k+1}} \lambda(\tau) \, d\tau\right),$$

$$\Sigma_k = q \int_{t_k}^{t_{k+1}} \exp\left(-2 \int_\tau^{t_{k+1}} \lambda(\tau') \, d\tau'\right) d\tau, \tag{6.15}$$

together with $u_k = 0$. These quantities give the (weak) solution at given discretization points $t_1, t_2, \ldots$ as

$$x(t_{k+1}) = a_k x(t_k) + q_k, \qquad q_k \sim N(0, \Sigma_k). \tag{6.16}$$

6.2 Linear Time-Invariant SDEs

In the case of a linear time-invariant (LTI) SDE

$$d\mathbf{x} = \mathbf{F}\,\mathbf{x}\,dt + \mathbf{L}\,d\boldsymbol{\beta}, \tag{6.17}$$

where $\mathbf{F}$ and $\mathbf{L}$ are constant, the mean and covariance are also given by Equation (6.2), which now takes the form

$$
\begin{aligned}
\frac{d\mathbf{m}}{dt} &= \mathbf{F}\,\mathbf{m}, \\
\frac{d\mathbf{P}}{dt} &= \mathbf{F}\,\mathbf{P} + \mathbf{P}\,\mathbf{F}^\mathsf{T} + \mathbf{L}\,\mathbf{Q}\,\mathbf{L}^\mathsf{T}.
\end{aligned}
\tag{6.18}
$$

Thus the only differences are that the matrices $\mathbf{F}$ and $\mathbf{L}$ are constant, and there is no input. In this LTI SDE case, the transition matrix is the matrix exponential function $\boldsymbol{\Psi}(t, \tau) = \exp(\mathbf{F}\,(t - \tau))$ and the solutions to the differential equations can be obtained by a substitution of it to Equations (6.3) and (6.4):

$$\mathbf{m}(t) = \exp(\mathbf{F}\,(t - t_0))\,\mathbf{m}(t_0), \tag{6.19}$$

$$
\begin{aligned}
\mathbf{P}(t) = {}& \exp(\mathbf{F}\,(t - t_0))\,\mathbf{P}(t_0)\,\exp(\mathbf{F}\,(t - t_0))^\mathsf{T} \\
&+ \int_{t_0}^{t} \exp(\mathbf{F}\,(t - \tau))\,\mathbf{L}\,\mathbf{Q}\,\mathbf{L}^\mathsf{T}\,\exp(\mathbf{F}\,(t - \tau))^\mathsf{T}\,d\tau.
\end{aligned}
\tag{6.20}
$$

The transition density is then given as

$$p(\mathbf{x}(t) \mid \mathbf{x}(s)) = N(\mathbf{x}(t) \mid \mathbf{m}(t \mid s), \mathbf{P}(t \mid s)), \tag{6.21}$$

where

$$
\begin{aligned}
\mathbf{m}(t \mid s) &= \exp(\mathbf{F}\,(t - s))\,\mathbf{x}(s), \\
\mathbf{P}(t \mid s) &= \int_{s}^{t} \exp(\mathbf{F}\,(t - \tau))\,\mathbf{L}\,\mathbf{Q}\,\mathbf{L}^\mathsf{T}\,\exp(\mathbf{F}\,(t - \tau))^\mathsf{T}\,d\tau.
\end{aligned}
\tag{6.22}
$$

The corresponding equivalent discrete system now takes the form

$$\mathbf{x}(t_{k+1}) = \mathbf{A}_k\,\mathbf{x}(t_k) + \mathbf{q}_k, \qquad \mathbf{q}_k \sim N(\mathbf{0}, \boldsymbol{\Sigma}_k), \tag{6.23}$$

where $\Delta t_k = t_{k+1} - t_k$ and

$$\mathbf{A}_k \triangleq \mathbf{A}(\Delta t_k) = \exp(\mathbf{F}\,\Delta t_k), \tag{6.24}$$

$$\boldsymbol{\Sigma}_k \triangleq \boldsymbol{\Sigma}(\Delta t_k) = \int_0^{\Delta t_k} \exp(\mathbf{F}\,(\Delta t_k - \tau))\,\mathbf{L}\,\mathbf{Q}\,\mathbf{L}^\mathsf{T}\,\exp(\mathbf{F}\,(\Delta t_k - \tau))^\mathsf{T}\,\mathrm{d}\tau. \tag{6.25}$$

These expressions can sometimes (actually quite often) be evaluated in closed form. There also exists a number of computational methods that can be used for computing them numerically. A review of available methods can be found in Axelsson and Gustafsson (2015). In the next section, we discuss one particularly useful method for the computation of the preceding covariance using the matrix fraction decomposition.

In terms of the preceding quantities, the transition density has the representation

$$p(\mathbf{x}(t) \mid \mathbf{x}(s)) = \mathrm{N}(\mathbf{x}(t) \mid \mathbf{A}(t - s)\,\mathbf{x}(s),\,\boldsymbol{\Sigma}(t - s)). \tag{6.26}$$

An example of computing these quantities is shown in the next example.

Example 6.2 (Discretized Ornstein–Uhlenbeck process). *For the (LTI) Ornstein–Uhlenbeck process*

$$\mathrm{d}x = -\lambda\,x\,\mathrm{d}t + \mathrm{d}\beta, \tag{6.27}$$

with $\lambda > 0$, and with the Brownian motion having the diffusion constant q, we get the transition density

$$p(x(t) \mid x(s)) = \mathrm{N}(x(t) \mid m(t \mid s),\, P(t \mid s)), \tag{6.28}$$

where

$$m(t \mid s) = \exp(-\lambda\,(t - s))\,x(s),$$
$$P(t \mid s) = \frac{q}{2\lambda}\,[1 - \exp(-2\lambda\,(t - s))]. \tag{6.29}$$

The corresponding discretization matrices (scalars) are then given by

$$a_k = \exp(-\lambda\,\Delta t_k),$$
$$\Sigma_k = \frac{q}{2\lambda}\,[1 - \exp(-2\lambda\,\Delta t_k)], \tag{6.30}$$

where $\Delta t_k = t_{k+1} - t_k$, which indeed agree with Equation (6.15), and the equivalently discretized system again has the form shown in Equation (6.16).

Equations (6.24) and (6.25) can often be found in tracking literature (Bar-Shalom et al., 2001; Grewal and Andrews, 2001; Särkkä, 2006, 2013), because they are useful in converting continuous-discrete Kalman filtering problems into equivalent discrete-time Kalman filtering problems. A typical example of a model in that context is the following.

Example 6.3 (Discretized Wiener velocity model). *The Wiener velocity model (see, e.g., Bar-Shalom et al., 2001; Särkkä, 2006) is a typical model found in target tracking, where the velocity (the first derivative of the process) is modeled as a Wiener process, that is, as a Brownian motion. In white noise interpretation, this means that the acceleration (i.e., the second derivative) is a white noise process with spectral density q:*

$$\frac{d^2 x(t)}{dt^2} = w(t). \tag{6.31}$$

In more rigorous Itô SDE form, this model can be written as

$$\begin{pmatrix} dx_1 \\ dx_2 \end{pmatrix} = \underbrace{\begin{pmatrix} 0 & 1 \\ 0 & 0 \end{pmatrix}}_{\mathbf{F}} \begin{pmatrix} x_1 \\ x_2 \end{pmatrix} dt + \underbrace{\begin{pmatrix} 0 \\ 1 \end{pmatrix}}_{\mathbf{L}} d\beta, \tag{6.32}$$

where $\beta(t)$ is a Brownian motion with diffusion coefficient q, $x_1(t) \triangleq x(t)$ is the actual process, and $x_2(t)$ is its derivative.

Now the matrices of the equivalent discrete-time model are given as follows (notice that $\mathbf{F}$ is a nilpotent matrix such that $\mathbf{F}^n = \mathbf{0}$ for $n > 1$):

$$\mathbf{A}(\Delta t) = \exp(\mathbf{F} \, \Delta t) = \mathbf{I} + \mathbf{F} \, \Delta + \underbrace{\frac{1}{2!} \mathbf{F}^2 \, \Delta t^2 + \cdots}_{=0} = \begin{pmatrix} 1 & \Delta t \\ 0 & 1 \end{pmatrix},$$

$$\mathbf{\Sigma}(\Delta t) = \int_0^{\Delta t} \begin{pmatrix} 1 & \Delta t - \tau \\ 0 & 1 \end{pmatrix} \begin{pmatrix} 0 & 0 \\ 0 & q \end{pmatrix} \begin{pmatrix} 1 & \Delta t - \tau \\ 0 & 1 \end{pmatrix}^{\mathsf{T}} d\tau$$

$$= \begin{pmatrix} \frac{1}{3} \Delta t^3 & \frac{1}{2} \Delta t^2 \\ \frac{1}{2} \Delta t^2 & \Delta t \end{pmatrix} q. \tag{6.33}$$

The transition density of the Wiener velocity model is thus given as

$$p(\mathbf{x}(t) \mid \mathbf{x}(s))$$
$$= \mathrm{N}\left(\begin{pmatrix} x_1(t) \\ x_2(t) \end{pmatrix} \Bigg| \begin{pmatrix} x_1(s) + x_2(s)(t - s) \\ x_2(s) \end{pmatrix}, \begin{pmatrix} \frac{q}{3}(t - s)^3 & \frac{q}{2}(t - s)^2 \\ \frac{q}{2}(t - s)^2 & q(t - s) \end{pmatrix} \right). \tag{6.34}$$

Example 6.4 (Discretized car model). *A two-dimensional version of the* *above Wiener velocity model was already presented in Example 3.4 for the purpose of modeling the movement of a car. The same model was also used in a Kalman filtering and smoothing context, for example, in Särkkä (2013). The corresponding discrete-time model matrices now become the following:*

$$\mathbf{A}(\Delta t) = \begin{pmatrix} 1 & 0 & \Delta t & 0 \\ 0 & 1 & 0 & \Delta t \\ 0 & 0 & 1 & 0 \\ 0 & 0 & 0 & 1 \end{pmatrix}, \quad \boldsymbol{\Sigma}(\Delta t) = \begin{pmatrix} \frac{q_1 \Delta t^3}{3} & 0 & \frac{q_1 \Delta t^2}{2} & 0 \\ 0 & \frac{q_2 \Delta t^3}{3} & 0 & \frac{q_2 \Delta t^2}{2} \\ \frac{q_1 \Delta t^2}{2} & 0 & q_1 \Delta t & 0 \\ 0 & \frac{q_2 \Delta t^2}{2} & 0 & q_2 \Delta t \end{pmatrix},$$

(6.35)

where q_1 and q_2 are the diffusion coefficients of the driving Brownian motions.

6.3 Matrix Fraction Decomposition

A convenient numerical method for solving the covariance $\mathbf{P}(t)$ from Equation (6.2) or (6.18) is by using matrix fractions (see, e.g., Stengel, 1994; Grewal and Andrews, 2001; Särkkä, 2006; Axelsson and Gustafsson, 2015). If we define matrices $\mathbf{C}(t)$ and $\mathbf{D}(t)$ such that $\mathbf{P}(t) = \mathbf{C}(t)\mathbf{D}^{-1}(t)$, then $\mathbf{P}(t)$ solves the matrix Lyapunov differential equation

$$\frac{d\mathbf{P}}{dt} = \mathbf{F}(t)\mathbf{P} + \mathbf{P}\mathbf{F}^{\mathsf{T}}(t) + \mathbf{L}(t)\mathbf{Q}\mathbf{L}^{\mathsf{T}}(t),$$

(6.36)

if the matrices $\mathbf{C}(t)$ and $\mathbf{D}(t)$ solve the differential equation

$$\begin{pmatrix} d\mathbf{C}/dt \\ d\mathbf{D}/dt \end{pmatrix} = \begin{pmatrix} \mathbf{F}(t) & \mathbf{L}(t)\mathbf{Q}\mathbf{L}^{\mathsf{T}}(t) \\ \mathbf{0} & -\mathbf{F}^{\mathsf{T}}(t) \end{pmatrix} \begin{pmatrix} \mathbf{C} \\ \mathbf{D} \end{pmatrix},$$

(6.37)

and $\mathbf{P}(t_0) = \mathbf{C}(t_0)\mathbf{D}^{-1}(t_0)$. We can select, for example,

$$\mathbf{C}(t_0) = \mathbf{P}(t_0) \quad \text{and} \quad \mathbf{D}(t_0) = \mathbf{I}.$$

(6.38)

The convenience of the matrix fraction decomposition comes from the fact that Equation (6.37) is a "normal" linear differential equation in the sense that we have a matrix multiplying the unknown from the left, whereas the Lyapunov differential equation (6.36) has a matrix multiplication from right. For this reason, the covariance solutions in Equations (6.4) and (6.20) have more complicated forms than the mean solutions in Equations (6.3) and (6.19).

The matrix fraction decomposition is especially useful in the case of linear time-variant systems. If the differential equation (6.37) is time-invariant, that is, if $\mathbf{F}$ and $\mathbf{L}$ are constant, then it can be solved using the matrix exponential function:

$$\begin{pmatrix} \mathbf{C}(t) \\ \mathbf{D}(t) \end{pmatrix} = \exp\left[\begin{pmatrix} \mathbf{F} & \mathbf{L}\,\mathbf{Q}\,\mathbf{L}^\mathsf{T} \\ \mathbf{0} & -\mathbf{F}^\mathsf{T} \end{pmatrix} (t - t_0)\right] \begin{pmatrix} \mathbf{C}(t_0) \\ \mathbf{D}(t_0) \end{pmatrix}. \tag{6.39}$$

The final solution is then given as $\mathbf{P}(t) = \mathbf{C}(t)\,\mathbf{D}^{-1}(t)$. This is useful, because now both the mean and covariance can be solved via a simple matrix exponential function computation, which allows for easy numerical treatment. These matrix exponentials can often also be computed in closed form using, for example, Laplace transforms.

In filtering and smoothing (see Chapter 10), we are often interested in forming the matrices $\mathbf{A}(\Delta t)$ and $\mathbf{\Sigma}(\Delta t)$ in Equations (6.24) and (6.25) by numerical means. This is because these numerical matrices can then be directly used in a discrete-time Kalman filter to infer the state of the SDE at a discrete set of time instants. The matrix fraction decomposition is useful for this task as well. First note that the numerical computation of $\mathbf{A}(\Delta t)$ is easy, because it is just a matrix exponential for which good numerical computation methods are available. However, the integral expression for $\mathbf{\Sigma}(\Delta t)$ is more problematic from a numerical point of view.

It turns out that the matrix fractions can also be used for reducing the computations of the matrix $\mathbf{\Sigma}(\Delta t)$ to a simple matrix exponential (see, e.g., Särkkä, 2006; Axelsson and Gustafsson, 2015). The trick is that the matrix is also the solution to the differential equation

$$\frac{d\mathbf{\Sigma}(t)}{dt} = \mathbf{F}\,\mathbf{\Sigma}(t) + \mathbf{\Sigma}(t)\,\mathbf{F}^\mathsf{T} + \mathbf{L}\,\mathbf{Q}\,\mathbf{L}^\mathsf{T}, \quad \mathbf{\Sigma}(0) = \mathbf{0}. \tag{6.40}$$

Thus we can now use the matrix fractions to solve $\mathbf{\Sigma}(\Delta t) = \mathbf{C}_\Sigma(\Delta t)\,\mathbf{D}_\Sigma^{-1}(\Delta t)$, where

$$\begin{pmatrix} \mathbf{C}_\Sigma(\Delta t) \\ \mathbf{D}_\Sigma(\Delta t) \end{pmatrix} = \underbrace{\exp\left[\begin{pmatrix} \mathbf{F} & \mathbf{L}\,\mathbf{Q}\,\mathbf{L}^\mathsf{T} \\ \mathbf{0} & -\mathbf{F}^\mathsf{T} \end{pmatrix} \Delta t\right]}_{\mathbf{\Phi}(\Delta t)} \begin{pmatrix} \mathbf{0} \\ \mathbf{I} \end{pmatrix}. \tag{6.41}$$

It is also useful to remark (see, e.g., Axelsson and Gustafsson, 2015) that the preceding matrix exponential $\mathbf{\Phi}(\Delta t)$ always has the form

$$\mathbf{\Phi}(\Delta t) = \begin{pmatrix} \mathbf{A}(\Delta t) & \mathbf{\Sigma}(\Delta t)\,\mathbf{A}(\Delta t)^{-\mathsf{T}} \\ \mathbf{0} & \mathbf{A}(\Delta t)^{-\mathsf{T}} \end{pmatrix}, \tag{6.42}$$

where $\mathbf{A}(\Delta t) = \exp(\mathbf{F}\,\Delta t)$. This also implies that we always have

$D_\Sigma^{-1}(\Delta t) = A^T(\Delta t)$ and thus the matrix inversion does not actually need to be computed.

Example 6.5 (Matrix fractions solution to the Wiener velocity model). *For the Wiener velocity model (6.32), we get*

$$\begin{pmatrix} C_\Sigma(\Delta t) \\ D_\Sigma(\Delta t) \end{pmatrix} = \exp\left[\underbrace{\begin{pmatrix} 0 & 1 & 0 & 0 \\ 0 & 0 & 0 & q \\ 0 & 0 & 0 & 0 \\ 0 & 0 & -1 & 0 \end{pmatrix}}_{G} \Delta t \right] \begin{pmatrix} \mathbf{0} \\ \mathbf{I} \end{pmatrix}. \tag{6.43}$$

By manual or symbolic computation, we can find out that the matrix is nilpotent of degree 4, *that is,* $G^n = 0$ *when* $n \geq 4$. *Hence we have*

$$\exp(G\,\Delta t) = I + G\,\Delta t + G^2 \frac{\Delta t^2}{2} + G^3 \frac{\Delta t^3}{6}$$

$$= \begin{pmatrix} 1 & \Delta t & -\frac{\Delta t^3 q}{6} & \frac{\Delta t^2 q}{2} \\ 0 & 1 & -\frac{\Delta t^2 q}{2} & \Delta t\, q \\ 0 & 0 & 1 & 0 \\ 0 & 0 & -\Delta t & 1 \end{pmatrix}. \tag{6.44}$$

This then leads to

$$C_\Sigma(\Delta t) = \begin{pmatrix} -\frac{\Delta t^3 q}{6} & \frac{\Delta t^2 q}{2} \\ -\frac{\Delta t^2 q}{2} & \Delta t\, q \end{pmatrix} \quad and \quad D_\Sigma^{-1}(\Delta t) = \begin{pmatrix} 1 & 0 \\ \Delta t & 1 \end{pmatrix}, \tag{6.45}$$

and further

$$\Sigma(\Delta) = C_\Sigma(\Delta t)\, D_\Sigma^{-1}(\Delta t) = \begin{pmatrix} \frac{1}{3}\Delta t^3 & \frac{1}{2}\Delta t^2 \\ \frac{1}{2}\Delta t^2 & \Delta t \end{pmatrix} q, \tag{6.46}$$

which agrees with the solution obtained in (6.33).

Obviously, the matrix G is not always nilpotent, and often a symbolic computation of the matrix exponential is not feasible by other means either. However, fortunately, commonly used software packages such as MATLAB® or Python already provide functions for efficient numerical computation of matrix exponentials. These functions can be easily used for computing the matrix fraction decompositions and hence equivalent discretizations numerically. This is illustrated in the next example.

Example 6.6 (Discretization of the spring model). *Recall the spring model* *in Example 3.10, which in proper SDE interpretation has the form*

$$\underbrace{\begin{pmatrix} dx_1 \\ dx_2 \end{pmatrix}}_{d\mathbf{x}} = \underbrace{\begin{pmatrix} 0 & 1 \\ -v^2 & -\gamma \end{pmatrix}}_{\mathbf{F}} \underbrace{\begin{pmatrix} x_1 \\ x_2 \end{pmatrix}}_{\mathbf{x}} dt + \underbrace{\begin{pmatrix} 0 \\ 1 \end{pmatrix}}_{\mathbf{L}} d\beta. \tag{6.47}$$

If we now wish to discretize this model, we encounter the problem that already the matrix exponential for $\mathbf{A}(\Delta t)$ is fairly complicated and we cannot hope to compute the integral for $\mathbf{\Sigma}(\Delta t)$ in Equation (6.33) in closed form. Hence, numerical integration would be needed, which can be cumbersome. However, for given values of the parameters, we can numerically use the matrix fraction decomposition and do the discretization as follows.

Assume, for example, that we the parameters $v = 1$, $\gamma = 1/10$, and $\Delta t = 1/2$. Then the numerical $\mathbf{G}$ matrix and the matrix exponential of $\mathbf{G}\,\Delta t$ are given as

$$\mathbf{G} = \begin{pmatrix} 0 & 1.0000 & 0 & 0 \\ -1.0000 & -0.1000 & 0 & 1.0000 \\ 0 & 0 & 0 & 1.0000 \\ 0 & 0 & -1.0000 & 0.1000 \end{pmatrix},$$

$$\exp(\mathbf{G}\,\Delta t) \approx \begin{pmatrix} 0.8796 & 0.4676 & -0.0203 & 0.1199 \\ -0.4676 & 0.8328 & -0.1199 & 0.4593 \\ 0 & 0 & 0.8755 & 0.4916 \\ 0 & 0 & -0.4916 & 0.9247 \end{pmatrix}. \tag{6.48}$$

The preceding matrix exponential was computed using MATLAB®*'s built-in* expm *function. The matrix exponential*

$$\mathbf{A}(\Delta t) = \exp(\mathbf{F}\,\Delta t) \approx \begin{pmatrix} 0.8796 & 0.4676 \\ -0.4676 & 0.8328 \end{pmatrix} \tag{6.49}$$

can be found in the top-left corner of the second matrix in Equation (6.48). The discretized covariance is then given as

$$\mathbf{\Sigma}(\Delta t) \approx \underbrace{\begin{pmatrix} -0.0203 & 0.1199 \\ -0.1199 & 0.4593 \end{pmatrix}}_{\mathbf{C}_\Sigma(\Delta t)} \underbrace{\begin{pmatrix} 0.8796 & -0.4676 \\ 0.4676 & 0.8328 \end{pmatrix}}_{\mathbf{D}_\Sigma^{-1}(\Delta t) = \mathbf{A}^{\mathsf{T}}(\Delta t)}$$

$$\approx \begin{pmatrix} 0.0382 & 0.1093 \\ 0.1093 & 0.4386 \end{pmatrix}, \tag{6.50}$$

where we have truncated the display of the intermediate quantities to four decimals although the result was computed by directly extracting the blocks numerically from the matrix exponential in the second equation in (6.48).

Remark 6.7 (LTI SDE with constant input). *Note that by the first glance the preceding method does not seem to directly work for discretization of LTI SDEs with a constant input* **u***:*

$$d\mathbf{x} = \mathbf{F}\,\mathbf{x}\,dt + \mathbf{u}\,dt + \mathbf{L}\,d\boldsymbol{\beta}, \tag{6.51}$$

but it turns out that it actually does. This is because we can rewrite the equation as

$$d\mathbf{x} = \mathbf{F}\,\mathbf{x}\,dt + \mathbf{u}\,dt + \mathbf{L}\,d\boldsymbol{\beta},$$
$$d\mathbf{u} = \mathbf{0}. \tag{6.52}$$

The discretization can now be done to the joint state-space $(\mathbf{x}, \mathbf{u})$, *which then gives one additional coefficient* $\mathbf{B}(\Delta t)$ *for the discretization:*

$$\mathbf{x}(t_{k+1}) = \mathbf{A}(\Delta t_k)\,\mathbf{x}(t_k) + \mathbf{B}(\Delta t_k)\,\mathbf{u} + \mathbf{q}_k. \tag{6.53}$$

If the input is some time-dependent $\mathbf{u}(t)$, *we can also directly use this result to form a zeroth-order-hold (ZOH) approximation to the input contribution. However, with additional augmentation tricks, we can also construct higher-order approximations (nth order holds) with respect to the input.*

6.4 Covariance Functions of Linear SDEs

So far, we have only considered the computation of time-marginal statistics of the state, that is, the statistics of $\mathbf{x}(t)$ for single t at a time. However, sometimes we are interested in the interaction of states at different times t and s. This kind of interaction (up to the second order) is characterized by the covariance function

$$\mathbf{C}(t, s) = \mathrm{E}[(\mathbf{x}(t) - \mathbf{m}(t))\,(\mathbf{x}(s) - \mathbf{m}(s))^{\mathsf{T}}], \tag{6.54}$$

where $\mathbf{m}(t) = \mathrm{E}[\mathbf{x}(t)]$ and $\mathbf{m}(s) = \mathrm{E}[\mathbf{x}(s)]$. The covariance function is a matrix-valued function that gives the cross-covariance matrix between the vectors $\mathbf{x}(t)$ and $\mathbf{x}(s)$. By evaluating the covariance function at $t = s$, we get $\mathbf{P}(t) = \mathbf{C}(t, t)$, which is the marginal covariance of $\mathbf{x}(t)$.

Let us now compute the covariance function of

$$d\mathbf{x} = \mathbf{F}(t)\,\mathbf{x}\,dt + \mathbf{L}(t)\,d\boldsymbol{\beta}, \quad \mathbf{x}(t_0) \sim \mathrm{N}(\mathbf{m}(t_0), \mathbf{P}(t_0)), \tag{6.55}$$

where again the initial condition is assumed to be independent of the Brownian motion. It is worth noting at this point that we have left out the input $\mathbf{u}(t)$ for notational convenience. However, the covariance function that we derive will be valid also for a linear SDE with deterministic input (6.1), because a deterministic input does not affect the covariance function.

The solution to the SDE can be expressed as (by (4.28))

$$\mathbf{x}(t) = \mathbf{\Psi}(t, t_0) \mathbf{x}(t_0) + \int_{t_0}^{t} \mathbf{\Psi}(t, \tau) \mathbf{L}(\tau) \, d\boldsymbol{\beta}(\tau), \qquad (6.56)$$

where $\mathbf{\Psi}$ is the transition matrix defined in Equation (2.34). Taking the expectation gives the mean

$$\mathbf{m}(t) = \mathrm{E}[\mathbf{\Psi}(t, t_0) \mathbf{x}(t_0)] = \mathbf{\Psi}(t, t_0) \mathbf{m}(t_0). \qquad (6.57)$$

By using the independence of initial conditions from the Brownian motion, we now get for the covariance function

$$\begin{aligned}
\mathbf{C}(t, s) &= \mathrm{E}[(\mathbf{x}(t) - \mathbf{m}(t)) (\mathbf{x}(s) - \mathbf{m}(s))^{\mathsf{T}}] \\
&= \mathrm{E}\left[\left(\mathbf{\Psi}(t, t_0) \mathbf{x}(t_0) + \int_{t_0}^{t} \mathbf{\Psi}(t, \tau) \mathbf{L}(\tau) \, d\boldsymbol{\beta}(\tau) - \mathbf{\Psi}(t, t_0) \mathbf{m}(t_0) \right) \right. \\
&\qquad \left. \times \left(\mathbf{\Psi}(s, t_0) \mathbf{x}(t_0) + \int_{t_0}^{s} \mathbf{\Psi}(s, \sigma) \mathbf{L}(\sigma) \, d\boldsymbol{\beta}(\sigma) - \mathbf{\Psi}(s, t_0) \mathbf{m}(t_0) \right)^{\mathsf{T}} \right] \\
&= \mathrm{E}\left[\mathbf{\Psi}(t, t_0) (\mathbf{x}(t_0) - \mathbf{m}(t_0)) (\mathbf{x}(t_0) - \mathbf{m}(t_0))^{\mathsf{T}} \mathbf{\Psi}^{\mathsf{T}}(s, t_0) \right] \\
&\qquad + \mathrm{E}\left[\left(\int_{t_0}^{t} \mathbf{\Psi}(t, \tau) \mathbf{L}(\tau) \, d\boldsymbol{\beta}(\tau) \right) \left(\int_{t_0}^{s} \mathbf{\Psi}(s, \sigma) \mathbf{L}(\sigma) \, d\boldsymbol{\beta}(\sigma) \right)^{\mathsf{T}} \right].
\end{aligned}$$

$$(6.58)$$

It is now convenient to split the evaluation into two parts:

- If $t < s$, then for the first term we get the following:

$$\begin{aligned}
&\mathrm{E}[\mathbf{\Psi}(t, t_0) (\mathbf{x}(t_0) - \mathbf{m}(t_0)) (\mathbf{x}(t_0) - \mathbf{m}(t_0))^{\mathsf{T}} \mathbf{\Psi}^{\mathsf{T}}(s, t_0)] \\
&= \mathbf{\Psi}(t, t_0) \mathbf{P}(t_0) \mathbf{\Psi}^{\mathsf{T}}(s, t_0) \\
&= \mathbf{\Psi}(t, t_0) \mathbf{P}(t_0) \mathbf{\Psi}^{\mathsf{T}}(t, t_0) \mathbf{\Psi}^{\mathsf{T}}(s, t). \qquad (6.59)
\end{aligned}$$

The second term then gives the following:

$$
\mathrm{E}\left[\left(\int_{t_0}^{t} \boldsymbol{\Psi}(t,\tau)\,\mathbf{L}(\tau)\,\mathrm{d}\boldsymbol{\beta}(\tau)\right)\left(\int_{t_0}^{s} \boldsymbol{\Psi}(s,\sigma)\,\mathbf{L}(\sigma)\,\mathrm{d}\boldsymbol{\beta}(\sigma)\right)^{\mathsf{T}}\right]
$$

$$
= \int_{t_0}^{t} \boldsymbol{\Psi}(t,\tau)\,\mathbf{L}(\tau)\,\mathbf{Q}\,\mathbf{L}^{\mathsf{T}}(\tau)\,\boldsymbol{\Psi}^{\mathsf{T}}(s,\tau)\,\mathrm{d}\tau
$$

$$
= \int_{t_0}^{t} \boldsymbol{\Psi}(t,\tau)\,\mathbf{L}(\tau)\,\mathbf{Q}\,\mathbf{L}^{\mathsf{T}}(\tau)\,[\boldsymbol{\Psi}(s,t)\,\boldsymbol{\Psi}(t,\tau)]^{\mathsf{T}}\,\mathrm{d}\tau
$$

$$
= \left[\int_{t_0}^{t} \boldsymbol{\Psi}(t,\tau)\,\mathbf{L}(\tau)\,\mathbf{Q}\,\mathbf{L}^{\mathsf{T}}(\tau)\,\boldsymbol{\Psi}^{\mathsf{T}}(t,\tau)\,\mathrm{d}\tau\right]\boldsymbol{\Psi}^{\mathsf{T}}(s,t), \tag{6.60}
$$

where $\mathbf{P}(t)$ is the marginal covariance. Thus when $t < s$, we have

$$
\mathbf{C}(t,s) = \left[\boldsymbol{\Psi}(t,t_0)\,\mathbf{P}(t_0)\,\boldsymbol{\Psi}^{\mathsf{T}}(t,t_0)\right.
$$

$$
\left. + \int_{t_0}^{t} \boldsymbol{\Psi}(t,\tau)\,\mathbf{L}(\tau)\,\mathbf{Q}\,\mathbf{L}^{\mathsf{T}}(\tau)\,\boldsymbol{\Psi}^{\mathsf{T}}(t,\tau)\,\mathrm{d}\tau\right]\boldsymbol{\Psi}^{\mathsf{T}}(s,t)
$$

$$
= \mathbf{P}(t)\,\boldsymbol{\Psi}^{\mathsf{T}}(s,t), \tag{6.61}
$$

where we have recalled the expression of $\mathbf{P}(t)$ given in Equation (6.4).

- If $t \geq s$, then we similarly get

$$
\mathrm{E}[\boldsymbol{\Psi}(t,t_0)\,(\mathbf{x}(t_0) - \mathbf{m}(t_0))\,(\mathbf{x}(t_0) - \mathbf{m}(t_0))^{\mathsf{T}}\,\boldsymbol{\Psi}^{\mathsf{T}}(s,t_0)]
$$

$$
= \boldsymbol{\Psi}(t,s)\,\boldsymbol{\Psi}(s,t_0)\,\mathbf{P}(t_0)\,\boldsymbol{\Psi}^{\mathsf{T}}(s,t_0), \tag{6.62}
$$

and

$$
\mathrm{E}\left[\left(\int_{t_0}^{t} \boldsymbol{\Psi}(t,\tau)\,\mathbf{L}(\tau)\,\mathrm{d}\boldsymbol{\beta}(\tau)\right)\left(\int_{t_0}^{s} \boldsymbol{\Psi}(s,\sigma)\,\mathbf{L}(\sigma)\,\mathrm{d}\boldsymbol{\beta}(\sigma)\right)^{\mathsf{T}}\right]
$$

$$
= \boldsymbol{\Psi}(t,s)\int_{t_0}^{s} \boldsymbol{\Psi}(s,\tau)\,\mathbf{L}(\tau)\,\mathbf{Q}\,\mathbf{L}^{\mathsf{T}}(\tau)\,\boldsymbol{\Psi}^{\mathsf{T}}(s,\tau)\,\mathrm{d}\tau. \tag{6.63}
$$

Thus when $t \geq s$, we get the covariance

$$
\mathbf{C}(t,s) = \boldsymbol{\Psi}(t,s)\,\mathbf{P}(s). \tag{6.64}
$$

From the preceding derivation, we get that the covariance function can be written as

$$
\mathbf{C}(t,s) = \begin{cases} \mathbf{P}(t)\,\boldsymbol{\Psi}^{\mathsf{T}}(s,t), & \text{if } t < s, \\ \boldsymbol{\Psi}(t,s)\,\mathbf{P}(s), & \text{if } t \geq s. \end{cases} \tag{6.65}
$$

In the LTI case, we have $\Psi(t, s) = \exp((t - s)\, \mathbf{F})$ and thus

$$\mathbf{C}(t, s) = \begin{cases} \mathbf{P}(t)\, \exp((s - t)\, \mathbf{F})^\mathsf{T}, & \text{if } t < s, \\ \exp((t - s)\, \mathbf{F})\, \mathbf{P}(s), & \text{if } t \geq s. \end{cases} \tag{6.66}$$

6.5 Steady-State Solutions of Linear SDEs

In this section, we consider the steady-state solutions of SDEs. When the SDE is time-invariant and stable in a suitable sense, the solution approaches a steady state when $t \to \infty$. Alternatively, we can think that the process has been started from $t_0 \to -\infty$ and thus the solution has reached its steady state already at time t. We already saw in Section 5.2 that the steady state can be solved easily from the FPK if the drift has a special gradient-of-potential form. Here the aim is to consider the steady-state solutions of LTI SDEs of the form

$$\mathrm{d}\mathbf{x} = \mathbf{F}\, \mathbf{x}\, \mathrm{d}t + \mathbf{L}\, \mathrm{d}\boldsymbol{\beta}, \tag{6.67}$$

where the gradient condition does not need to hold. It turns out that such (nonsingular) stationary solution can only exist if the matrix $\mathbf{F}$ is stable (i.e., a Hurwitz matrix), that is, if the real parts of its eigenvalues are strictly negative.

As the solution to a linear SDE is characterized by the mean and covariance, our aim is now to solve the stationary mean and covariance. We now notice that at the steady state, the time derivatives of mean and covariance should be zero:

$$\begin{aligned} \frac{\mathrm{d}\mathbf{m}}{\mathrm{d}t} &= \mathbf{F}\, \mathbf{m} = \mathbf{0}, \\ \frac{\mathrm{d}\mathbf{P}}{\mathrm{d}t} &= \mathbf{F}\, \mathbf{P} + \mathbf{P}\, \mathbf{F}^\mathsf{T} + \mathbf{L}\, \mathbf{Q}\, \mathbf{L}^\mathsf{T} = \mathbf{0}. \end{aligned} \tag{6.68}$$

The first equation implies that the stationary mean should be identically zero $\mathbf{m}_\infty = \mathbf{0}$. Here we use the subscript ∞ to mean the steady-state value, which in a sense corresponds to the value after an infinite duration of time. The second equation leads to the so-called Lyapunov equation, which is a special case of the so-called algebraic Riccati equations (AREs):

$$\mathbf{F}\, \mathbf{P}_\infty + \mathbf{P}_\infty\, \mathbf{F}^\mathsf{T} + \mathbf{L}\, \mathbf{Q}\, \mathbf{L}^\mathsf{T} = \mathbf{0}. \tag{6.69}$$

The steady-state covariance $\mathbf{P}_\infty$ can be algebraically solved from the preceding equation. Note that although the equation is linear in $\mathbf{P}_\infty$, it cannot be solved via simple matrix inversion, because the matrix $\mathbf{F}$ appears on the left- and right-hand sides of the covariance. However, most mathematical software packages (e.g., MATLAB®) have built-in routines for solving this type of equation numerically.

It is worth noting that provided that we have computed the matrix $\mathbf{P}_\infty$ for an LTI SDE, forming the equivalent discretization becomes a bit easier. Note that the equivalent discrete-time version of the Lyapunov equation is

$$\mathbf{P}_\infty = \mathbf{A}(\Delta t)\, \mathbf{P}_\infty \, \mathbf{A}^\mathsf{T}(\Delta t) + \mathbf{\Sigma}(\Delta t), \tag{6.70}$$

where $\mathbf{A}(\Delta t)$ and $\mathbf{\Sigma}(\Delta t)$ are given by Equations (6.24) and (6.25), respectively. This implies that once we have computed $\mathbf{P}_\infty$ and $\mathbf{A}(\Delta t)$, we can easily compute

$$\mathbf{\Sigma}(\Delta t) = \mathbf{P}_\infty - \mathbf{A}(\Delta t)\, \mathbf{P}_\infty \, \mathbf{A}^\mathsf{T}(\Delta t), \tag{6.71}$$

and the use of matrix fractions for computing $\mathbf{\Sigma}(\Delta t)$ is not necessary.

By substituting the steady-state covariance function $\mathbf{P}(t) = \mathbf{P}_\infty$ to Equation (6.66), we can readily derive the covariance function at the steady state. It now turns out that the covariance function does not explicitly depend on times t and s, but instead only on the difference $s - t$. Thus, by letting $\tau = s - t$, the steady-state covariance function becomes

$$\mathbf{C}(\tau) = \begin{cases} \mathbf{P}_\infty \exp(\tau\, \mathbf{F})^\mathsf{T}, & \text{if } \tau > 0, \\ \exp(-\tau\, \mathbf{F})\, \mathbf{P}_\infty, & \text{if } \tau \leq 0, \end{cases} \tag{6.72}$$

which we have written as function of a single variable τ only. We will continue to use this slight abuse of notation in the next sections. We also refer this function, which is sometimes called the *autocovariance function*, simply as the *covariance function*.

Example 6.8 (Steady-state solution of the Ornstein–Uhlenbeck model). *Consider the steady-state solution of the Ornstein–Uhlenbeck SDE*

$$\mathrm{d}x = -\lambda\, x\, \mathrm{d}t + \mathrm{d}\beta, \tag{6.73}$$

where $\lambda > 0$ and the Brownian motion β has a diffusion constant $q > 0$. The steady-state variance can be obtained from

$$\frac{\mathrm{d}P}{\mathrm{d}t} = -2\lambda\, P + q = 0, \tag{6.74}$$

which gives $P = \frac{q}{2\lambda}$, and the covariance function is given as

$$C(\tau) = \begin{cases} \frac{q}{2\lambda} \exp(-\lambda\, \tau), & \text{if } \tau \geq 0, \\ \exp(\lambda\, \tau)\, \frac{q}{2\lambda}, & \text{if } \tau < 0, \end{cases} \tag{6.75}$$

which can also be written simply as

$$C(\tau) = \frac{q}{2\lambda} \exp(-\lambda |\tau|). \tag{6.76}$$

6.6 Fourier Analysis of LTI SDEs

Another way to study the steady-state solutions of LTI SDEs is to use the Fourier domain. In that case, a useful quantity is the spectral density, which is the expectation of the squared absolute value of the Fourier transform of the process. For example, if the Fourier transform of a (real-valued) scalar process $x(t)$ is $X(i\,\omega)$, then its spectral density is

$$S_x(\omega) = \mathrm{E}\left[|X(i\,\omega)|^2\right] = \mathrm{E}\left[X(i\,\omega)\,X(-i\,\omega)\right]. \tag{6.77}$$

In the case of a vector process $\mathbf{x}(t)$, we have the spectral density matrix

$$\mathbf{S_x}(\omega) = \mathrm{E}\left[\mathbf{X}(i\,\omega)\,\mathbf{X}^\mathsf{T}(-i\,\omega)\right]. \tag{6.78}$$

Now if $\mathbf{w}(t)$ is a white noise process with spectral density $\mathbf{Q}$, it really means that the expected squared absolute value of the Fourier transform is $\mathbf{Q}$:

$$\mathbf{S_w}(\omega) = \mathrm{E}\left[\mathbf{W}(i\,\omega)\,\mathbf{W}^\mathsf{T}(-i\,\omega)\right] = \mathbf{Q}. \tag{6.79}$$

However, one needs to be extra careful when using this, because the Fourier transform of a white noise process is defined only as a kind of limit of smooth processes. Fortunately, as long as we only work with linear systems, this definition indeed works. It also provides a useful tool for determining covariance functions of stochastic differential equations.

As we already saw in the previous section, the covariance function of a zero mean stationary stochastic process $\mathbf{x}(t)$ can be defined as

$$\mathbf{C_x}(\tau) = \mathrm{E}[\mathbf{x}(t)\,\mathbf{x}^\mathsf{T}(t + \tau)]. \tag{6.80}$$

This function is independent of t, because we have assumed that the process is *stationary*. This means that formally we think that the process has been started at time $t_0 = -\infty$ and it has reached its stationary state such that its statistics no longer depend on the absolute time t, but only the difference of time steps τ.

The Wiener–Khinchin theorem says that the covariance function is the inverse Fourier transform of the spectral density:

$$\mathbf{C_x}(\tau) = \mathfrak{F}^{-1}[\mathbf{S_x}(\omega)]. \tag{6.81}$$

For the white noise process, we get

$$\mathbf{C_w}(\tau) = \mathfrak{F}^{-1}[\mathbf{Q}] = \mathbf{Q}\,\mathfrak{F}^{-1}[1] = \mathbf{Q}\,\delta(\tau), \tag{6.82}$$

as expected.

Now consider the stochastic differential equation

$$\frac{d\mathbf{x}}{dt} = \mathbf{F}\,\mathbf{x} + \mathbf{L}\,\mathbf{w}, \tag{6.83}$$

and assume that it has already reached its stationary state and hence it also has zero mean. Recall from the previous section that the stationary stage can only exist of the matrix $\mathbf{F}$ corresponds to a *stable* system, which means that all its eigenvalues have negative real parts. Let us now assume that it is indeed the case.

Similarly, as in Section 2.4 we get the following solution for the Fourier transform $\mathbf{X}(i\,\omega)$ of $\mathbf{x}(t)$:

$$\mathbf{X}(i\,\omega) = ((i\,\omega)\,\mathbf{I} - \mathbf{F})^{-1}\,\mathbf{L}\,\mathbf{W}(i\,\omega), \tag{6.84}$$

where $\mathbf{W}(i\,\omega)$ is the *formal* Fourier transform of white noise $\mathbf{w}(t)$. Note that this transform does not strictly exist, because a white noise process is not square-integrable, but let us now pretend that it does.

The spectral density of $\mathbf{x}(t)$ is now given by the matrix

$$\begin{aligned}
\mathbf{S_x}(\omega) &= (\mathbf{F} - (i\,\omega)\,\mathbf{I})^{-1}\,\mathbf{L}\,\mathrm{E}\left[\mathbf{W}(i\,\omega)\,\mathbf{W}^\mathsf{T}(-i\,\omega)\right]\mathbf{L}^\mathsf{T}\,(\mathbf{F} + (i\,\omega)\,\mathbf{I})^{-\mathsf{T}} \\
&= (\mathbf{F} - (i\,\omega)\,\mathbf{I})^{-1}\,\mathbf{L}\,\mathbf{Q}\,\mathbf{L}^\mathsf{T}\,(\mathbf{F} + (i\,\omega)\,\mathbf{I})^{-\mathsf{T}}. \tag{6.85}
\end{aligned}$$

Thus the covariance function is

$$\mathbf{C_x}(\tau) = \mathfrak{F}^{-1}[(\mathbf{F} - (i\,\omega)\,\mathbf{I})^{-1}\,\mathbf{L}\,\mathbf{Q}\,\mathbf{L}^\mathsf{T}\,(\mathbf{F} + (i\,\omega)\,\mathbf{I})^{-\mathsf{T}}]. \tag{6.86}$$

Even though this looks complicated, it provides useful means of computing the covariance function of a solution to stochastic differential equation without first explicitly solving the equation.

Note that because

$$\mathbf{C_x}(0) = \mathbf{P}_\infty, \tag{6.87}$$

by Equation (6.72), where $\mathbf{P}_\infty$ is the stationary solution considered in the previous section, we also get the following identity:

$$\mathbf{P}_\infty = \frac{1}{2\pi}\int_{-\infty}^{\infty}(\mathbf{F} - (i\,\omega)\,\mathbf{I})^{-1}\,\mathbf{L}\,\mathbf{Q}\,\mathbf{L}^\mathsf{T}\,(\mathbf{F} + (i\,\omega)\,\mathbf{I})^{-\mathsf{T}}\,d\omega, \tag{6.88}$$

which can sometimes be used for computing solutions to stationary (algebraic) Lyapunov equations.

From the covariance function, we can also extract the discretization matrices as follows using Equation (6.72):

$$\mathbf{A}(\Delta t) = \mathbf{C}_\mathbf{x}^\mathsf{T}(\Delta t)\,\mathbf{P}_\infty^{-1}, \tag{6.89}$$

and $\boldsymbol{\Sigma}(\Delta t)$ can be further computed from (6.71). Thus, provided that we know the full covariance function of a linear time-invariant Itô process, we can form its equivalent discretization without forming its continuous-time state-space form.

Example 6.9 (Spectrum and covariance of the Ornstein–Uhlenbeck process). *Consider the Ornstein–Uhlenbeck SDE in its white noise form:*

$$\frac{dx}{dt} = -\lambda\,x + w, \tag{6.90}$$

where $\lambda > 0$ and the spectral density of white noise $w(t)$ is q. Taking formal Fourier transform from both sides yields

$$(i\,\omega)\,X(i\,\omega) = -\lambda\,X(i\,\omega) + W(i\,\omega), \tag{6.91}$$

and solving for $X(i\,\omega)$ gives

$$X(i\,\omega) = \frac{W(i\,\omega)}{(i\,\omega) + \lambda}. \tag{6.92}$$

Thus we get the following spectral density:

$$S_x(\omega) = \frac{\mathrm{E}\left[|W(i\,\omega)|^2\right]}{|(i\,\omega) + \lambda|^2} = \frac{q}{\omega^2 + \lambda^2}, \tag{6.93}$$

where q is the spectral density of the white noise input process $w(t)$. The Fourier transform then leads to the covariance function

$$C_x(\tau) = \frac{q}{2\lambda}\,\exp(-\lambda\,|\tau|), \tag{6.94}$$

which agrees with the result in (6.76). Furthermore, we get

$$P_\infty = \frac{1}{2\pi}\int_{-\infty}^{\infty}\frac{q}{\lambda^2 + \omega^2}\,d\omega = \frac{q}{2\,\lambda}, \tag{6.95}$$

which is the solution to the stationary Lyapunov equation in (6.74).

As pointed out previously, the analysis in this section has not been entirely rigorous, because we had to resort to computation of the Fourier transform of white noise

$$W(i\,\omega) = \int_{-\infty}^{\infty} w(t)\,\exp(-i\,\omega\,t)\,dt, \tag{6.96}$$

which is not well defined as an ordinary integral. The obvious substitution $d\beta = w(t)\,dt$ will not help us either, because we would still have trouble in defining what is meant by this resulting highly oscillatory stochastic process.

The problem can be solved by using the integrated Fourier transform as follows. It can be shown (see, e.g., Van Trees, 1968) that every stationary Gaussian process $x(t)$ has a representation of the form

$$x(t) = \int_{-\infty}^{\infty} \exp(i\,\omega\,t)\,d\zeta(i\,\omega), \tag{6.97}$$

where $\omega \mapsto \zeta(i\,\omega)$ is some complex-valued Gaussian process with independent increments. Then the mean squared difference $E[|\zeta(\omega_{k+1}) - \zeta(\omega_k)|^2]$ roughly corresponds to the mean power on the interval $[\omega_k, \omega_{k+1}]$. The spectral density then corresponds to a function $S(\omega)$ such that

$$E[|\zeta(\omega_{k+1}) - \zeta(\omega_k)|^2] = \frac{1}{\pi} \int_{\omega_k}^{\omega_{k+1}} S(\omega)\,d\omega, \tag{6.98}$$

where the constant factor results from two-sidedness of $S(\omega)$ and from the constant factor $(2\pi)^{-1}$ in the inverse Fourier transform.

By replacing the Fourier transform in the preceding analysis with the integrated Fourier transform, it is possible derive the spectral densities of covariance functions of LTI SDEs without resorting to the formal Fourier transform of white noise. However, the results remain exactly the same. For more information on this procedure, see, for example, Van Trees (1968).

Another way to treat the problem is to recall that the solution of a LTI ODE of the form

$$\frac{d\mathbf{x}}{dt} = \mathbf{F}\,\mathbf{x} + \mathbf{L}\,\mathbf{u}, \tag{6.99}$$

where $\mathbf{u}(t)$ is a smooth process, approaches the solution of the corresponding LTI SDE in the Stratonovich sense when the correlation length of $\mathbf{u}(t)$ goes to zero. Thus we can start by replacing the formal white noise process with a Gaussian process with covariance function

$$\mathbf{C_u}(\tau; \Delta t) = \mathbf{Q}\,\frac{1}{\sqrt{2\pi\,\Delta t^2}}\,\exp\left(-\frac{1}{2\,\Delta t^2}\tau^2\right), \tag{6.100}$$

which in the limit $\Delta t \to 0$ gives the white noise:

$$\lim_{\Delta t \downarrow 0} \mathbf{C_u}(\tau; \Delta t) = \mathbf{Q}\,\delta(\tau). \tag{6.101}$$

If we now carry out the derivation in the beginning of this section, we end up with the following spectral density:

$$\mathbf{S_x}(\omega; \Delta t) = (\mathbf{F} - (\mathrm{i}\,\omega)\,\mathbf{I})^{-1}\,\mathbf{L}\,\mathbf{Q}\,\exp\left(-\frac{\Delta t^2}{2}\,\omega^2\right)\,\mathbf{L}^\mathsf{T}\,(\mathbf{F} + (\mathrm{i}\,\omega)\,\mathbf{I})^{-\mathsf{T}}.$$

$$(6.102)$$

We can now compute the limit $\Delta t \to 0$ to get the spectral density corresponding to the white noise input:

$$\mathbf{S_x}(\omega) = \lim_{\Delta t \to 0}\,\mathbf{S_x}(\omega; \Delta t) = (\mathbf{F} - (\mathrm{i}\,\omega)\,\mathbf{I})^{-1}\,\mathbf{L}\,\mathbf{Q}\,\mathbf{L}^\mathsf{T}\,(\mathbf{F} + (\mathrm{i}\,\omega)\,\mathbf{I})^{-\mathsf{T}},$$

$$(6.103)$$

which agrees with the result obtained in Section 6.6. This also implies that the covariance function of $\mathbf{x}$ is indeed

$$\mathbf{C_x}(\tau) = \mathfrak{F}^{-1}[(\mathbf{F} - (\mathrm{i}\,\omega)\,\mathbf{I})^{-1}\,\mathbf{L}\,\mathbf{Q}\,\mathbf{L}^\mathsf{T}\,(\mathbf{F} + (\mathrm{i}\,\omega)\,\mathbf{I})^{-\mathsf{T}}]. \qquad (6.104)$$

6.7 Exercises

6.1   Consider a "Wiener acceleration model", where the model matrices of the linear time-invariant SDE are

$$\mathbf{F} = \begin{pmatrix} 0 & 1 & 0 \\ 0 & 0 & 1 \\ 0 & 0 & 0 \end{pmatrix} \quad \text{and} \quad \mathbf{L} = \begin{pmatrix} 0 \\ 0 \\ 1 \end{pmatrix},$$

which corresponds to $\mathrm{d}^3 x(t)/\mathrm{d}t^3 = w(t)$. Provide the discrete-time solution in terms of $\mathbf{A}(\Delta t)$ and $\boldsymbol{\Sigma}(\Delta t)$.

6.2   Provide the discrete-time solution in terms of $\mathbf{A}(\Delta t)$ and $\boldsymbol{\Sigma}(\Delta t)$ to the linear SDE given for smartphone orientation tracking in Example 3.5:

$$\mathrm{d}\mathbf{g}_\mathrm{L} = -\boldsymbol{\omega}_\mathrm{L} \times \mathbf{g}_\mathrm{L}\,\mathrm{d}t + \mathrm{d}\boldsymbol{\beta},$$

where $\mathbf{g}_\mathrm{L}$ represents the locally seen gravitation, $\boldsymbol{\omega}_\mathrm{L}$ angular velocity (here assumed to be constant), and $\boldsymbol{\beta}(t)$ is a Brownian motion with diffusion matrix $q\,\mathbf{I}$. Recall that the matrix exponential was already solved in Exercise 3.6 using the Rodrigues formula.

6.3   Calculate the matrix fraction decomposition for the Ornstein–Uhlenbeck process

$$\mathrm{d}x = -\lambda\,x\,\mathrm{d}t + \mathrm{d}\beta,$$

where $\beta(t)$ is a Brownian motion with diffusion constant q.

6.4   Derive the expression for $\mathbf{B}(\Delta t_k)$ based on Remark 6.7.

6.5   Derive the covariance function for the Wiener velocity model

$$\mathrm{d}\mathbf{x} = \begin{pmatrix} 0 & 1 \\ 0 & 0 \end{pmatrix}\mathbf{x}\,\mathrm{d}t + \begin{pmatrix} 0 \\ 1 \end{pmatrix}\mathrm{d}\beta,$$

where $\lambda > 0$, $\beta(t)$ is a Brownian motion with diffusion q and the initial condition is $\mathbf{x}(0) = \mathbf{0}$.

6.6　Derive the covariance function of the nonstationary model

$$d\mathbf{x} = \begin{pmatrix} 0 & 1 \\ 0 & 0 \end{pmatrix} \mathbf{x}\, dt,$$

where $\mathbf{x}(0) \sim N(\mathbf{0}, \mathrm{diag}(0, \sigma^2))$.

6.7　Theorem 5.6 tells that we can form a stationary solution to the FPK equation in form of Equation (5.26) when the drift can be written as a gradient of a potential function. In case of a linear SDE

$$d\mathbf{x} = \mathbf{F}\,\mathbf{x}\, dt + \mathbf{L}\, d\boldsymbol{\beta},$$

what kinds of conditions does this imply for the matrix $\mathbf{F}$? Can you use this result to obtain a class of solutions to Equation (6.69)?

6.8　Derive the steady-state mean $\mathbf{m}_\infty$ and covariance $\mathbf{P}_\infty$ of the Matérn ($\nu = 3/2$) model

$$d\mathbf{x} = \begin{pmatrix} 0 & 1 \\ -\lambda^2 & -2\lambda \end{pmatrix} \mathbf{x}\, dt + \begin{pmatrix} 0 \\ 1 \end{pmatrix} d\beta, \tag{6.105}$$

by solving the stationary states of the mean and covariance differential equations. The parameter $\lambda > 0$ and $\beta(t)$ is a Brownian motion with diffusion q. Also derive the covariance function of the process.

6.9　Derive the steady-state solution and covariance function of the Matérn ($\nu = 3/2$) model (6.105) using the Fourier transform similarly as in Example 6.9.

7

Useful Theorems and Formulas for SDEs

This chapter contains a set of useful results and theorems that are somewhat on the edge of more advanced analysis and methods for SDEs. Although the chapter contains important topics, it can be safely skipped during the first reading, because it is a bit of a detour from the main storyline of the book. We start with Lamperti transforms for solving multiplicative noise SDEs and then proceed to different constructions and characteristics of Brownian motion, including the definition of the Wiener measure. We then introduce and discuss the Girsanov theorem that can be used for transforming measures and finding weak solutions to SDEs. The Doob's h-transform is then introduced as a method for conditioning SDEs on their end point, and finally we discuss path integrals and the Feynman–Kac formulae that connect SDEs to quantum mechanics and partial differential equations.

7.1 Lamperti Transform

The Lamperti transform (see, e.g., Iacus, 2008; Møller and Madsen, 2010) is a change of variables $\mathbf{y} = \mathbf{h}(\mathbf{x}, t)$ in an SDE

$$\mathrm{d}\mathbf{x} = \mathbf{f}(\mathbf{x}, t)\,\mathrm{d}t + \mathbf{L}(\mathbf{x}, t)\,\mathrm{d}\boldsymbol{\beta}, \tag{7.1}$$

such that it transforms into

$$\mathrm{d}\mathbf{y} = \mathbf{g}(\mathbf{y}, t)\,\mathrm{d}t + \mathrm{d}\boldsymbol{\beta}. \tag{7.2}$$

That is, the multiplicative noise is transformed into additive noise. This transformation is useful, for example, for numerical treatment of SDEs, because numerically solving an additive-noise SDE (7.2) is usually easier than solving a general SDE of the form (7.1). Even though the Lamperti transform can be, in a sense, extended to a multivariate setting, it is essentially a one-dimensional transform (Møller and Madsen, 2010).

To illustrate the idea of the Lamperti transform, let us consider a scalar

SDE

$$dx = f(x,t)\, dt + L(x,t)\, d\beta, \tag{7.3}$$

where $\beta(t)$ is a standard Brownian motion. We can then define a transform as follows:

$$y = h(x,t) = \int_\xi^x \frac{1}{L(u,t)}\, du, \tag{7.4}$$

where ξ is an arbitrary point. Applying the Itô formula on h gives

$$
\begin{aligned}
dy &= \frac{\partial h(x,t)}{\partial t}\, dt + \left(\frac{1}{L(x,t)}\right) dx + \frac{1}{2}\left(-\frac{\partial L(x,t)/\partial x)}{L^2(x,t)}\right) dx^2 \\
&= \left[\frac{\partial}{\partial t}\int_\xi^x \frac{1}{L(u,t)}\, du\right] dt + \left(\frac{1}{L(x,t)}\right) \left(f(x,t)\, dt + L(x,t)\, d\beta\right) \\
&\quad + \frac{1}{2}\left(-\frac{\partial L(x,t)/\partial x)}{L^2(x,t)}\right) L^2(x,t)\, dt \\
&= \left(\frac{\partial}{\partial t}\int_\xi^x \frac{1}{L(u,t)}\, du + \frac{f(x,t)}{L(x,t)} - \frac{1}{2}\frac{\partial L(x,t)}{\partial x}\right) dt + d\beta. \tag{7.5}
\end{aligned}
$$

Provided that $x = h^{-1}(y,t)$ exists, we can express the SDE in terms of y by

$$dy = \underbrace{\left(\frac{\partial}{\partial t}\int_\xi^x \frac{1}{L(u,t)}\, du + \frac{f(x,t)}{L(x,t)} - \frac{1}{2}\frac{\partial L(x,t)}{\partial x}\right)\Bigg|_{x=h^{-1}(y,t)}}_{g(y,t)} dt + d\beta. \tag{7.6}$$

The transformation generalizes to a multivariate setting if $\mathbf{L}(\mathbf{x},t)$ is diagonal such that each $L_{ii}(\mathbf{x},t)$ only depends on the component x_i. Then we can apply the Lamperti transform to each component separately. More generally the transformation does not have a simple formula, but instead the transformation can only be defined as a solution to a system of partial differential equations (Møller and Madsen, 2010).

Example 7.1 (Lamperti transform). *Let us consider again the SDE in Example 4.9, which was given as*

$$dx = \left(\alpha x \log x + \frac{1}{2}x\right) dt + x\, d\beta. \tag{7.7}$$

The corresponding Lamperti transform is

$$h(x) = \int_\xi^x \frac{1}{x}\, dx = \log x - \log \xi, \tag{7.8}$$

where we can, for example, choose $\xi = 1$, which then gives $h^{-1}(y) = \exp(y)$. The transformed SDE is then given as

$$dy = \left. \left(\frac{\alpha x \log x + \frac{1}{2} x}{x} - \frac{1}{2} \right) \right|_{x=\exp(y)} dt + d\beta$$

$$= \alpha \, y \, dt + d\beta, \tag{7.9}$$

which is a linear SDE and thus easily solvable. Solving this equation and transforming back leads to the result (4.47) that we obtained earlier.

7.2 Constructions of Brownian Motion and the Wiener Measure

In this section, we discuss some constructions of Brownian motion. We only consider one-dimensional standard Brownian motion with $q = 1$ – which is also often referred to as the Wiener process – but all the ideas generalize to multivariate and nonstandard Brownian motions with other initial conditions. We can already see as a special case of Definition 4.1 that a standard Brownian motion is a process with independent Gaussian increments of the form

$$\beta(t_{k+1}) - \beta(t_k) \sim N(0, t_{k+1} - t_k), \quad t_{k+1} > t_k. \tag{7.10}$$

From the preceding we can also deduce that Brownian motion is nowhere differentiable and that the quadratic variation of Brownian motion is $[b, b]_t = t$, where the quadratic variation is defined as the limit

$$[\beta, \beta]_t = \lim_{t_{k+1} - t_k \to 0} \sum_{t_k \le t} |\beta(t_{k+1}) - \beta(t_k)|^2. \tag{7.11}$$

It turns out that the preceding already characterizes the Brownian motion provided that we also demand that it is a continuous process with this property. This is called the Lévy characterization theorem, which states the following (see, e.g., Karatzas and Shreve, 1991).

Theorem 7.2 (Lévy's characterization of Brownian motion). *Let $\{\beta(t) \mid t \in [0, \infty)\}$ be a continuous martingale with $\beta(0) = 0$ and $[\beta, \beta]_t = t$, then $\beta(t)$ is a standard Brownian motion.*

Based on Equation (7.10), we can also proceed to write down the probability density of an arbitrary finite-dimensional distribution $p(\beta(t_1), \ldots, \beta(t_T))$ as follows:

$$p(\beta(t_1), \ldots, \beta(t_T)) = \prod_{k=0}^{T-1} N(\beta(t_{k+1}) \mid \beta(t_k), t_{k+1} - t_k). \tag{7.12}$$

Because this is valid for any collection of time points, it defines a probability measure for the whole process. The probability measure is called the *Wiener measure* (Wiener, 1923). Nowadays, the Wiener measure is typically defined implicitly as the probability measure of the process that has the preceding finite-dimensional distributions. The corresponding measure can be shown to be unique, have a continuous version, and other useful properties (see, e.g., Karatzas and Shreve, 1991; Rogers and Williams, 2000a).

In physics literature (see, e.g., Chaichian and Demichev, 2001a), the Wiener measure is often not defined via its finite-dimensional distributions, but in connection with path integrals, which is also close to the original exposition by Wiener (1923). If we plug in the density of the normal distribution to the preceding equation, we get

$$p(\beta(t_1), \ldots, \beta(t_T))$$

$$= \prod_{k=0}^{T-1} \frac{1}{\sqrt{2\pi (t_{k+1} - t_k)}} \exp\left(-\frac{1}{2(t_{k+1} - t_k)} (\beta(t_{k+1}) - \beta(t_k))^2\right)$$

$$= \exp\left(-\frac{1}{2} \sum_{k=0}^{T-1} \frac{(\beta(t_{k+1}) - \beta(t_k))^2}{(t_{k+1} - t_k)^2} (t_{k+1} - t_k)\right) \prod_{k=0}^{T-1} \frac{1}{\sqrt{2\pi (t_{k+1} - t_k)}}.$$

$$(7.13)$$

The probability of a set $\mathcal{B}_T$ of discrete paths is then given as

$$P((\beta(t_1), \ldots, \beta(t_T)) \in \mathcal{B}_T)$$

$$= \int_{\mathcal{B}_T} p(\beta(t_1), \ldots, \beta(t_T)) \, d\beta(t_1) \times \cdots \times d\beta(t_T)$$

$$= \int_{\mathcal{B}_T} \exp\left(-\frac{1}{2} \sum_{k=0}^{T-1} \frac{(\beta(t_{k+1}) - \beta(t_k))^2}{(t_{k+1} - t_k)^2} (t_{k+1} - t_k)\right)$$

$$\times \prod_{k=0}^{T-1} \frac{d\beta(t_{k+1})}{\sqrt{2\pi (t_{k+1} - t_k)}}.$$

$$(7.14)$$

We can now formally take the limit $(\beta(t_{k+1}) - \beta(t_k))^2 / (t_{k+1} - t_k)^2 \to \dot\beta^2(t)$, where $\dot\beta \triangleq d\beta/dt$ and thus formally define the Wiener measure P_W of a set of functions $\beta \in \mathcal{B}$ on interval $[0, t]$ as

$$P_W(\beta \in \mathcal{B}) = \int_{\mathcal{B}} \exp\left(-\frac{1}{2} \int_0^t \dot\beta^2(\tau) \, d\tau\right) \prod_{\tau=0}^t \frac{d\beta(\tau)}{\sqrt{2\pi \, d\tau}}, \qquad (7.15)$$

which is a "path integral" in physics terms, that is, an integral over the paths of $\beta(t)$. Note that the product the end of the expression is actually a notational shorthand for a "continuous product" over the span $[\tau, t]$.

This representation of the Wiener measure should be considered a formal expression, because Brownian motion is not differentiable and thus the derivative $\dot{\beta}(t)$ is not well defined. However, this representation still works well in practice in physics.

It is also possible to construct Brownian motion as a limit of a random walk, which is taking integer steps up and down. Note that the construction (7.10) can also be seen as a random walk

$$\beta_{k+1} = \beta_k + q_k, \qquad (7.16)$$

where q_k is a zero-mean Gaussian random variable with variance $\sigma_k^2 = t_{k+1} - t_k$. Now the central limit theorem states that the limit of a sum of identically distributed random variables ξ_i with zero mean and variance σ^2 tends to a Gaussian random variable in the following sense:

$$\frac{\xi_1 + \xi_2 + \cdots + \xi_n}{\sqrt{n}} \to N(0, \sigma^2), \qquad (7.17)$$

when $n \to \infty$. Let us write

$$s_k = \sum_{i=1}^{k} \xi_i \qquad (7.18)$$

and select the random variable ξ_i such that it takes values $\{-1, 1\}$ with equal probabilities. Now consider the process

$$S_n(t) = \frac{s_{[nt]}}{\sqrt{n}}, \qquad (7.19)$$

where $[\bullet]$ denotes the integer part of its argument. Then it turns out that

$$S_n(t_{k+1}) - S_n(t_k) = \frac{s_{[nt_{k+1}]}}{\sqrt{n}} - \frac{s_{[nt_k]}}{\sqrt{n}} \to N(0, t_{k+1} - t_k), \qquad (7.20)$$

when $n \to \infty$. Thus we can construct Brownian motion as the limit of the integer-step random walk:

$$\beta(t) = \lim_{n \to \infty} S_n(t). \qquad (7.21)$$

Yet another way of looking at Brownian motion is to see it as a zero-mean Gaussian process with the covariance function

$$C(t, t') = \min(t, t'). \qquad (7.22)$$

This also directly results from the property (7.10).

Let us consider a certain finite interval $t \in [0, T]$. The so-called Mercer's theorem (Mercer, 1909; Loève, 1963) tells that we can expand the preceding covariance function as

$$C(t, t') = \sum_{n=1}^{\infty} \lambda_n \phi_n(t) \phi_n(t'), \tag{7.23}$$

where λ_n and ϕ_n are the eigenvalues and eigenfunctions of the covariance function, respectively, in this case defined as follows:

$$\int_0^T C(t, t') \phi_n(t') \, dt' = \lambda_n \phi_n(t). \tag{7.24}$$

It turns out that the eigenvalues and eigenfunctions of standard Brownian motion are given as

$$\lambda_n = \left(\frac{2T}{(2n-1)\pi} \right)^2,$$

$$\phi_n(t) = \left(\frac{2}{T} \right)^{1/2} \sin \left(\frac{(2n-1)\pi}{2T} t \right). \tag{7.25}$$

We can then construct Brownian motion through its Karhunen–Loeve expansion (Loève, 1963):

$$\beta(t) = \sum_{n=1}^{\infty} z_n \phi_n(t), \tag{7.26}$$

where $z_n \sim N(0, \lambda_n)$ are independent Gaussian random variables.

Example 7.3 (Karhunen–Loeve series of Brownian motion). *It is possible to approximate Brownian motion by using a finite truncation of its Karhunen–Loeve series. The convergence of the series to Brownian motion is illustrated in Figure 7.1.*

We can also take any orthonormal basis $\{\varphi_n(t)\}$ of the Hilbert space with the inner product

$$\langle f, g \rangle = \int_0^T f(\tau) g(\tau) \, d\tau \tag{7.27}$$

and expand Brownian motion on $t \in [0, T]$ as

$$\beta(t) = \sum_{n=1}^{\infty} z_n \int_0^t \varphi_n(\tau) \, d\tau, \tag{7.28}$$

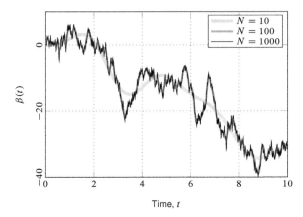

Figure 7.1 Illustration of the Karhunen–Loeve expansion of Brownian motion with 10, 100, and 1000 terms in the truncated series. With increasing N the process becomes rougher and eventually non-differentiable in the limit.

where $z_n \sim N(0, 1)$ are independent (see, e.g., Luo, 2006). An interesting choice is, for example, to use Haar functions, which correspond to the Lévy–Ciesielski construction of Brownian motion (see, e.g., Rogers and Williams, 2000a).

Note that the Karhunen–Loeve expansion corresponds to the selection

$$\varphi_n(t) = \left(\frac{2}{T} \right)^{1/2} \cos \left(\frac{(2n-1)\,\pi}{2T} t \right) \tag{7.29}$$

in the preceding representation.

7.3 Girsanov Theorem

The Girsanov theorem is an important theorem in stochastics that can be used for transforming the probability measures of SDEs. It is often used for removing or transforming drift functions of SDEs, and it can also be used for finding weak solutions to SDEs. The original Girsanov theorem is given as Theorem 7.8, and it is due to Girsanov (1960). In addition to the original article, its derivation can be found, for example, in Karatzas and Shreve (1991) (see also Øksendal, 2003). The derivation of Theorem 7.4 from the Girsanov theorem can be found, for example, in Särkkä and Sottinen (2008). Here we proceed backward from Theorem 7.4 to Theorem 7.8. Some intuition behind the theorem is provided in the next section.

As in Chapter 5, we denote the whole path of the Itô process $\mathbf{x}(t)$ on a time interval $[0, t]$ as follows:

$$\mathcal{X}_t = \{\mathbf{x}(\tau) \mid 0 \le \tau \le t\}. \tag{7.30}$$

Let $\mathbf{x}(t)$ be the solution to

$$d\mathbf{x} = \mathbf{f}(\mathbf{x}, t)\, dt + d\boldsymbol{\beta}. \tag{7.31}$$

Here we have set $\mathbf{L}(\mathbf{x}, t) = \mathbf{I}$ for notational simplicity. In fact, the Girsanov theorem can be used for general time-varying $\mathbf{L}(t)$ provided that $\mathbf{L}(t)$ is invertible for each t. This invertibility requirement can also be relaxed in some situations (cf. Särkkä and Sottinen, 2008).

For any finite n, the joint probability density $p(\mathbf{x}(t_1), \mathbf{x}(t_2), \ldots, \mathbf{x}(t_n))$ exists (provided that certain technical conditions are met) for an arbitrary finite collection of times $t_1, t_2, \ldots, t_n$. We will now formally define the probability density of the whole path as

$$p(\mathcal{X}_t) = \lim_{n \to \infty} p(\mathbf{x}(t_1), \mathbf{x}(t_2), \ldots, \mathbf{x}(t_n)), \tag{7.32}$$

where the times $t_1, t_2, \ldots, t_n$ need to be selected such that they become dense in the limit. In fact, this density is not normalizable, but we can still define the density through the ratio between the joint probability density of $\mathbf{x}$ and another process $\mathbf{y}$:

$$\frac{p(\mathcal{X}_t)}{p(\mathcal{Y}_t)} = \lim_{n \to \infty} \frac{p(\mathbf{x}(t_1), \mathbf{x}(t_2), \ldots, \mathbf{x}(t_n))}{p(\mathbf{y}(t_1), \mathbf{y}(t_2), \ldots, \mathbf{y}(t_n))}. \tag{7.33}$$

This is a finite quantity with a suitable choice of $\mathbf{y}$. We can also denote the expectation of a functional $h(\mathcal{X}_t)$ of the path as follows:

$$E[h(\mathcal{X}_t)] = \int h(\mathcal{X}_t)\, p(\mathcal{X}_t)\, d\mathcal{X}_t. \tag{7.34}$$

In physics these kinds of integrals are called path integrals (Chaichian and Demichev, 2001a,b, see also Section 7.6). Note that this notation is purely formal, because the density $p(\mathcal{X}_t)$ is actually an infinite quantity. However, the expectation is indeed finite. Let us now compute the ratio of probability densities for a pair of processes.

Theorem 7.4 (Likelihood ratio of Itô processes). *Consider the Itô processes*

$$\begin{aligned} d\mathbf{x} &= \mathbf{f}(\mathbf{x}, t)\, dt + d\boldsymbol{\beta}, & \mathbf{x}(0) &= \mathbf{x}_0, \\ d\mathbf{y} &= \mathbf{g}(\mathbf{y}, t)\, dt + d\boldsymbol{\beta}, & \mathbf{y}(0) &= \mathbf{x}_0, \end{aligned} \tag{7.35}$$

where $\mathbf{x}(t), \mathbf{y}(t) \in \mathbb{R}^D$ and the Brownian motion $\boldsymbol{\beta}(t) \in \mathbb{R}^D$ has a non-singular diffusion matrix $\mathbf{Q}$. Then the ratio of the probability laws of $\mathcal{X}_t$ and $\mathcal{Y}_t$ is given as

$$\frac{p(\mathcal{X}_t)}{p(\mathcal{Y}_t)} = Z(t), \tag{7.36}$$

where

$$Z(t) = \exp\left(-\frac{1}{2}\int_0^t [\mathbf{f}(\mathbf{y}, \tau) - \mathbf{g}(\mathbf{y}, \tau)]^\mathsf{T}\, \mathbf{Q}^{-1}\, [\mathbf{f}(\mathbf{y}, \tau) - \mathbf{g}(\mathbf{y}, \tau)]\, d\tau \right.$$
$$\left. + \int_0^t [\mathbf{f}(\mathbf{y}, \tau) - \mathbf{g}(\mathbf{y}, \tau)]^\mathsf{T}\, \mathbf{Q}^{-1}\, d\boldsymbol{\beta}(\tau) \right) \tag{7.37}$$

in the sense that for an arbitrary functional $h(\bullet)$ of the path from 0 to t, we have

$$\mathrm{E}[h(\mathcal{X}_t)] = \mathrm{E}[Z(t)\, h(\mathcal{Y}_t)], \tag{7.38}$$

where the expectation is over the randomness induced by the Brownian motion. Furthermore, under the probability measure defined through the transformed probability density

$$\tilde{p}(\mathcal{X}_t) = Z(t)\, p(\mathcal{X}_t), \tag{7.39}$$

the process

$$\tilde{\boldsymbol{\beta}} = \boldsymbol{\beta} - \int_0^t [\mathbf{f}(\mathbf{y}, \tau) - \mathbf{g}(\mathbf{y}, \tau)]\, d\tau \tag{7.40}$$

is a Brownian motion with diffusion matrix $\mathbf{Q}$.

Note that in the preceding theorem, we have required that the states and the Brownian motion have the same dimensionality (i.e., $S = D$), but as discussed in Särkkä and Sottinen (2008), it can be easily relaxed.

Remark 7.5. *We need to have*

$$\mathrm{E}\left[\exp\left(\int_0^t \mathbf{f}(\mathbf{y}, \tau)^\mathsf{T}\, \mathbf{Q}^{-1}\, \mathbf{f}(\mathbf{y}, \tau)\, d\tau\right)\right] < \infty,$$
$$\mathrm{E}\left[\exp\left(\int_0^t \mathbf{g}(\mathbf{y}, \tau)^\mathsf{T}\, \mathbf{Q}^{-1}\, \mathbf{g}(\mathbf{y}, \tau)\, d\tau\right)\right] < \infty, \tag{7.41}$$

because otherwise $Z(t)$ will be zero.

From Theorem 7.4, we also get the following useful special case by setting $\mathbf{g} = \mathbf{0}$.

Corollary 7.6 (Likelihood ratio of Itô process and Brownian motion). *The likelihood ratio of the law of the process X_t as defined in Equation (7.35), to the law of the driving Brownian motion is*

$$Z(t) = \exp\left(-\frac{1}{2}\int_0^t \mathbf{f}^\mathsf{T}(\boldsymbol{\beta},\tau)\,\mathbf{Q}^{-1}\,\mathbf{f}(\boldsymbol{\beta},\tau)\,d\tau\right.$$
$$\left. + \int_0^t \mathbf{f}^\mathsf{T}(\boldsymbol{\beta},\tau)\,\mathbf{Q}^{-1}\,d\boldsymbol{\beta}(\tau)\right). \quad (7.42)$$

Now we write a slightly more abstract version of Theorem 7.4, which is roughly equivalent to the actual Girsanov theorem in the form that it is usually found in stochastics literature.

Theorem 7.7 (Girsanov I). *Let $\boldsymbol{\theta}(t)$ be a process that is driven by a standard Brownian motion $\boldsymbol{\beta}(t)$ such that*

$$\mathrm{E}\left[\int_0^t \boldsymbol{\theta}^\mathsf{T}(\tau)\,\boldsymbol{\theta}(\tau)\,d\tau\right] < \infty, \quad (7.43)$$

then under the measure defined by the formal probability density

$$\tilde{p}(\Theta_t) = Z(t)\,p(\Theta_t), \quad (7.44)$$

where $\Theta_t = \{\boldsymbol{\theta}(\tau) \mid 0 \le \tau \le t\}$ and

$$Z(t) = \exp\left(\int_0^t \boldsymbol{\theta}^\mathsf{T}(\tau)\,d\boldsymbol{\beta} - \frac{1}{2}\int_0^t \boldsymbol{\theta}^\mathsf{T}(\tau)\,\boldsymbol{\theta}(\tau)\,d\tau\right), \quad (7.45)$$

the following process is a standard Brownian motion:

$$\tilde{\boldsymbol{\beta}}(t) = \boldsymbol{\beta}(t) - \int_0^t \boldsymbol{\theta}(\tau)\,d\tau. \quad (7.46)$$

Derivation   Select $\boldsymbol{\theta}(t) = \mathbf{f}(\mathbf{y},t) - \mathbf{g}(\mathbf{y},t)$ and $\mathbf{Q} = \mathbf{I}$ in Theorem 7.4.
□

In fact, the preceding derivation does not yet guarantee that any selected $\boldsymbol{\theta}(t)$ can be constructed like this. But still, it reveals the link between the likelihood ratio and the Girsanov theorem. Despite the limited derivation, Theorem 7.7 is generally true. The detailed technical conditions can be found in the original article of Girsanov (1960).

However, Theorem 7.7 is still in the heuristic notation in terms of the formal probability densities of paths. In the proper formulation of the theorem $\boldsymbol{\theta}$ being "driven" by Brownian motion actually means that the process $\boldsymbol{\theta}$ is *adapted* to the Brownian motion. To be more explicit in notation, it is also common to write down the event space element $\omega \in \Omega$ as the argument of

$\boldsymbol{\beta}(t, \omega)$. The processes $\boldsymbol{\theta}(t, \omega)$ and $Z(t, \omega)$ should then be functions of the event space element as well. In fact, $\boldsymbol{\theta}(t, \omega)$ should be nonanticipative (not looking into the future) functional of Brownian motion, that is, adapted to the natural filtration $\mathcal{F}_t$ (Øksendal, 2003) of the Brownian motion. Furthermore, the ratio of probability densities is in fact the Radon–Nikodym derivative of the measure $\tilde{\mathbb{P}}(\omega)$ with respect to the other measure $\mathbb{P}(\omega)$. With these notations the Girsanov theorem looks like the following, which is roughly the format found in stochastics books.

Theorem 7.8 (Girsanov II). *Let $\boldsymbol{\beta}(t, \omega)$ be a standard S-dimensional Brownian motion under the probability measure $\mathbb{P}$. Let $\boldsymbol{\theta} : \mathbb{R}_+ \times \Omega \mapsto \mathbb{R}^S$ be an adapted process such that the process Z defined as*

$$Z(t, \omega) = \exp\left\{\int_0^t \boldsymbol{\theta}^\top(t, \omega) \, \mathrm{d}\boldsymbol{\beta}(t, \omega) - \frac{1}{2}\int_0^t \boldsymbol{\theta}^\top(t, \omega) \, \boldsymbol{\theta}(t, \omega) \, \mathrm{d}t\right\}$$
(7.47)

satisfies $\mathrm{E}[Z(t, \omega)] = 1$. Then the process

$$\tilde{\boldsymbol{\beta}}(t, \omega) = \boldsymbol{\beta}(t, \omega) - \int_0^t \boldsymbol{\theta}(\omega, \tau) \, \mathrm{d}\tau$$
(7.48)

is a standard Brownian motion under the probability measure $\tilde{\mathbb{P}}$ defined via the relation

$$\mathrm{E}\left[\frac{\mathrm{d}\tilde{\mathbb{P}}}{\mathrm{d}\mathbb{P}}(\omega) \,\middle|\, \mathcal{F}_t\right] = Z(t, \omega),$$
(7.49)

where $\mathcal{F}_t$ is the natural filtration of the Brownian motion $\boldsymbol{\beta}(\omega, t)$.

The Girsanov theorem can be used for eliminating the drift functions by changing the measure suitably (see, e.g., Øksendal, 2003). The basic idea in drift removal is to define $\boldsymbol{\theta}(t)$ in terms of the drift function suitably such that in the transformed SDE the drift cancels out.

A useful application of the theorem is in finding weak solutions to SDEs. Construction of weak solutions is based on the result that the process $\tilde{\boldsymbol{\beta}}(t)$ is a Brownian motion under the transformed measure. We can select $\boldsymbol{\theta}(t)$ such that there is another easily constructed process which then serves as the corresponding $\tilde{\mathbf{x}}(t)$ that solves the SDE driven by this new Brownian motion. This is illustrated in the following corollary.

Corollary 7.9 (Weak solution of SDE). *Assume that the process $\mathbf{x}(t)$ solves the stochastic differential equation*

$$\mathrm{d}\mathbf{x} = \mathbf{f}(\mathbf{x}, t) \, \mathrm{d}t + \mathrm{d}\boldsymbol{\beta}, \qquad \mathbf{x}(0) = \mathbf{x}_0.$$
(7.50)

If we now define

$$Z(t) = \exp\left(-\frac{1}{2}\int_0^t \mathbf{f}^\mathsf{T}(\mathbf{x}_0 + \boldsymbol{\beta}(\tau), \tau)\, \mathbf{Q}^{-1}\, \mathbf{f}(\mathbf{x}_0 + \boldsymbol{\beta}(\tau), \tau)\, d\tau \right.$$
$$\left. + \int_0^t \mathbf{f}^\mathsf{T}(\mathbf{x}_0 + \boldsymbol{\beta}(\tau), \tau)\, \mathbf{Q}^{-1}\, d\boldsymbol{\beta}(\tau) \right), \quad (7.51)$$

then the expectation of any function $\mathbf{h}(\bullet)$ *over the SDE solution can be expressed as*

$$E[\mathbf{h}(\mathbf{x}(t))] = E[Z(t)\,\mathbf{h}(\mathbf{x}_0 + \boldsymbol{\beta}(t))]. \quad (7.52)$$

Furthermore, the pair $(\tilde{\mathbf{x}}, \tilde{\boldsymbol{\beta}})$, *where*

$$\tilde{\mathbf{x}}(t) = \mathbf{x}_0 + \boldsymbol{\beta}(t),$$
$$\tilde{\boldsymbol{\beta}}(t) = \boldsymbol{\beta}(t) - \int_0^t \mathbf{f}(\mathbf{x}_0 + \boldsymbol{\beta}(\tau))\, d\tau, \quad (7.53)$$

is a weak solution to the SDE under the transformed probability density $\tilde{p}(\mathcal{X}_t) = Z(t)\, p(\mathcal{X}_t)$.

Proof If we select $\mathbf{g} = \mathbf{0}$, then by Theorem 7.4 $\tilde{\boldsymbol{\beta}}(t)$ is a Brownian motion under the transformed measure. Substituting $\tilde{\mathbf{x}}(t)$ to the equation of $\tilde{\boldsymbol{\beta}}(t)$ and rearranging then gives

$$\tilde{\mathbf{x}}(t) = \mathbf{x}_0 + \int_0^t \mathbf{f}(\tilde{\mathbf{x}}(t))\, d\tau + \tilde{\boldsymbol{\beta}}(t), \quad (7.54)$$

which indeed shows that the pair $(\tilde{\mathbf{x}}, \tilde{\boldsymbol{\beta}})$ is a weak solution to the original SDE. $\qquad\square$

The use of the preceding result is illustrated in the following example, where we find the transition density of the SDE.

Example 7.10 (Weak solution of the Beneš SDE). *Consider the Itô process*

$$dx = \tanh(x)\, dt + d\beta, \quad x(0) = x_0, \quad (7.55)$$

where $\beta(t)$ *is a standard Brownian motion. The likelihood ratio is now*

$$Z(t) = \exp\left(\int_0^t \tanh(x_0 + \beta(t))\, d\beta(t) - \frac{1}{2}\int_0^t \tanh^2(x_0 + \beta(t))\, dt\right). \quad (7.56)$$

The first integral can be evaluated by applying the Itô formula:

$$d(\log(\cosh(x_0 + \beta(t))))$$
$$= \tanh(x_0 + \beta(t)) \, d\beta(t) + \frac{1}{2}(1 - \tanh^2(x_0 + \beta(t))) \, dt, \quad (7.57)$$

that is,

$$\int_0^t \tanh(x_0 + \beta(t)) \, d\beta(t) - \frac{1}{2} \int_0^t \tanh^2(x_0 + \beta(t)) \, dt$$
$$= \log(\cosh(x_0 + \beta(t))) - \log(\cosh(x_0)) - \int_0^t \frac{1}{2} \, dt, \quad (7.58)$$

which gives

$$Z(t) = \exp\left(\log(\cosh(x_0 + \beta(t))) - \log(\cosh(x_0)) - \int_0^t \frac{1}{2} \, dt \right)$$
$$= \frac{\cosh(x_0 + \beta(t))}{\cosh(x_0)} \exp\left(-\frac{1}{2}t\right). \quad (7.59)$$

Because this term depends only on the state of Brownian motion at time t, the probability density can be computed explicitly. The probability density of $\tilde{x}(t) = x_0 + \beta(t)$ is

$$N(\tilde{x}(t) \mid 0, t) = \frac{1}{\sqrt{2\pi t}} \exp\left(-\frac{1}{2t}(\tilde{x}(t) - x_0)^2\right) \quad (7.60)$$

and thus the probability density of $x(t)$ is

$$p(x(t)) = \frac{1}{\sqrt{2\pi t}} \frac{\cosh(x(t))}{\cosh(x_0)} \exp\left(-\frac{1}{2}t\right) \exp\left(-\frac{1}{2t}(x(t) - x_0)^2\right). \quad (7.61)$$

Because this is true for any initial condition, it implies that the transition density is given as

$$p(x(t_{k+1}) \mid x(t_k)) = \frac{1}{\sqrt{2\pi \Delta t_k}} \frac{\cosh(x(t_{k+1}))}{\cosh(x(t_k))} \exp\left(-\frac{1}{2}\Delta t_k\right)$$
$$\times \exp\left(-\frac{1}{2\Delta t_k}(x(t_{k+1}) - x(t_k))^2\right), \quad (7.62)$$

where $\Delta t_k = t_{k+1} - t_k$.

The Girsanov theorem is also important in stochastic filtering theory (see Chapter 10). The theorem can be used as the starting point of deriving the so-called Kallianpur–Striebel formula (Bayes' rule in continuous time). From this, we can derive the whole stochastic filtering theory. The formula can also be used to form Monte Carlo (particle) methods for approximating filtering solutions. For details, see Crisan and Rozovskiĭ (2011). In so-called continuous-discrete filtering (continuous-time dynamics, discrete-time measurements), the theorem has turned out to be useful in constructing importance sampling and exact sampling methods for conditioned SDEs (Beskos et al., 2006a; Särkkä and Sottinen, 2008).

7.4 Some Intuition on the Girsanov Theorem

The purpose of this section is to give some intuition to the Girsanov theorem presented in the previous section. We will not attempt to derive the general theorem, but sketch a derivation of a discrete-time analogue of the theorem. Let us start by considering the ratio of the probability densities of a discrete-time Brownian motion and Brownian motion with a drift. Let $\beta(t_k)$ be a Brownian motion with diffusion coefficient q, which has been sampled at times $t_0 = 0, t_1 = \Delta t, t_2 = 2\Delta t, \ldots$. The sampled Brownian motion is then a discrete-time process of the form

$$\beta(t_k) = \beta(t_{k-1}) + \Delta\beta_k, \quad \Delta\beta_k \sim N(0, \Delta t\, q), \tag{7.63}$$

and its joint probability density can be expressed as

$$p(\beta(t_1), \ldots, \beta(t_n))$$
$$= \prod_{k=1}^{n} \left[\frac{1}{\sqrt{2\pi\, \Delta t\, q}} \exp\left(-\frac{(\beta(t_k) - \beta(t_{k-1}))^2}{2\, \Delta t\, q} \right) \right]$$
$$= \prod_{k=1}^{n} \left[\frac{1}{\sqrt{2\pi\, \Delta t\, q}} \exp\left(-\frac{\Delta\beta_k^2}{2\, \Delta t\, q} \right) \right]$$
$$= \frac{1}{\left(\sqrt{2\pi\, \Delta t\, q} \right)^n} \exp\left(-\frac{1}{2\, \Delta t\, q} \sum_{k=1}^{n} \Delta\beta_k^2 \right). \tag{7.64}$$

Let us now assume that we have another process defined as

$$x(t_k) = x(t_{k-1}) + f_k\, \Delta t + \Delta\beta_k, \quad \Delta\beta_k \sim N(0, \Delta t\, q), \tag{7.65}$$

where f_k is some given mean sequence. We have included the factor Δt so that this can be seen as a discrete-time analogue of the equation $dx = f(t)\, dt + d\beta$. The process x is thus a *drifted* version of the process β.

The probability density of the latter process is

$$p(x(t_1), x(t_2), \ldots, x(t_n))$$

$$= \prod_{k=1}^{n} \left[\frac{1}{\sqrt{2\pi \, \Delta t \, q}} \exp\left(-\frac{(x(t_k) - x(t_{k-1}) - f_k \, \Delta t)^2}{2 \, \Delta t \, q} \right) \right]$$

$$= \prod_{k=1}^{n} \left[\frac{1}{\sqrt{2\pi \, \Delta t \, q}} \exp\left(-\frac{(\Delta \beta_k - f_k \, \Delta t)^2}{2 \, \Delta t \, q} \right) \right]$$

$$= \frac{1}{(\sqrt{2\pi \, \Delta t \, q})^n} \exp\left(-\frac{1}{2 \, \Delta t \, q} \sum_{k=1}^{n} \Delta \beta_k^2 + \frac{1}{q} \sum_{k=1}^{n} f_k \, \Delta \beta_k \right.$$

$$\left. - \frac{1}{2q} \sum_{k=1}^{n} f_k^2 \, \Delta t \right). \tag{7.66}$$

The ratio of the probability densities now has the interesting form

$$\frac{p(x(t_1), \ldots, x(t_k))}{p(\beta(t_1), \ldots, \beta(t_n))} = \exp\left(\frac{1}{q} \sum_{k=1}^{n} f_k \, \Delta \beta_k - \frac{1}{2q} \sum_{k=1}^{n} f_k^2 \, \Delta t \right)$$

$$\triangleq Z_n(\beta(t_1), \ldots, \beta(t_n)), \quad (7.67)$$

which explicitly is only a function of the driving random variables $\Delta \beta_k$ (or equivalently $\beta(t_k)$) and the biases f_k, not of the values, of $x(t_k)$. Another interesting thing is that the expression seems to stay meaningful even when we take the limit $n \to \infty$. In fact, we would expect that

$$\exp\left(\frac{1}{q} \sum_{k=1}^{n} f_k \, \Delta \beta_k - \frac{1}{2q} \sum_{k=1}^{n} f_k^2 \, \Delta t \right)$$

$$\to \exp\left(\frac{1}{q} \int_0^t f(t) \, d\beta - \frac{1}{2q} \int_0^t f^2(t) \, dt \right), \quad (7.68)$$

which in fact *is* true and is the essence of the Girsanov theorem. It also turns out that we can indeed let f to depend on the process β and the formula still works. This corresponds to the results of Corollary 7.6. Similarly, we could also derive a likelihood ratio of processes with drifts g and f, which would lead to a result corresponding to Theorem 7.4.

Because Z_n is just the ratio of the densities, we have the following property for expectations of an arbitrary function h:

$$E[h(x(t_1), \ldots, x(t_n))] = E[Z_k(\beta(t_1), \ldots, \beta(t_n)) \, h(\beta(t_1), \ldots, \beta(t_n))]. \tag{7.69}$$

We can now think that Z_n defines a *transformed* probability distribution (or measure) via

$$\tilde{p}(\beta(t_1), \ldots, \beta(t_n)) = Z_n(\beta(t_1), \ldots, \beta(t_k))\, p(\beta(t_1), \ldots, \beta(t_n)). \quad (7.70)$$

There is yet another interesting process that can be defined by

$$\tilde{\beta}(t_k) = \beta(t_k) - \sum_k f_k\, \Delta t. \quad (7.71)$$

It turns out that this process is a Brownian motion under the distribution defined by $\tilde{p}$ in the sense that

$$\mathrm{E}[Z_k(\beta(t_1), \ldots, \beta(t_n))\, h(\tilde{\beta}(t_1), \ldots, \tilde{\beta}(t_n))] = \mathrm{E}[h(\beta(t_1), \ldots, \beta(t_n))]$$
$$(7.72)$$

for an arbitrary function h.

7.5 Doob's h-Transform

The Doob's h-transform (e.g., Rogers and Williams, 2000a,b) is a method that can be used, for example, for deriving the SDE, which is obtained by conditioning another SDE at its end point. It can also be used for removing the drift from an SDE as well as for analyzing hitting times (the time when the process reaches a certain subset of the state space) and excursions of SDEs (Rogers and Williams, 2000b). The main idea in the h-transform is that we multiply the transition density of the original SDE with a suitable term (involving the h-function) such that we can construct an SDE corresponding to the transformed transition density.

Let $p(\mathbf{y}, t' \mid \mathbf{x}, t) \triangleq p(\mathbf{y}(t') \mid \mathbf{x}(t))$ denote a transition density of an SDE. Let a function $h(t, \mathbf{x})$ be defined via the space–time regularity property

$$h(t, \mathbf{x}) = \int p(\mathbf{y}, t + s \mid \mathbf{x}, t)\, h(t + s, \mathbf{y})\, \mathrm{d}\mathbf{y}. \quad (7.73)$$

We can now define another Markov process with the transition kernel $p^h(\mathbf{y}, t' \mid \mathbf{x}, t) \triangleq p(\mathbf{y}(t') \mid \mathbf{x}(t))$ via

$$p^h(\mathbf{y}, t + s \mid \mathbf{x}, t) = p(\mathbf{y}, t + s \mid \mathbf{x}, t)\, \frac{h(t + s, \mathbf{y})}{h(t, \mathbf{x})}. \quad (7.74)$$

Due to the property (7.73), this is a genuine probability density and thus defines a Markov process:

$$\int p^h(\mathbf{y}, t + s \mid \mathbf{x}, t)\, d\mathbf{y}$$
$$= \int p(\mathbf{y}, t + s \mid \mathbf{x}, t) \frac{h(t + s, \mathbf{y})}{h(t, \mathbf{x})}\, d\mathbf{y} = \frac{h(t, \mathbf{x})}{h(t, \mathbf{x})} = 1. \quad (7.75)$$

Assume that our SDE has the form

$$d\mathbf{x} = \mathbf{f}(\mathbf{x}, t)\, dt + \mathbf{L}(\mathbf{x}, t)\, d\boldsymbol{\beta}. \quad (7.76)$$

Note that (7.73) implies that $h(t, \mathbf{x})$ obeys $\mathcal{A}_t h = 0$, where $\mathcal{A}_t$ is the generalized generator (Definition 5.3) of the SDE at hand. Using this property, the generator of p^h can be computed as

$$\mathcal{A}^h \phi = \lim_{s \downarrow 0} \frac{E^h[\phi(\mathbf{x}(t + s))] - \phi(\mathbf{x}(t))}{s}$$

$$= \lim_{s \downarrow 0} \frac{E[\phi(\mathbf{x}(t + s))\, h(t + s, \mathbf{y})] - \phi(\mathbf{x}(t))\, h(\mathbf{x}, t)}{s\, h(t, \mathbf{x})}$$

$$= \frac{1}{h(t, \mathbf{x})}\, \mathcal{A}_t \{h(t, \mathbf{x})\, \phi(\mathbf{x})\}$$

$$= \frac{1}{h(t, \mathbf{x})} \left\{ \frac{\partial h(t, \mathbf{x})}{\partial t} \phi + \sum_i \left[\frac{\partial h(t, \mathbf{x})}{\partial x_i} \phi(\mathbf{x}) + h(t, \mathbf{x}) \frac{\partial \phi(\mathbf{x})}{\partial x_i} \right] f_i(\mathbf{x}, t) \right.$$

$$\left. + \frac{1}{2} \sum_{i,j} \frac{\partial^2 [h(t, \mathbf{x})\, \phi(\mathbf{x})]}{\partial x_i\, \partial x_j} \, [\mathbf{L}(\mathbf{x}, t)\, \mathbf{Q}\, \mathbf{L}^\mathsf{T}(\mathbf{x}, t)]_{ij} \right\}$$

$$= \frac{1}{h(t, \mathbf{x})} \left\{ \underbrace{\left[\frac{\partial h(t, \mathbf{x})}{\partial t} + \sum_i \frac{\partial h(t, \mathbf{x})}{\partial x_i} f_i(\mathbf{x}, t) \right.}_{} \right.$$

$$\underbrace{\left. + \frac{1}{2} \sum_{i,j} \frac{\partial^2 h(t, \mathbf{x})}{\partial x_i\, \partial x_j} \, [\mathbf{L}(\mathbf{x}, t)\, \mathbf{Q}\, \mathbf{L}^\mathsf{T}(\mathbf{x}, t)]_{ij} \right]}_{\mathcal{A}_t h = 0} \phi(\mathbf{x})$$

$$+ \sum_i h(t, \mathbf{x}) \frac{\partial \phi(\mathbf{x})}{\partial x_i} f_i(\mathbf{x}, t) + \frac{1}{2} \sum_{i,j} \left[\frac{\partial h(t, \mathbf{x})}{\partial x_j} \frac{\partial \phi(\mathbf{x})}{\partial x_i} \right.$$

$$\left. + \frac{\partial h(t, \mathbf{x})}{\partial x_i} \frac{\partial \phi(\mathbf{x})}{\partial x_j} + h(t, \mathbf{x}) \frac{\partial^2 \phi(\mathbf{x})}{\partial x_i\, \partial x_j} \right] [\mathbf{L}(\mathbf{x}, t)\, \mathbf{Q}\, \mathbf{L}^\mathsf{T}(\mathbf{x}, t)]_{ij} \right\}$$

$$= \sum_i \left[f_i(\mathbf{x}, t) + \mathbf{L}(\mathbf{x}, t) \, \mathbf{Q} \, \mathbf{L}^\mathsf{T}(\mathbf{x}, t) \, \frac{\nabla h(t, \mathbf{x})}{h(t, \mathbf{x})} \right] \frac{\partial \phi(\mathbf{x})}{\partial x_i}$$

$$+ \frac{1}{2} \sum_{i,j} \frac{\partial^2 \phi(\mathbf{x})}{\partial x_i \, \partial x_j} \, [\mathbf{L}(\mathbf{x}, t) \, \mathbf{Q} \, \mathbf{L}^\mathsf{T}(\mathbf{x}, t)]_{ij}, \tag{7.77}$$

which shows that the new, h-transformed, SDE is

$$\begin{aligned} d\mathbf{x} &= \left[\mathbf{f}(\mathbf{x}, t) + \mathbf{L}(\mathbf{x}, t) \, \mathbf{Q} \, \mathbf{L}^\mathsf{T}(\mathbf{x}, t) \, \frac{\nabla h(t, \mathbf{x})}{h(t, \mathbf{x})} \right] dt + \mathbf{L} \, d\boldsymbol{\beta} \\ &= \left[\mathbf{f}(\mathbf{x}, t) + \mathbf{L}(\mathbf{x}, t) \, \mathbf{Q} \, \mathbf{L}^\mathsf{T}(\mathbf{x}, t) \, \nabla \log h(t, \mathbf{x}) \right] dt + \mathbf{L} \, d\boldsymbol{\beta}. \end{aligned} \tag{7.78}$$

We can use the preceding result to condition an SDE on its end point.

Theorem 7.11. *Assume that we have an SDE of form (7.76) and we wish to condition its solution to hit $\mathbf{x}(T)$ at time $t = T$. The h-transform now gives the following SDE for the end-point-conditioned process:*

$$\begin{aligned} d\mathbf{x} = \left[\mathbf{f}(\mathbf{x}, t) + \mathbf{L}(\mathbf{x}, t) \, \mathbf{Q} \, \mathbf{L}^\mathsf{T}(\mathbf{x}, t) \, \nabla \log p(\mathbf{x}(T) \mid \mathbf{x}(t)) \right] dt \\ + \mathbf{L}(\mathbf{x}, t) \, d\boldsymbol{\beta}, \quad (7.79) \end{aligned}$$

where ∇ is understood to act on the variable $\mathbf{x}(t)$.

Proof   The conditioning of an SDE on an end point can be achieved by Bayes' rule:

$$\begin{aligned} p(\mathbf{x}(t + s) \mid \mathbf{x}(t), \mathbf{x}(T)) &= \frac{p(\mathbf{x}(T) \mid \mathbf{x}(t + s), \mathbf{x}(t)) \, p(\mathbf{x}(t + s) \mid \mathbf{x}(t))}{p(\mathbf{x}(T) \mid \mathbf{x}(t))} \\ &= \frac{p(\mathbf{x}(T) \mid \mathbf{x}(t + s)) \, p(\mathbf{x}(t + s) \mid \mathbf{x}(t))}{p(\mathbf{x}(T) \mid \mathbf{x}(t))}. \end{aligned}$$
$$\tag{7.80}$$

We can now put

$$h(t, \mathbf{x}) = p(\mathbf{x}(T) \mid \mathbf{x}(t)). \tag{7.81}$$

We can now check that this h is also valid in the sense that the space–time regularity property (7.73) is satisfied:

$$p(\mathbf{x}(T) \mid \mathbf{x}(t)) = \int p(\mathbf{x}(t + s) \mid \mathbf{x}(t)) \, p(\mathbf{x}(T) \mid \mathbf{x}(t + s)) \, d\mathbf{x}(t + s). \tag{7.82}$$

Substituting (7.81) to (7.78) then leads to the result in the theorem.   □

Example 7.12 (Conditioned Ornstein–Uhlenbeck process). *Assume that the original process is an Ornstein–Uhlenbeck process:*

$$dx = -\lambda\, x\, dt + d\beta, \quad x(0) = 0, \tag{7.83}$$

and we wish to condition on $x(T) = x_T$ with $x_T = 5$ and $T = 1$. Then we have

$$h(t, x) = \mathrm{N}(x_T \mid a(t)\, x, \sigma^2(t)), \tag{7.84}$$

where

$$a(t) = \exp(-\lambda\,(T - t)),$$
$$\sigma^2(t) = \frac{q}{2\lambda}\, [1 - \exp(-2\lambda\,(T - t))]. \tag{7.85}$$

The h-transform now gives

$$\frac{\partial}{\partial x} \log h(t, x) = \frac{\partial}{\partial x}\left[-\frac{1}{2}\log(2\pi\,\sigma^2(t)) - \frac{1}{2\sigma^2(t)}\,(x_T - a(t)\,x)^2\right]$$
$$= \frac{a(t)}{\sigma^2(t)}\,(x_T - a(t)x), \tag{7.86}$$

which leads to

$$dx = \left[-\lambda\, x + \frac{q\,a(t)}{\sigma^2(t)}\,(x_T - a(t)\,x)\right] dt + d\beta. \tag{7.87}$$

The solution of this SDE is illustrated in Figure 7.2.

We can also use the h-transform to remove a drift from an SDE. This can be done by constructing a space–time regular h such that

$$\mathbf{L}(\mathbf{x}, t)\,\mathbf{Q}\,\mathbf{L}^\mathsf{T}(\mathbf{x}, t)\,\nabla \log h(t, \mathbf{x}) = -\mathbf{f}(\mathbf{x}, t), \tag{7.88}$$

which then cancels out the drift. For more applications of the transform, see Rogers and Williams (2000b).

7.6 Path Integrals

Given the Wiener measure in the path integral form (7.15), we can also define the integral (i.e., expectation) of a functional $F[\beta]$ with respect to this measure. The expectation can then be formally written as a path integral

$$\mathrm{E}[F] = \int F[\beta]\, \exp\left(-\frac{1}{2}\int_0^t \dot{\beta}^2(\tau)\, d\tau\right) \prod_{\tau=0}^t \frac{d\beta(\tau)}{\sqrt{2\pi\, d\tau}}. \tag{7.89}$$

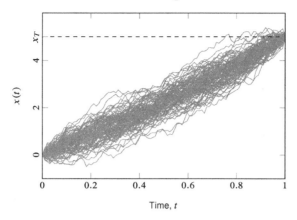

Figure 7.2 Sample trajectories drawn from the conditioned Ornstein–Uhlenbeck model using the Euler–Maruyama method (see Example 7.12).

Methods for directly computing this kind of integrals and their connection with SDEs can be found, for example, in the books by Chaichian and Demichev (2001a) and Zinn-Justin (2002). Quantum mechanics can also be formulated in terms of special kinds of path integrals, not with respect to a Wiener measure, but with respect to suitable oscillating measures (see, e.g., Feynman and Hibbs, 2010; Chaichian and Demichev, 2001a; Zinn-Justin, 2002).

As the (Wiener) path integral is just the expectation of the functional, we can sometimes compute or approximate its value by using SDE methods. It is possible when we can construct an SDE such that the functional can be expressed as a function (not functional) of its solution at a certain time point. The path integral then reduces to computing an expectation of the SDE solution, which can be done numerically using the methods presented in this book. This is illustrated in the following example.

Example 7.13 (Path integral). *Let us compute path integral of the functional*

$$F[\beta] = \exp\left(\int_0^t \beta(s)\,\mathrm{d}s\right). \tag{7.90}$$

Now $x_1(t) = \int_0^t \beta(s)\,\mathrm{d}s$ can be constructed as the solution to

$$\begin{aligned} \mathrm{d}x_1 &= x_2\,\mathrm{d}t, \\ \mathrm{d}x_2 &= \mathrm{d}\beta, \end{aligned} \tag{7.91}$$

which gives $x_1 = \int_0^t \beta(s)\,dt \sim N(0, t^3/3)$. *Thus the path integral of the functional has the value*

$$E[F] = \int \exp(x_1)\,N(x_1 \mid 0, t^3/3)\,dx_1 = \exp\left(\frac{t^3}{6}\right). \tag{7.92}$$

In practice, path integrals are often computed with respect to more general measures than the Wiener measure. As already mentioned, the whole quantum mechanics can be formulated in terms of path integrals with respect to suitably defined measures. Path integral methods for computing the moments of SDEs are discussed, for example, in Chow and Buice (2015).

7.7 Feynman–Kac Formula

The Feynman–Kac formula (see, e.g., Øksendal, 2003; Pardoux and Răşcanu, 2014) gives a link between solutions of partial differential equations (PDEs) and certain expected values of SDE solutions. In this section, we present the general idea by deriving the scalar Feynman–Kac formulae and give the more general multidimensional results in the algorithms. Some recent generalizations to the formulae can be found, for example, in Simon (2015) and Piiroinen and Simon (2016).

Let us start by considering the following PDE for a function $u(x,t)$:

$$\frac{\partial u}{\partial t} + f(x)\frac{\partial u}{\partial x} + \frac{1}{2}L^2(x)\frac{\partial^2 u}{\partial x^2} = 0, \tag{7.93}$$
$$u(x,T) = \Psi(x),$$

where $f(x)$, $L(x)$, and $\Psi(x)$ are some given functions and T is a fixed time instant. We define a process $x(t)$ on the interval $[t', T]$ as follows:

$$dx = f(x)\,dt + L(x)\,d\beta, \qquad x(t') = x', \tag{7.94}$$

that is, the process starts from a given x' at time t'. Using the Itô formula for $u(x, t)$ and recalling that it solves the PDE (7.93) gives

$$
\begin{aligned}
du &= \frac{\partial u}{\partial t}\, dt + \frac{\partial u}{\partial x}\, dx + \frac{1}{2}\frac{\partial^2 u}{\partial x^2}\, dx^2 \\
&= \frac{\partial u}{\partial t}\, dt + \frac{\partial u}{\partial x}\, f(x)\, dt + \frac{\partial u}{\partial x} L(x)\, d\beta + \frac{1}{2}\frac{\partial^2 u}{\partial x^2} L^2(x)\, dt \\
&= \underbrace{\left[\frac{\partial u}{\partial t} + f(x)\frac{\partial u}{\partial x} + \frac{1}{2}L^2(x)\frac{\partial^2 u}{\partial x^2}\right]}_{=0} dt + \frac{\partial u}{\partial x} L(x)\, d\beta \\
&= \frac{\partial u}{\partial x} L(x)\, d\beta.
\end{aligned}
\tag{7.95}
$$

Integrating from t' to T now gives

$$
u(x(T), T) - u(x(t'), t') = \int_{t'}^{T} \frac{\partial u}{\partial x} L(x)\, d\beta,
\tag{7.96}
$$

and by substituting the initial and terminal terms, we get the following:

$$
\Psi(x(T)) - u(x', t') = \int_{t'}^{T} \frac{\partial u}{\partial x} L(x)\, d\beta.
\tag{7.97}
$$

We can now take expectations of both sides and recall that the expectation of any Itô integral is zero. Thus after rearranging, we get

$$
u(x', t') = \mathrm{E}[\Psi(x(T))].
\tag{7.98}
$$

This means that we can solve the value of $u(x', t')$ for arbitrary x' and t' by starting the process in Equation (7.94) from x' and time t', and letting it run until time T. The solution is then the expected value of $\Psi(x(T))$ over the process realizations.

This idea can also be generalized to equations of the form

$$
\begin{aligned}
\frac{\partial u}{\partial t} + f(x)\frac{\partial u}{\partial x} + \frac{1}{2}L^2(x)\frac{\partial^2 u}{\partial x^2} - r\, u &= 0, \\
u(x, T) &= \Psi(x),
\end{aligned}
\tag{7.99}
$$

where r is a positive constant. The corresponding SDE will be the same, but we need to apply the Itô formula to $\exp(-r\, t)\, u(x, t)$ instead of $u(x, t)$. The resulting *Feynman–Kac equation* is

$$
u(x', t') = \exp(-r\, (T - t'))\, \mathrm{E}[\Psi(x(T))].
\tag{7.100}
$$

In practice, this result means that we can solve the PDE (7.99) by simulating paths from the SDE (7.94) and computing the empirical average

(7.100). We can generalize the result a lot, but here we restrict the generalization to the multivariate case in order to keep the result quite clean but useful. For more general results, the reader is referred to Pardoux and Răşcanu (2014), Simon (2015), and Piiroinen and Simon (2016). We get the following algorithm.

Algorithm 7.14 (Solving backward PDEs with SDE simulation). *An approximation of the solution $u(\mathbf{x}, t)$ to the following backward PDE*

$$\frac{\partial u(\mathbf{x}, t)}{\partial t} + \sum_i \frac{\partial u(\mathbf{x}, t)}{\partial x_i} f_i(\mathbf{x}, t)$$

$$+ \frac{1}{2} \sum_{i,j} \left(\frac{\partial^2 u(\mathbf{x}, t)}{\partial x_i \partial x_j} \right) [\mathbf{L}(\mathbf{x}, t) \mathbf{Q} \mathbf{L}^\mathsf{T}(\mathbf{x}, t)]_{ij} - r \, u(\mathbf{x}, t) = 0,$$

$$(7.101)$$

where $\mathbf{x} \in \mathbb{R}^D$ and $\mathbf{Q} \in \mathbb{R}^{S \times S}$, with the boundary condition

$$u(\mathbf{x}, T) = \Psi(\mathbf{x}) \qquad (7.102)$$

for any fixed $(\mathbf{x}', t')$ can be computed as follows:

1. *Simulate N solutions to the following SDE from time t' to time T:*

$$\mathrm{d}\mathbf{x} = \mathbf{f}(\mathbf{x}, t) \, \mathrm{d}t + \mathbf{L}(\mathbf{x}, t) \, \mathrm{d}\boldsymbol{\beta}, \quad \mathbf{x}(t') = \mathbf{x}', \qquad (7.103)$$

   where $\mathbf{x}(t) \in \mathbb{R}^D$ and the Brownian motion $\boldsymbol{\beta}(t) \in \mathbb{R}^S$ has the diffusion matrix $\mathbf{Q}$.
2. *Estimate the following expected value from the simulations:*

$$u(\mathbf{x}', t') = \exp(-r \, (T - t')) \, \mathrm{E}[\Psi(\mathbf{x}(T))]. \qquad (7.104)$$

When the system is time-invariant, we can put $u(\mathbf{x}, t) \leftarrow u(\mathbf{x}, T - t)$, and we get the following algorithm for forward PDEs (see, e.g., Pardoux and Răşcanu, 2014, for details and generalizations).

Algorithm 7.15 (Solving forward PDEs with SDE simulation). *The solution $u(\mathbf{x}, t)$ to the following forward PDE with $\mathbf{x} \in \mathbb{R}^D$, $\mathbf{Q} \in \mathbb{R}^{S \times S}$*

$$\frac{\partial u(\mathbf{x}, t)}{\partial t} = \sum_i \frac{\partial u(\mathbf{x}, t)}{\partial x_i} f_i(\mathbf{x}, t)$$

$$+ \frac{1}{2} \sum_{i,j} \left(\frac{\partial^2 u(\mathbf{x}, t)}{\partial x_i \partial x_j} \right) [\mathbf{L}(\mathbf{x}, t) \mathbf{Q} \mathbf{L}^\mathsf{T}(\mathbf{x}, t)]_{ij} - r \, u(\mathbf{x}, t)$$

$$(7.105)$$

with the boundary condition

$$u(\mathbf{x}, 0) = \Psi(\mathbf{x}) \tag{7.106}$$

for any fixed $(\mathbf{x}', t')$ *can be computed as follows:*

1. *Simulate N solutions to the following SDE from time 0 to time t':*

$$d\mathbf{x} = \mathbf{f}(\mathbf{x})\, dt + \mathbf{L}(\mathbf{x})\, d\boldsymbol{\beta}, \quad \mathbf{x}(0) = \mathbf{x}', \tag{7.107}$$

   where $\mathbf{x}(t) \in \mathbb{R}^D$ and $\boldsymbol{\beta}(t) \in \mathbb{R}^S$ has the diffusion matrix $\mathbf{Q}$.
2. *Estimate the following expected value from the simulations:*

$$u(\mathbf{x}', t') = \exp(-r\, t')\, \mathrm{E}[\Psi(\mathbf{x}(t'))]. \tag{7.108}$$

The Feynman–Kac equation can also be used for computing solutions to boundary value problems, which do not include time variables at all (see, e.g., Øksendal, 2003; Pardoux and Răşcanu, 2014; Simon, 2015; Piiroinen and Simon, 2016). In the following derivation, we only consider the scalar case, but analogous derivation works for the multidimensional case as well. Furthermore, proper derivation of the results in this section would need us to define the concept of random stopping time, which we have not done and thus in this sense the derivation is quite heuristic.

Consider the following boundary value problem for a function $u(x)$ defined on some finite domain Ω with boundary $\partial\Omega$:

$$f(x)\frac{\partial u}{\partial x} + \frac{1}{2}L^2(x)\frac{\partial^2 u}{\partial x^2} = 0,$$
$$u(x) = \Psi(x), \quad x \in \partial\Omega. \tag{7.109}$$

Again, let us define a process $x(t)$ in the same way as in Equation (7.94). Further, the application of the Itô formula to $u(x)$ gives

$$\begin{aligned}
du &= \frac{\partial u}{\partial x}\, dx + \frac{1}{2}\frac{\partial^2 u}{\partial x^2}\, dx^2 \\
&= \frac{\partial u}{\partial x}\, f(x)\, dt + \frac{\partial u}{\partial x}\, L(x)\, d\beta + \frac{1}{2}\frac{\partial^2 u}{\partial x^2}\, L^2(x)\, dt \\
&= \underbrace{\left[f(x)\frac{\partial u}{\partial x} + \frac{1}{2}L^2(x)\frac{\partial^2 u}{\partial x^2} \right]}_{=0} dt + \frac{\partial u}{\partial x}\, L(x)\, d\beta \\
&= \frac{\partial u}{\partial x}\, L(x)\, d\beta. \tag{7.110}
\end{aligned}$$

Let T_e be the first exit time of the process $x(t)$ from the domain Ω. Integration from t' to T_e gives

$$u(x(T_e)) - u(x(t')) = \int_{t'}^{T_e} \frac{\partial u}{\partial x} L(x) \, d\beta. \tag{7.111}$$

The value of $u(x)$ on the boundary is $\Psi(x)$ and $x(t') = x'$, and thus we have

$$\Psi(x(T_e)) - u(x') = \int_{t'}^{T_e} \frac{\partial u}{\partial x} L(x) \, d\beta. \tag{7.112}$$

Taking expectation and rearranging then gives

$$u(x') = E[\Psi(x(T_e))]. \tag{7.113}$$

That is, the value $u(x')$ with arbitrary x' can be obtained by starting the process $x(t)$ from $x(t') = x'$ in Equation (7.94) at arbitrary time t' and computing the expectation of $\Psi(x(T_e))$ over the first exit points of the process $x(t)$ from the domain Ω.

Again, we can generalize the derivation to equations of the form

$$f(x) \frac{\partial u}{\partial x} + \frac{1}{2} L^2(x) \frac{\partial^2 u}{\partial x^2} - r u = 0, \tag{7.114}$$
$$u(x) = \Psi(x), \quad x \in \partial\Omega,$$

which gives

$$u(x') = \exp(-r \, (T_e - t')) \, E[\Psi(x(T_e))]. \tag{7.115}$$

By extending this to multiple dimensions we get the following algorithm (see Pardoux and Răşcanu, 2014, for further generalizations).

Algorithm 7.16 (Solving boundary value problems with SDE simulation). *An approximation of the solution $u(\mathbf{x})$ to the following elliptic PDE defined on some domain $\mathbf{x} \in \Omega \subset \mathbb{R}^D$*

$$\sum_i \frac{\partial u(\mathbf{x})}{\partial x_i} f_i(\mathbf{x})$$
$$+ \frac{1}{2} \sum_{i,j} \left(\frac{\partial^2 u(\mathbf{x})}{\partial x_i \partial x_j} \right) [\mathbf{L}(\mathbf{x}) \mathbf{Q} \mathbf{L}^\mathsf{T}(\mathbf{x})]_{ij} - r u(\mathbf{x}, t) = 0, \tag{7.116}$$

where $\mathbf{Q} \in \mathbb{R}^{S \times S}$, with the boundary condition

$$u(\mathbf{x}) = \Psi(\mathbf{x}), \quad \mathbf{x} \in \partial\Omega, \tag{7.117}$$

on the edge $\partial\Omega$ of the domain Ω, for any fixed $\mathbf{x}'$, can be computed as follows:

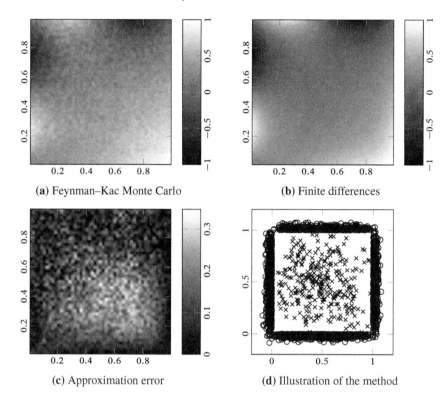

(a) Feynman–Kac Monte Carlo

(b) Finite differences

(c) Approximation error

(d) Illustration of the method

Figure 7.3 Subfigure (a) shows the Monte Carlo solution to the elliptic PDE in Example 7.17, (b) is the finite-differences solution, and (c) shows the error between finite-differences and Monte Carlo solutions. Subfigure (d) illustrates the method: a set of Ornstein–Uhlenbeck processes is simulated until they exit the rectangle.

1. Simulate N solutions to the following SDE from time t' (e.g. $t' = 0$) to the first exit time T_e from the domain Ω:

$$\mathrm{d}\mathbf{x} = \mathbf{f}(\mathbf{x})\,\mathrm{d}t + \mathbf{L}(\mathbf{x})\,\mathrm{d}\boldsymbol{\beta}, \quad \mathbf{x}(t') = \mathbf{x}'. \tag{7.118}$$

In the SDE, $\mathbf{x}(t) \in \mathbb{R}^D$ and the Brownian motion $\boldsymbol{\beta}(t) \in \mathbb{R}^S$ has the diffusion matrix $\mathbf{Q}$.

2. Estimate the following expected value from the simulations:

$$u(\mathbf{x}') = \exp(-r\,(T_e - t'))\,\mathrm{E}[\Psi(\mathbf{x}(T_e))]. \tag{7.119}$$

The use of Algorithm 7.16 is illustrated in the following example.

Example 7.17 (Solution of an elliptic PDE using SDE simulation). *We consider the following PDE*

$$f_1(x, y) \frac{\partial u(x, y)}{\partial x} + f_2(x, y) \frac{\partial u(x, y)}{\partial y}$$

$$+ \frac{q}{2} \left[\frac{\partial^2 u(x, y)}{\partial x^2} + \frac{\partial^2 u(x, y)}{\partial y^2} \right] - r\, u(x, y) = 0 \quad (7.120)$$

with $f_1(x, y) = -x/10$, $f_2(x, y) = -y/10$ *and the following boundary conditions on the square* $[0, L]^2$:

$$
\begin{aligned}
u(x, 0) &= \sin(2\pi\, x), & u(x, L) &= 1 - x, \\
u(0, y) &= y, & u(L, y) &= \sin(2\pi\, y).
\end{aligned}
\quad (7.121)
$$

The parameter values were selected as $L = 1$, $q = 1/10$, *and* $r = 1$. *The result of the Feynman–Kac–based Monte Carlo solution is compared to a finite-differences approximation in Figure 7.3. The Monte Carlo solution was computed by starting an Ornstein–Uhlenbeck process 100 from each point on a* 50×50 *grid and simulating it using the exact transition density until it hit one of the boundaries.*

7.8 Exercises

7.1   Consider the modified Cox–Ingersoll–Ross model (5.69) being transformed by $y(t) = h(x(t))$ using the Lamperti transform

$$h(x) = \int_0^x \frac{1}{L(u)} \, du = \int_0^x \frac{1}{\theta_2 \sqrt{1 + u^2}} \, du. \quad (7.122)$$

Rewrite the original process in terms of the $y(t)$ process.

7.2   Derive the weak solution to the Ornstein–Uhlenbeck process following the techniques in Example 7.10.

7.3   Derive the weak solution to the SDE model

$$dx = \sin(x) \, dt + d\beta,$$

where $\beta(t)$ is a standard Brownian motion, by using the Girsanov theorem.

7.4   Consider the solution given in Example 7.12:

  (a) Write down the mean and variance expressions for the solution as a function of t.

  (b) Check numerically using the Euler–Maruyama method that the solution matches your mean and variance expressions.

7.5   Derive the conditioned SDE (similar to Example 7.12) for the Beneš model $dx = \tanh(x) \, dt + d\beta$ conditioned on $x(T) = x_T$.

7.6   Solve the following PDE by using finite differences and SDE simulation and check that the results match. The PDE is

$$\frac{\partial^2 u(x, y)}{\partial x^2} + \frac{\partial^2 u(x, y)}{\partial y^2} = 0$$

with boundary conditions

$$u(x, -1) = \cos(\pi x/2), \quad u(x, 1) = -\cos(\pi x/2),$$
$$u(-1, y) = \cos(\pi y/2), \quad u(1, y) = -\cos(\pi y/2).$$

8

Numerical Simulation of SDEs

This chapter is concerned with solving – or simulating from – stochastic differential equations by numerical approximations. The presentation in this chapter tries to link numerical solution concepts familiar from ordinary differential equations to their stochastic equivalents for SDEs. We start by going through Taylor series approximations of ODEs and then extend them to strong and weak Itô–Taylor series approximations of SDEs. After that, our focus will be on deterministic *Runge–Kutta methods*, which are an important family of explicit and implicit methods for numerical solving of differential equations. Then we discuss strong stochastic Runge–Kutta methods and thereafter weak methods. We will try to provide general tools, but it is worth noting that often in the case of stochastic differential equations special structure of the problem can have a large impact on the complexity of the solution method. In the end of the chapter, we briefly discuss Verlet methods and the Exact algorithm, of which neither strictly fits under the frameworks of Itô–Taylor and Runge–Kutta schemes. Some of the methods presented in the next chapter can also be used for numerical simulation and hence numerical solving of SDEs.

8.1 Taylor Series of ODEs

One way to find approximate solutions of ODEs is by using Taylor series expansions in time direction. Even though this method as a practical ODE numerical approximation method is quite much superseded by the Runge–Kutta type of derivative-free methods, it still is an important theoretical tool for finding and analyzing numerical schemes (e.g., the theory of Runge–Kutta methods is based on the Taylor series). In the case of SDEs, the corresponding Itô–Taylor series solutions provide a useful basis for numerical methods for SDEs. However, we cannot simply apply the ODE-based numerical schemes to SDEs. An inconvenient fact is that in the stochastic case, Runge–Kutta methods are not as easy to use as in the case of ODEs.

In this section, we derive the Taylor series–based solutions of ODEs in detail, because the derivation of the Itô–Taylor series can be done in an analogous way. As the idea is the same, by first going through the deterministic case, it is easy to see the essential things behind the technical details in the stochastic case.

We start by considering the following differential equation:

$$\frac{d\mathbf{x}(t)}{dt} = \mathbf{f}(\mathbf{x}(t), t), \qquad \mathbf{x}(t_0) = \text{given}, \tag{8.1}$$

which can be integrated to give

$$\mathbf{x}(t) = \mathbf{x}(t_0) + \int_{t_0}^{t} \mathbf{f}(\mathbf{x}(\tau), \tau) \, d\tau. \tag{8.2}$$

If the function $\mathbf{f}$ is differentiable, we can also write $t \mapsto \mathbf{f}(\mathbf{x}(t), t)$ as the solution to the differential equation

$$\frac{d\mathbf{f}(\mathbf{x}(t), t)}{dt} = \frac{\partial}{\partial t}\mathbf{f}(\mathbf{x}(t), t) + \sum_{i} f_i(\mathbf{x}(t), t) \frac{\partial}{\partial x_i}\mathbf{f}(\mathbf{x}(t), t), \tag{8.3}$$

where $\mathbf{f}(\mathbf{x}(t_0), t_0)$ is the given initial condition. The integral form of this is

$$\mathbf{f}(\mathbf{x}(t), t) = \mathbf{f}(\mathbf{x}(t_0), t_0)$$
$$+ \int_{t_0}^{t} \left[\frac{\partial}{\partial t}\mathbf{f}(\mathbf{x}(\tau), \tau) + \sum_{i} f_i(\mathbf{x}(\tau), \tau) \frac{\partial}{\partial x_i}\mathbf{f}(\mathbf{x}(\tau), \tau) \right] d\tau. \tag{8.4}$$

At this point, it is convenient to define the linear operator

$$\mathcal{L}(\bullet) = \frac{\partial}{\partial t}(\bullet) + \sum_{i} f_i \frac{\partial}{\partial x_i}(\bullet) \tag{8.5}$$

and rewrite the integral equation as

$$\mathbf{f}(\mathbf{x}(t), t) = \mathbf{f}(\mathbf{x}(t_0), t_0) + \int_{t_0}^{t} \mathcal{L}\mathbf{f}(\mathbf{x}(\tau), \tau) \, d\tau. \tag{8.6}$$

Substituting this into Equation (8.2) gives

$$\mathbf{x}(t) = \mathbf{x}(t_0) + \int_{t_0}^{t} \left[\mathbf{f}(\mathbf{x}(t_0), t_0) + \int_{t_0}^{\tau} \mathcal{L}\mathbf{f}(\mathbf{x}(\tau'), \tau') \, d\tau' \right] d\tau$$
$$= \mathbf{x}(t_0) + \mathbf{f}(\mathbf{x}(t_0), t_0)(t - t_0) + \int_{t_0}^{t} \int_{t_0}^{\tau} \mathcal{L}\mathbf{f}(\mathbf{x}(\tau'), \tau') \, d\tau' \, d\tau.$$
$$\tag{8.7}$$

The term in the integrand on the right can again be defined as the solution to the differential equation

$$\frac{d[\mathcal{L}\mathbf{f}(\mathbf{x}(t),t)]}{dt} = \frac{\partial[\mathcal{L}\mathbf{f}(\mathbf{x}(t),t)]}{\partial t} + \sum_i f_i(\mathbf{x}(t),t)\frac{\partial[\mathcal{L}\mathbf{f}(\mathbf{x}(t),t)]}{\partial x_i}$$

$$= \mathcal{L}^2\mathbf{f}(\mathbf{x}(t),t), \tag{8.8}$$

which in integral form is

$$\mathcal{L}\mathbf{f}(\mathbf{x}(t),t) = \mathcal{L}\mathbf{f}(\mathbf{x}(t_0),t_0) + \int_{t_0}^{t}\mathcal{L}^2\mathbf{f}(\mathbf{x}(\tau),\tau)\,d\tau. \tag{8.9}$$

Substituting into the equation of $\mathbf{x}(t)$ then gives

$$\mathbf{x}(t) = \mathbf{x}(t_0) + \mathbf{f}(\mathbf{x}(t_0),t)(t-t_0)$$

$$+ \int_{t_0}^{t}\int_{t_0}^{\tau}\left[\mathcal{L}\mathbf{f}(\mathbf{x}(t_0),t_0) + \int_{t_0}^{\tau'}\mathcal{L}^2\mathbf{f}(\mathbf{x}(\tau''),\tau'')\,d\tau''\right]d\tau'\,d\tau$$

$$= \mathbf{x}(t_0) + \mathbf{f}(\mathbf{x}(t_0),t_0)(t-t_0) + \frac{1}{2}\mathcal{L}\mathbf{f}(\mathbf{x}(t_0),t_0)(t-t_0)^2$$

$$+ \int_{t_0}^{t}\int_{t_0}^{\tau}\int_{t_0}^{\tau'}\mathcal{L}^2\mathbf{f}(\mathbf{x}(\tau''),\tau'')\,d\tau''\,d\tau'\,d\tau. \tag{8.10}$$

If we continue this procedure ad infinitum, we obtain the following Taylor series expansion for the solution of the ODE:

$$\mathbf{x}(t) = \mathbf{x}(t_0) + \mathbf{f}(\mathbf{x}(t_0),t_0)(t-t_0) + \frac{1}{2!}\mathcal{L}\mathbf{f}(\mathbf{x}(t_0),t_0)(t-t_0)^2$$

$$+ \frac{1}{3!}\mathcal{L}^2\mathbf{f}(\mathbf{x}(t_0),t_0)(t-t_0)^3 + \cdots \tag{8.11}$$

From the preceding derivation, we also get the result that if we truncate the series at the nth term, the residual error is

$$\mathbf{r}_n(t) = \int_{t_0}^{t}\cdots\int_{t_0}^{\tau}\mathcal{L}^n\mathbf{f}(\mathbf{x}(\tau),\tau)\,d\tau^{n+1}, \tag{8.12}$$

which could be further simplified via integration by parts and by using the mean value theorem. To derive the series expansion for an arbitrary function $\mathbf{x}(t)$, we can define it as the solution to the trivial differential equation

$$\frac{d\mathbf{x}(t)}{dt} = \mathbf{f}(t), \qquad \mathbf{x}(t_0) = \text{given}, \tag{8.13}$$

where $\mathbf{f}(t) = d\mathbf{x}(t)/dt$. Because $\mathbf{f}$ is independent of $\mathbf{x}$, we have

$$\mathcal{L}^n \mathbf{f}(t) = \frac{d^{n+1}\mathbf{x}(t)}{dt^{n+1}}. \qquad (8.14)$$

Thus the corresponding series becomes the classical Taylor series:

$$\mathbf{x}(t) = \mathbf{x}(t_0) + \frac{d\mathbf{x}}{dt}(t_0)(t - t_0) + \frac{1}{2!}\frac{d^2\mathbf{x}}{dt^2}(t_0)(t - t_0)^2$$
$$+ \frac{1}{3!}\frac{d^3\mathbf{x}}{dt^3}(t_0)(t - t_0)^3 + \cdots \qquad (8.15)$$

8.2 Itô–Taylor Series–Based Strong Approximations of SDEs

Itô–Taylor series (see Kloeden et al., 1994; Kloeden and Platen, 1999) are an extension of the Taylor series of ODEs to SDEs. The derivation is basically identical to the Taylor series solution in the previous section except that we replace the time derivative computations with application of the Itô formula.

Let us consider the following SDE:

$$d\mathbf{x} = \mathbf{f}(\mathbf{x}, t)\, dt + \mathbf{L}(\mathbf{x}, t)\, d\boldsymbol{\beta}, \qquad \mathbf{x}(t_0) \sim p(\mathbf{x}(t_0)), \qquad (8.16)$$

where $\mathbf{x}(t) \in \mathbb{R}^D$ and the Brownian motion $\boldsymbol{\beta}(t) \in \mathbb{R}^S$ has the diffusion matrix $\mathbf{Q}$. We also assume that the initial condition is independent of the Brownian motion. In integral form, this SDE can be expressed as

$$\mathbf{x}(t) = \mathbf{x}(t_0) + \int_{t_0}^{t} \mathbf{f}(\mathbf{x}(\tau), \tau)\, d\tau + \int_{t_0}^{t} \mathbf{L}(\mathbf{x}(\tau), \tau)\, d\boldsymbol{\beta}(\tau). \qquad (8.17)$$

Applying the Itô formula to terms $\mathbf{f}(\mathbf{x}(t), t)$ and $\mathbf{L}(\mathbf{x}(t), t)$ gives the following for the drift:

$$d\mathbf{f}(\mathbf{x}(t), t) = \frac{\partial \mathbf{f}(\mathbf{x}(t), t)}{\partial t}\, dt + \sum_i \frac{\partial \mathbf{f}(\mathbf{x}(t), t)}{\partial x_i} f_i(\mathbf{x}(t), t)\, dt$$
$$+ \sum_i \frac{\partial \mathbf{f}(\mathbf{x}(t), t)}{\partial x_i} [\mathbf{L}(\mathbf{x}(t), t)\, d\boldsymbol{\beta}(\tau)]_i$$
$$+ \frac{1}{2}\sum_{i,j} \frac{\partial^2 \mathbf{f}(\mathbf{x}(t), t)}{\partial x_i\, \partial x_j} [\mathbf{L}(\mathbf{x}(t), t)\, \mathbf{Q}\, \mathbf{L}^{\mathsf{T}}(\mathbf{x}(t), t)]_{ij}\, dt, \qquad (8.18)$$

and the following for the dispersion matrix:

$$dL(\mathbf{x}(t), t) = \frac{\partial L(\mathbf{x}(t), t)}{\partial t} dt + \sum_i \frac{\partial L(\mathbf{x}(t), t)}{\partial x_i} f_i(\mathbf{x}(t), t) dt$$

$$+ \sum_i \frac{\partial L(\mathbf{x}(t), t)}{\partial x_i} [L(\mathbf{x}(t), t) d\boldsymbol{\beta}(\tau)]_i$$

$$+ \frac{1}{2} \sum_{i,j} \frac{\partial^2 L(\mathbf{x}(t), t)}{\partial x_i \partial x_j} [L(\mathbf{x}(t), t) Q L^\mathsf{T}(\mathbf{x}(t), t)]_{ij} dt. \quad (8.19)$$

In integral form, these can be written as

$$\mathbf{f}(\mathbf{x}(t), t) = \mathbf{f}(\mathbf{x}(t_0), t_0) + \int_{t_0}^t \frac{\partial \mathbf{f}(\mathbf{x}(\tau), \tau)}{\partial t} d\tau$$

$$+ \int_{t_0}^t \sum_i \frac{\partial \mathbf{f}(\mathbf{x}(\tau), \tau)}{\partial x_i} f_i(\mathbf{x}(\tau), \tau) d\tau$$

$$+ \int_{t_0}^t \sum_i \frac{\partial \mathbf{f}(\mathbf{x}(\tau), \tau)}{\partial x_i} [L(\mathbf{x}(\tau), \tau) d\boldsymbol{\beta}(\tau)]_i$$

$$+ \int_{t_0}^t \frac{1}{2} \sum_{i,j} \frac{\partial^2 \mathbf{f}(\mathbf{x}(\tau), \tau)}{\partial x_i \partial x_j} [L(\mathbf{x}(\tau), \tau) Q L^\mathsf{T}(\mathbf{x}(\tau), \tau)]_{ij} d\tau,$$

$$(8.20)$$

and

$$L(\mathbf{x}(t), t) = L(\mathbf{x}(t_0), t_0) + \int_{t_0}^t \frac{\partial L(\mathbf{x}(\tau), \tau)}{\partial t} d\tau$$

$$+ \int_{t_0}^t \sum_i \frac{\partial L(\mathbf{x}(\tau), \tau)}{\partial x_i} f_i(\mathbf{x}(\tau), \tau) d\tau$$

$$+ \int_{t_0}^t \sum_i \frac{\partial L(\mathbf{x}(\tau), \tau)}{\partial x_i} [L(\mathbf{x}(\tau), \tau) d\boldsymbol{\beta}(\tau)]_i$$

$$+ \int_{t_0}^t \frac{1}{2} \sum_{i,j} \frac{\partial^2 L(\mathbf{x}(\tau), \tau)}{\partial x_i \partial x_j} [L(\mathbf{x}(\tau), \tau) Q L^\mathsf{T}(\mathbf{x}(\tau), \tau)]_{ij} d\tau. \quad (8.21)$$

If we define the following two operators

$$\mathcal{L}_t(\bullet) = \frac{\partial(\bullet)}{\partial t} + \sum_i \frac{\partial(\bullet)}{\partial x_i} f_i + \frac{1}{2} \sum_{i,j} \frac{\partial^2(\bullet)}{\partial x_i \partial x_j} [L Q L^\mathsf{T}]_{ij},$$

$$(8.22)$$

$$\mathcal{L}_{\beta,j}(\bullet) = \sum_i \frac{\partial(\bullet)}{\partial x_i} L_{ij}, \quad \text{for } j = 1, 2, \ldots, S,$$

then we can conveniently write

$$
\mathbf{f}(\mathbf{x}(t), t) = \mathbf{f}(\mathbf{x}(t_0), t_0) + \int_{t_0}^{t} \mathcal{L}_t \mathbf{f}(\mathbf{x}(\tau), \tau) \, d\tau
$$

$$
+ \sum_j \int_{t_0}^{t} \mathcal{L}_{\beta,j} \mathbf{f}(\mathbf{x}(\tau), \tau) \, d\beta_j(\tau), \qquad (8.23)
$$

$$
\mathbf{L}(\mathbf{x}(t), t) = \mathbf{L}(\mathbf{x}(t_0), t_0) + \int_{t_0}^{t} \mathcal{L}_t \mathbf{L}(\mathbf{x}(\tau), \tau) \, d\tau
$$

$$
+ \sum_j \int_{t_0}^{t} \mathcal{L}_{\beta,j} \mathbf{L}(\mathbf{x}(\tau), \tau) \, d\beta_j(\tau). \qquad (8.24)
$$

If we now substitute these into the expression of $\mathbf{x}(t)$ in Equation (8.17), we get

$$
\mathbf{x}(t) = \mathbf{x}(t_0) + \mathbf{f}(\mathbf{x}(t_0), t_0)(t - t_0) + \mathbf{L}(\mathbf{x}(t_0), t_0)(\boldsymbol{\beta}(t) - \boldsymbol{\beta}(t_0))
$$

$$
+ \int_{t_0}^{t} \int_{t_0}^{\tau} \mathcal{L}_t \mathbf{f}(\mathbf{x}(\tau), \tau) \, d\tau \, d\tau + \sum_j \int_{t_0}^{t} \int_{t_0}^{\tau} \mathcal{L}_{\beta,j} \mathbf{f}(\mathbf{x}(\tau), \tau) \, d\beta_j(\tau) \, d\tau
$$

$$
+ \int_{t_0}^{t} \int_{t_0}^{\tau} \mathcal{L}_t \mathbf{L}(\mathbf{x}(\tau), \tau) \, d\tau \, d\boldsymbol{\beta}(\tau)
$$

$$
+ \sum_j \int_{t_0}^{t} \int_{t_0}^{\tau} \mathcal{L}_{\beta,j} \mathbf{L}(\mathbf{x}(\tau), \tau) \, d\beta_j(\tau) \, d\boldsymbol{\beta}(\tau). \qquad (8.25)
$$

This can be seen to have the form

$$
\mathbf{x}(t) = \mathbf{x}(t_0) + \mathbf{f}(\mathbf{x}(t_0), t_0)(t - t_0)
$$

$$
+ \mathbf{L}(\mathbf{x}(t_0), t_0)(\boldsymbol{\beta}(t) - \boldsymbol{\beta}(t_0)) + \mathbf{r}(t), \qquad (8.26)
$$

where the remainder $\mathbf{r}(t)$ consists of higher-order multiple stochastic integrals involving the function itself, the drift and diffusion, and their derivatives such that

$$
\mathbf{r}(t) = \int_{t_0}^{t} \int_{t_0}^{\tau} \mathcal{L}_t \mathbf{f}(\mathbf{x}(\tau), \tau) \, d\tau \, d\tau + \sum_j \int_{t_0}^{t} \int_{t_0}^{\tau} \mathcal{L}_{\beta,j} \mathbf{f}(\mathbf{x}(t), t) \, d\beta_j(\tau) \, d\tau
$$

$$
+ \int_{t_0}^{t} \int_{t_0}^{\tau} \mathcal{L}_t \mathbf{L}(\mathbf{x}(\tau), \tau) \, d\tau \, d\boldsymbol{\beta}(\tau)
$$

$$
+ \sum_j \int_{t_0}^{t} \int_{t_0}^{\tau} \mathcal{L}_{\beta,j} \mathbf{L}(\mathbf{x}(\tau), \tau) \, d\beta_j(\tau) \, d\boldsymbol{\beta}(\tau). \qquad (8.27)
$$

We can now form a first-order approximation to the solution by discarding the remainder term:

$$\mathbf{x}(t) \approx \mathbf{x}(t_0) + \mathbf{f}(\mathbf{x}(t_0), t_0)(t - t_0) + \mathbf{L}(\mathbf{x}(t_0), t_0)(\boldsymbol{\beta}(t) - \boldsymbol{\beta}(t_0)). \quad (8.28)$$

This leads to the Euler–Maruyama method already discussed in Section 3.4.

Algorithm 8.1 (Euler–Maruyama method). *Draw* $\hat{\mathbf{x}}(t_0) \sim p(\mathbf{x}(t_0))$ *and the divide time interval* $[t_0, t]$ *into* M *steps of length* Δt. *At each step* k, *do the following:*

1. *Draw random variable* $\Delta\boldsymbol{\beta}_k$ *from the distribution*

$$\Delta\boldsymbol{\beta}_k \sim \mathrm{N}(\mathbf{0}, \mathbf{Q}\,\Delta t). \quad (8.29)$$

2. *Compute*

$$\hat{\mathbf{x}}(t_{k+1}) = \hat{\mathbf{x}}(t_k) + \mathbf{f}(\hat{\mathbf{x}}(t_k), t_k)\,\Delta t + \mathbf{L}(\hat{\mathbf{x}}(t_k), t_k)\,\Delta\boldsymbol{\beta}_k. \quad (8.30)$$

The *strong order of convergence* of a stochastic numerical integration method can be roughly defined to be the largest exponent γ such that if we numerically solve an SDE using $M = 1/\Delta t$ steps of length Δt, then there exists a constant K such that

$$\mathrm{E}\left[|\mathbf{x}(t_M) - \hat{\mathbf{x}}(t_M)|\right] \le K\,\Delta t^{\gamma}. \quad (8.31)$$

For stochastic methods, there also exists a second type of convergence, namely *weak order of convergence*. This will be discussed in more detail in the next section.

It can be shown (Kloeden and Platen, 1999) that in the case of the Euler–Maruyama method (under assumptions of sufficient regularity), the strong order of convergence is $\gamma = 1/2$. However, as will be shown later on, it has the weak order of convergence $\alpha = 1$. The reason that the strong order of convergence is just $1/2$ is that the term with $\mathrm{d}\beta_j(\tau)\,\mathrm{d}\boldsymbol{\beta}(\tau)$ in the residual, when integrated, leaves us with a term with $\mathrm{d}\boldsymbol{\beta}(\tau)$, which is only of order $\mathrm{d}t^{1/2}$. Thus we can increase the strong order to one by expanding that term.

We can now do the same kind of expansion for the term $\mathscr{L}_{\beta,j}\mathbf{L}(\mathbf{x}(\tau), \tau)$ as we did in Equation (8.24), which leads to

$$\mathscr{L}_{\beta,j}\mathbf{L}(\mathbf{x}(t), t) = \mathscr{L}_{\beta,j}\mathbf{L}(\mathbf{x}(t_0), t_0) + \int_{t_0}^{t} \mathscr{L}_t\mathscr{L}_{\beta,j}\mathbf{L}(\mathbf{x}(t), t)\,\mathrm{d}t$$
$$+ \sum_j \int_{t_0}^{t} \mathscr{L}_{\beta,j}^2\mathbf{L}(\mathbf{x}(t), t)\,\mathrm{d}\beta_j(\tau). \quad (8.32)$$

Substituting this into the Equation (8.25) gives

$$\mathbf{x}(t) = \mathbf{x}(t_0) + \mathbf{f}(\mathbf{x}(t_0), t_0)(t - t_0) + \mathbf{L}(\mathbf{x}(t_0), t_0)(\boldsymbol{\beta}(t) - \boldsymbol{\beta}(t_0))$$
$$+ \sum_j \mathcal{L}_{\beta,j} \mathbf{L}(\mathbf{x}(t_0), t_0) \int_{t_0}^{t} \int_{t_0}^{\tau} \mathrm{d}\beta_j(\tau) \, \mathrm{d}\boldsymbol{\beta}(\tau) + \text{remainder.}$$

(8.33)

Now the important thing is to notice the *iterated Itô integral* appearing in the equation:

$$\int_{t_0}^{t} \int_{t_0}^{\tau} \mathrm{d}\beta_j(\tau) \, \mathrm{d}\boldsymbol{\beta}(\tau).$$

(8.34)

Computation of this kind of integrals and more general iterated Itô integrals turns out to be quite nontrivial. However, assuming that we can indeed compute the integral, as well as draw the corresponding Brownian increment (recall that the terms are not independent), we can form the following scheme known as the *Milstein method*.

Algorithm 8.2 (Milstein method). *Draw $\hat{\mathbf{x}}(t_0) \sim p(\mathbf{x}(t_0))$ and divide the time interval $[t_0, t]$ into M steps of length Δt. At each step k, do the following:*

1. *Jointly draw a Brownian motion increment and the iterated Itô integral of it:*

$$\Delta\boldsymbol{\beta}_k = \boldsymbol{\beta}(t_{k+1}) - \boldsymbol{\beta}(t_k),$$
$$\Delta\chi_{v,k} = \int_{t_k}^{t_{k+1}} \int_{t_k}^{\tau} \mathrm{d}\beta_j(\tau) \, \mathrm{d}\boldsymbol{\beta}(\tau).$$

(8.35)

2. *Compute*

$$\hat{\mathbf{x}}(t_{k+1}) = \hat{\mathbf{x}}(t_k) + \mathbf{f}(\hat{\mathbf{x}}(t_k), t_k)\,\Delta t + \mathbf{L}(\hat{\mathbf{x}}(t_k), t_k)\,\Delta\boldsymbol{\beta}_k$$
$$+ \sum_j \left[\sum_i \frac{\partial \mathbf{L}}{\partial x_i}(\hat{\mathbf{x}}(t_k), t_k)\, \mathbf{L}_{i,j}(\hat{\mathbf{x}}(t_k), t_k) \right] \Delta\chi_{v,k}. \quad (8.36)$$

The strong and weak orders of the preceding method are both one ($\gamma = \alpha = 1$). However, the difficulty is that drawing the iterated stochastic integral jointly with the Brownian motion is hard (cf. Kloeden and Platen, 1999). But if the noise is additive, that is, $\mathbf{L}(\mathbf{x}, t) = \mathbf{L}(t)$, then the Milstein method reduces to the Euler–Maruyama method. Thus in the additive noise case, the strong order of Euler–Maruyama is $\gamma = 1$ as well.

In the scalar case, we can compute the iterated stochastic integral:

$$\int_{t_0}^{t} \int_{t_0}^{\tau} d\beta(\tau)\, d\beta(\tau) = \frac{1}{2}\left[(\beta(t) - \beta(t_0))^2 - q\,(t - t_0)\right]. \qquad (8.37)$$

Thus in the scalar case, we can write down the Milstein method explicitly as follows.

Algorithm 8.3 (Scalar Milstein method). *Draw $\hat{x}(t_0) \sim p(x(t_0))$ and divide the time interval $[t_0, t]$ into M steps of length Δt. At each step k, do the following:*

1. *Draw random variable $\Delta\beta_k$ from the distribution*

$$\Delta\beta_k \sim N(0, q\,\Delta t). \qquad (8.38)$$

2. *Compute*

$$\hat{x}(t_{k+1}) = \hat{x}(t_k) + f(\hat{x}(t_k), t_k)\,\Delta t + L(\hat{x}(t_k), t_k)\,\Delta\beta_k$$
$$+ \frac{1}{2}\frac{\partial L(\hat{x}(t_k), t_k)}{\partial x}\,L(\hat{x}(t_k), t_k)\,(\Delta\beta_k^2 - q\,\Delta t). \quad (8.39)$$

We could now form even higher-order Itô–Taylor series expansions by including more terms into the series. However, if we try to derive higher-order methods than the Milstein method, we encounter higher-order iterated Itô integrals, which will turn out to be very difficult to compute. Fortunately, the additive noise case is much easier and often useful as well.

Now consider the case that $\mathbf{L}$ is in fact constant, which implies that $\mathcal{L}_t\mathbf{L} = \mathcal{L}_{\beta,j}\mathbf{L} = 0$. In that case, Equation (8.25) gives

$$\mathbf{x}(t) = \mathbf{x}(t_0) + \mathbf{f}(\mathbf{x}(t_0), t_0)\,(t - t_0) + \mathbf{L}\,(\boldsymbol{\beta}(t) - \boldsymbol{\beta}(t_0))$$
$$+ \int_{t_0}^{t}\int_{t_0}^{\tau}\mathcal{L}_t\mathbf{f}(\mathbf{x}(\tau), \tau)\,d\tau\,d\tau + \sum_j\int_{t_0}^{t}\int_{t_0}^{\tau}\mathcal{L}_{\beta,j}\mathbf{f}(\mathbf{x}(t), t)\,d\beta_j\,d\tau.$$
$$(8.40)$$

As the identities in Equation (8.24) are completely general, we can also apply them to $\mathcal{L}_t \mathbf{f}(\mathbf{x}(t), t)$ and $\mathcal{L}_{\beta,j} \mathbf{f}(\mathbf{x}(t), t)$, which gives

$$\mathcal{L}_t \mathbf{f}(\mathbf{x}(t), t) = \mathcal{L}_t \mathbf{f}(\mathbf{x}(t_0), t_0) + \int_{t_0}^t \mathcal{L}_t^2 \mathbf{f}(\mathbf{x}(t), t) \, dt$$

$$+ \sum_j \int_{t_0}^t \mathcal{L}_{\beta,j} \mathcal{L}_t \mathbf{f}(\mathbf{x}(t), t) \, d\beta_j, \qquad (8.41)$$

$$\mathcal{L}_{\beta,j} \mathbf{f}(\mathbf{x}(t), t) = \mathcal{L}_{\beta,j} \mathbf{f}(\mathbf{x}(t_0), t_0) + \int_{t_0}^t \mathcal{L}_t \mathcal{L}_{\beta,j} \mathbf{f}(\mathbf{x}(t), t) \, dt$$

$$+ \sum_j \int_{t_0}^t \mathcal{L}_{\beta,j}^2 \mathbf{f}(\mathbf{x}(t), t) \, d\beta_j. \qquad (8.42)$$

Substituting these identities into Equation (8.40) gives

$$\mathbf{x}(t) = \mathbf{x}(t_0) + \mathbf{f}(\mathbf{x}(t_0), t_0)(t - t_0) + \mathbf{L}(\boldsymbol{\beta}(t) - \boldsymbol{\beta}(t_0))$$

$$+ \mathcal{L}_t \mathbf{f}(\mathbf{x}(t_0), t_0) \frac{(t - t_0)^2}{2}$$

$$+ \sum_j \mathcal{L}_{\beta,j} \mathbf{f}(\mathbf{x}(t_0), t_0) \int_{t_0}^t [\beta_j(\tau) - \beta_j(t_0)] \, d\tau + \text{remainder.} \quad (8.43)$$

Thus the resulting approximation is

$$\mathbf{x}(t) \approx \mathbf{x}(t_0) + \mathbf{f}(\mathbf{x}(t_0), t_0)(t - t_0)$$

$$+ \mathbf{L}(\boldsymbol{\beta}(t) - \boldsymbol{\beta}(t_0)) + \mathcal{L}_t \mathbf{f}(\mathbf{x}(t_0), t_0) \frac{(t - t_0)^2}{2}$$

$$+ \sum_j \mathcal{L}_{\beta,j} \mathbf{f}(\mathbf{x}(t_0), t_0) \int_{t_0}^t [\beta_j(\tau) - \beta_j(t_0)] \, d\tau. \quad (8.44)$$

Note that the term $\boldsymbol{\beta}(t) - \boldsymbol{\beta}(t_0)$ and the integral $\int_{t_0}^t [\beta_j(\tau) - \beta_j(t_0)] \, d\tau$ really refer to the *same* Brownian motion and thus the terms are correlated. Fortunately, in this case both the terms are Gaussian and it is easy to compute their joint distribution:

$$\begin{pmatrix} \int_{t_0}^t [\boldsymbol{\beta}(\tau) - \boldsymbol{\beta}(t_0)] \, d\tau \\ \boldsymbol{\beta}(t) - \boldsymbol{\beta}(t_0) \end{pmatrix} \sim \mathrm{N}\left(\begin{pmatrix} \mathbf{0} \\ \mathbf{0} \end{pmatrix}, \begin{pmatrix} \mathbf{Q}(t - t_0)^3/3 & \mathbf{Q}(t - t_0)^2/2 \\ \mathbf{Q}(t - t_0)^2/2 & \mathbf{Q}(t - t_0) \end{pmatrix} \right).$$

$$(8.45)$$

From the resulting approximation, we get a strong order 1.5 Itô–Taylor expansion method, which has also been recently studied in the context of filtering theory (Arasaratnam et al., 2010; Särkkä and Solin, 2012).

Algorithm 8.4 (Strong order 1.5 Itô–Taylor for constant dispersion matrix). *Draw* $\hat{\mathbf{x}}(t_0) \sim p(\mathbf{x}(t_0))$ *and divide time interval* $[t_0, t]$ *into* M *steps of length* Δt. *At each step* k, *do the following:*

1. *Draw random variables* $\Delta \zeta_k$ *and* $\Delta \beta_k$ *from the joint distribution*

$$\begin{pmatrix} \Delta \zeta_k \\ \Delta \beta_k \end{pmatrix} \sim \mathrm{N} \left(\begin{pmatrix} \mathbf{0} \\ \mathbf{0} \end{pmatrix}, \begin{pmatrix} \mathbf{Q} \, \Delta t^3/3 & \mathbf{Q} \, \Delta t^2/2 \\ \mathbf{Q} \, \Delta t^2/2 & \mathbf{Q} \, \Delta t \end{pmatrix} \right). \qquad (8.46)$$

2. *Compute*

$$\hat{\mathbf{x}}(t_{k+1}) = \hat{\mathbf{x}}(t_k) + \mathbf{f}\left(\hat{\mathbf{x}}(t_k), t_k\right) \Delta t + \mathbf{L} \, \Delta \beta_k$$
$$+ \mathbf{a}_k\left(\hat{\mathbf{x}}(t_k)\right) \frac{\Delta t^2}{2} + \sum_j \mathbf{b}_{k,j}\left(\hat{\mathbf{x}}(t_k)\right) \Delta \zeta_k, \quad (8.47)$$

where

$$\mathbf{a}_k\left(\hat{\mathbf{x}}(t_k)\right) = \frac{\partial \mathbf{f}\left(\hat{\mathbf{x}}(t_k), t_k\right)}{\partial t} + \sum_i \frac{\partial \mathbf{f}\left(\hat{\mathbf{x}}(t_k), t_k\right)}{\partial x_i} f_i\left(\hat{\mathbf{x}}(t_k), t_k\right)$$
$$+ \frac{1}{2} \sum_{i,j} \frac{\partial^2 \mathbf{f}\left(\hat{\mathbf{x}}(t_k), t_k\right)}{\partial x_i \, \partial x_j} [\mathbf{L} \, \mathbf{Q} \, \mathbf{L}^\mathsf{T}]_{ij}, \qquad (8.48)$$
$$\mathbf{b}_{k,j}\left(\hat{\mathbf{x}}(t_k)\right) = \sum_i \frac{\partial \mathbf{f}\left(\hat{\mathbf{x}}(t_k), t_k\right)}{\partial x_i} L_{ij}.$$

As an interesting note on higher-order iterated Itô integrals, we point out a relation for the scalar case originally published by Itô himself (Itô, 1951). Applying the Itô formula for n times iterated Itô integrals leads to the following equation:

$$n! \int_{t_0}^t \int_{t_0}^{\tau_n} \cdots \int_{t_0}^{\tau_2} \mathrm{d}\beta(\tau_1) \, \mathrm{d}\beta(\tau_2) \ldots \mathrm{d}\beta(\tau_n)$$
$$= q^{n/2} \, (t - t_0)^{n/2} \, \mathrm{H}_n \left(\frac{\beta(t) - \beta(t_0)}{\sqrt{q \, (t - t_0)}} \right), \quad (8.49)$$

where $\mathrm{H}_n(t)$ denotes the probabilists' Hermite polynomials ($\mathrm{H}_0(t) = 1, \mathrm{H}_1(t) = t, \mathrm{H}_2(t) = t^2 - 1, \mathrm{H}_3(t) = t^3 - 3t, \ldots$). They are defined through the recursion $\mathrm{H}_{n+1}(t) = t \, \mathrm{H}_n(t) - \mathrm{d}/\mathrm{d}t \, \mathrm{H}_n(t)$. The result in Equation (8.37) can easily be verified from the preceding formula.

8.3 Weak Approximations of Itô–Taylor Series

The interest in solving SDEs is not always in the solution trajectories. Often, more interest is put into the distribution of the trajectories at a given time point rather than their paths. Thus we might be interested in forming approximations that describe accurately enough the probability distribution of the state trajectories. Weak approximations of Itô processes, that is, processes with approximately the same probability distribution, provide much more freedom in forming the approximations.

For instance, we can replace the distribution of $\mathbf{x}(t_0)$ with some other appropriate probability distribution, or more importantly we can replace the random increments $\Delta \boldsymbol{\beta}_k$ with more convenient approximations $\Delta \hat{\boldsymbol{\beta}}_k$ with similar moment properties.

The kind of approximations required here are much weaker than those required by the strong convergence criterion. The *weak order of convergence* can be defined to be the largest exponent α such that

$$| \mathrm{E}\left[g(\mathbf{x}(t_M))\right] - \mathrm{E}\left[g(\hat{\mathbf{x}}(t_M))\right] | \leq K \, \Delta t^\alpha \qquad (8.50)$$

for any polynomial function g. When the diffusion coefficient vanishes ($\mathbf{L}(\mathbf{x}, t) = \mathbf{0}$), this weak convergence criterion with $g(\mathbf{x}) = \mathbf{x}$ reduces to the usual deterministic convergence criterion for ordinary differential equations – as does the criterion for strong convergence.

For weak convergence, we only need to approximate the distribution of the Itô process $\mathbf{x}(t)$, so we can replace the Gaussian increments by other random variables with similar moment properties. Considering this, we can replace the increments in Algorithm 8.1 with approximate increments $\Delta \hat{\boldsymbol{\beta}}_k$. This leads to the *simplified weak Euler–Maruyama scheme*

$$\hat{\mathbf{x}}(t_{k+1}) = \hat{\mathbf{x}}(t_k) + \mathbf{f}(\hat{\mathbf{x}}(t_k), t_k) \, \Delta t + \mathbf{L}(\hat{\mathbf{x}}(t_k), t_k) \, \Delta \hat{\boldsymbol{\beta}}_k, \qquad (8.51)$$

where the $\Delta \hat{\beta}_k^{(j)}$, $j = 1, 2, \ldots, S$, must be independent random variables fulfilling suitable moment conditions. For example, we could use the following two-point distributed random variables (Kloeden and Platen, 1999)

$$\mathrm{P}(\Delta \hat{\beta}_k^{(j)} = \pm \sqrt{\Delta t}) = \frac{1}{2}. \qquad (8.52)$$

As we noticed in the previous section, multidimensional and higher-order Itô–Taylor approximations also involve additional random variables and iterated Itô integrals, which make them difficult to use in practice. The same applies to weak Itô–Taylor approximations, but handling them is much simpler than in the case of strong approximations. As a rule of thumb, Kloeden and Platen (1999) state that an Itô–Taylor approximation converges with any desired weak order $\alpha = 1.0, 2.0, \ldots$, when the number of stochastic integrals up to multiplicity α are included in the expansion. As an example, they give the following scalar time-invariant weak order $\alpha = 2.0$ scheme:

Algorithm 8.5 (Scalar weak order 2.0 Itô–Taylor method). *Draw $\hat{x}(t_0) \sim p(x(t_0))$ and divide the time interval $[t_0, t]$ into M steps of length Δt. At each step k, do the following:*

$$\hat{x}(t_{k+1}) = \hat{x}(t_k) + f(\hat{x}(t_k))\,\Delta t + L(\hat{x}(t_k))\,\Delta\hat{\beta}_k$$

$$+ \frac{1}{2}L(\hat{x}(t_k))\,\frac{\partial L(\hat{x}(t_k))}{\partial x}((\Delta\hat{\beta}_k)^2 - \Delta t)$$

$$+ \frac{\partial f(\hat{x}(t_k))}{\partial x}\,L(\hat{x}(t_k))\,\Delta\hat{\zeta}_k$$

$$+ \frac{1}{2}\left(f(\hat{x}(t_k))\,\frac{\partial f(\hat{x}(t_k))}{\partial x} + \frac{1}{2}\frac{\partial^2 f(\hat{x}(t_k))}{\partial x^2}\,L^2(\hat{x}(t_k))\right)(\Delta t)^2$$

$$+ \left(f(\hat{x}(t_k))\,\frac{\partial L(\hat{x}(t_k))}{\partial x} + \frac{1}{2}\frac{\partial^2 L(\hat{x}(t_k))}{\partial x^2}\,L^2(\hat{x}(t_k))\right)\left(\Delta\hat{\beta}_k\,\Delta t - \Delta\hat{\zeta}_k\right). \tag{8.53}$$

We can choose (see section 14.2 in Kloeden and Platen, 1999, and Eqs. 29–32)

$$\Delta\hat{\beta}_k = \Delta\beta_k \quad and \quad \Delta\hat{\zeta}_k = \frac{1}{2}\Delta\beta_k\,\Delta t \tag{8.54}$$

with $\Delta\beta_k \sim N(0, \Delta t)$, or instead of considering the normal increments, we could use

$$\Delta\hat{\beta}_k = (\Delta t)^{1/2}\,\theta_k \quad and \quad \Delta\hat{\zeta}_k = \frac{1}{2}(\Delta t)^{3/2}\,\theta_k, \tag{8.55}$$

where θ_ks are independent three-point distributed random variables with

$$P(\theta_k = \pm\sqrt{3}) = \frac{1}{6} \quad and \quad P(\theta_k = 0) = \frac{2}{3}. \tag{8.56}$$

As an example of an application of the preceding algorithm, we provide the following.

Example 8.6 (Simulating from a trigonometric nonlinear SDE). *Consider the nonlinear Itô stochastic differential equation model:*

$$dx = -\left(\frac{1}{10}\right)^2 \sin(x)\,\cos^3(x)\,dt + \frac{1}{10}\cos^2(x)\,d\beta, \qquad x(0) = x_0. \tag{8.57}$$

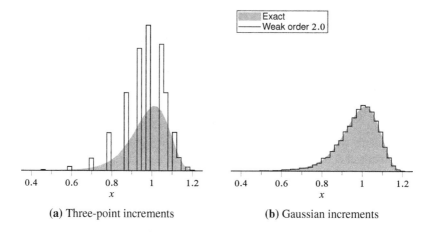

(a) Three-point increments   **(b)** Gaussian increments

Figure 8.1 The exact probability density at $t = 10$, and histograms of the 10,000 samples simulated by the weak order 2.0 scheme from Example 8.6 both using the three-point distributed and Gaussian increments. Both solutions (a) and (b) have similar moment properties.

This model has the solution $x(t) = \arctan(1/10\,\beta(t) + \tan(x_0))$ which we will use as ground truth, where $\beta(t)$ is a standard Brownian motion. The first and second derivatives of the drift $f(x)$ and diffusion $L(x)$ functions are:

$$\frac{\mathrm{d}f(x)}{\mathrm{d}x} = -\frac{1}{100}\cos^2(x)\,(2\cos(2x) - 1), \qquad \frac{\mathrm{d}L(x)}{\mathrm{d}x} = -\frac{1}{5}\sin(x)\,\cos(x),$$

$$\frac{\mathrm{d}^2 f(x)}{\mathrm{d}x^2} = \frac{1}{100}(\sin(2x) + 2\sin(4x)), \qquad \frac{\mathrm{d}^2 L(x)}{\mathrm{d}x^2} = -\frac{1}{5}\cos(2x).$$

We apply the weak order 2.0 scheme in Algorithm 8.5 to this problem, and characterize the solution at $t = 10$. We use a large step size $\Delta t = 1$ and simulate 10,000 trajectories. Figure 8.1 shows histograms of the values at $\hat{x}(10)$ both using the three-point distributed increments and Gaussian increments. Even though both solutions have similar moment properties (mean, variance, skewness, kurtosis, ...), the Gaussian increments appear neater in the visualization.

8.4 Ordinary Runge–Kutta Methods

Runge–Kutta (RK) methods are an important family of iterative methods for approximation of solutions of ordinary differential equations. The name stems from the German mathematicians Carl Runge and Martin Wilhelm Kutta, upon whose work many of the modern-day methods have been built.

The simplest Runge–Kutta method is the (forward) Euler scheme (see Algorithm 2.6), which is based on sequential linearization of the ODE system. This method is easy to understand and implement, but the global error of the method depends linearly on the step size Δt. The innovation Runge came up with was that subdivision of the integration interval into intermediate steps (as had earlier been done in quadrature methods, e.g., the *midpoint rule*, where the integrand was independent of t), can help build more efficient methods. Such higher-order methods can reach the same precision with fewer steps, which makes them appealing.

Consider the first-order nonlinear ODE from Chapter 2

$$\frac{d\mathbf{x}(t)}{dt} = \mathbf{f}(\mathbf{x}(t), t), \qquad \mathbf{x}(t_0) = \text{given}, \tag{8.58}$$

which can be integrated to give

$$\mathbf{x}(t) = \mathbf{x}(t_0) + \int_{t_0}^{t} \mathbf{f}(\mathbf{x}(\tau), \tau) \, d\tau. \tag{8.59}$$

Recall from the previous chapter that we used a *Taylor series expansion* for the solution of the ODE

$$\begin{aligned}
\mathbf{x}(t) = \mathbf{x}(t_0) &+ \mathbf{f}(\mathbf{x}(t_0), t_0)(t - t_0) \\
&+ \frac{1}{2!} \mathcal{L}\mathbf{f}(\mathbf{x}(t_0), t_0)(t - t_0)^2 \\
&+ \frac{1}{3!} \mathcal{L}^2 \mathbf{f}(\mathbf{x}(t_0), t_0)(t - t_0)^3 \\
&+ \cdots,
\end{aligned} \tag{8.60}$$

where we used the linear operator

$$\mathcal{L}(\bullet) = \frac{\partial}{\partial t}(\bullet) + \sum_i f_i \frac{\partial}{\partial x_i}(\bullet) \tag{8.61}$$

to come up with the convenient formulation of the series. Thus the series expansion is equal to

$$\mathbf{x}(t) = \mathbf{x}(t_0) + \mathbf{f}(\mathbf{x}(t_0), t_0)(t - t_0)$$
$$+ \frac{1}{2!} \left\{ \frac{\partial}{\partial t} \mathbf{f}(\mathbf{x}(t_0), t_0) + \sum_i f_i(\mathbf{x}(t_0), t_0) \frac{\partial}{\partial x_i} \mathbf{f}(\mathbf{x}(t_0), t_0) \right\} (t - t_0)^2$$
$$+ \frac{1}{3!} \left\{ \frac{\partial [\mathcal{L} \mathbf{f}(\mathbf{x}(t_0), t_0)]}{\partial t} + \sum_i f_i(\mathbf{x}(t_0), t_0) \frac{\partial [\mathcal{L} \mathbf{f}(\mathbf{x}(t_0), t_0)]}{\partial x_i} \right\} (t - t_0)^3$$
$$+ \cdots \tag{8.62}$$

If we were only to consider the terms up to Δt, we would recover the Euler method, which is a derivative-free Runge–Kutta scheme. However, here we wish to get hold of higher-order methods. For the sake of simplicity, we now stop at the term $(t - t_0)^2 = (\Delta t)^2$ and write

$$\mathbf{x}(t_0 + \Delta t) \approx \mathbf{x}(t_0) + \mathbf{f}(\mathbf{x}(t_0), t_0) \Delta t$$
$$+ \frac{1}{2} \left\{ \frac{\partial}{\partial t} \mathbf{f}(\mathbf{x}(t_0), t_0) + \sum_i f_i(\mathbf{x}(t_0), t_0) \frac{\partial}{\partial x_i} \mathbf{f}(\mathbf{x}(t_0), t_0) \right\} (\Delta t)^2. \tag{8.63}$$

This equation still contains derivatives, and we aim to get rid of them by writing the expression in terms of the function $\mathbf{f}(\bullet, \bullet)$ evaluated at various points. That is, we seek the following form:

$$\mathbf{x}(t_0 + \Delta t) \approx \mathbf{x}(t_0) + a \mathbf{f}(\mathbf{x}(t_0), t_0) \Delta t$$
$$+ b \mathbf{f}(\mathbf{x}(t_0) + c \mathbf{f}(\mathbf{x}(t_0), t_0) \Delta t, t_0 + d \Delta t) \Delta t, \tag{8.64}$$

where a, b, c, and d are unknown. The last term can be approximated with a truncated Taylor expansion (linearization) around $\mathbf{f}(\mathbf{x}(t_0), t_0)$ with the increments chosen as follows:

$$\mathbf{f}(\mathbf{x}(t_0) + c \mathbf{f}(\mathbf{x}(t_0), t_0) \Delta t, t_0 + d \Delta t) \approx \mathbf{f}(\mathbf{x}(t_0), t_0)$$
$$+ c \left(\sum_i f_i(\mathbf{x}(t_0), t_0) \frac{\partial}{\partial x_i} \mathbf{f}(\mathbf{x}(t_0), t_0) \right) \Delta t + d \frac{\partial \mathbf{f}(\mathbf{x}(t_0), t_0)}{\partial t} \Delta t.$$
$$\tag{8.65}$$

Combining the preceding two equations gives the following:

$$\mathbf{x}(t_0 + \Delta t) \approx \mathbf{x}(t_0) + (a + b) \mathbf{f}(\mathbf{x}(t_0), t_0) \Delta t$$
$$+ b \left\{ c \sum_i f_i(\mathbf{x}(t_0), t_0) \frac{\partial}{\partial x_i} \mathbf{f}(\mathbf{x}(t_0), t_0) + d \frac{\partial \mathbf{f}(\mathbf{x}(t_0), t_0)}{\partial t} \right\} (\Delta t)^2.$$
$$\tag{8.66}$$

If we now compare the preceding equation to the original truncated Taylor expansion in Equation (8.63), we get the following conditions for our coefficients:

$$a = \frac{1}{2}, \quad b = \frac{1}{2}, \quad c = 1, \quad \text{and} \quad d = 1. \tag{8.67}$$

Thus the approximative step given by this method can be written as

$$\hat{\mathbf{x}}(t_0 + \Delta t) = \mathbf{x}(t_0) + \frac{1}{2}\left[\mathbf{f}(\tilde{\mathbf{x}}_1, t_0) + \mathbf{f}(\tilde{\mathbf{x}}_2, t_0 + \Delta t)\right]\Delta t, \tag{8.68}$$

where the supporting values are given by

$$\begin{aligned} \tilde{\mathbf{x}}_1 &= \mathbf{x}(t_0), \\ \tilde{\mathbf{x}}_2 &= \mathbf{x}(t_0) + \mathbf{f}(\tilde{\mathbf{x}}_1, t_0)\,\Delta t. \end{aligned} \tag{8.69}$$

What we derived here is a two-stage method (actually the Heun method presented in Algorithm 2.7) with the finite differences determined by the choices we made in truncating the series expansion. The choices of how and what to truncate determine the number of terms in the expansion and thus also affect the number of equations to solve. Coming up with higher-order methods becomes increasingly complicated with the number of terms. The general principle, however, remains the same, and Runge–Kutta methods are constructed by evaluating the function $\mathbf{f}(\bullet, \bullet)$ at a number of points and weighting these evaluations.

We can write down a general s-stage algorithm for ordinary Runge–Kutta methods as follows:

Algorithm 8.7 (Runge–Kutta methods). *Start from $\hat{\mathbf{x}}(t_0) = \mathbf{x}(t_0)$ and divide the integration interval $[t_0, t]$ into M steps $t_0 < t_1 < t_2 < \ldots < t_M = t$ such that $\Delta t = t_{k+1} - t_k$. The integration method is defined by its Butcher tableau:*

$$\begin{array}{c|c} \mathbf{c} & \mathbf{A} \\ \hline & \boldsymbol{\alpha}^{\mathsf{T}} \end{array}. \tag{8.70}$$

On each step k, approximate the solution as follows:

$$\hat{\mathbf{x}}(t_{k+1}) = \hat{\mathbf{x}}(t_k) + \sum_{i=1}^{s} \alpha_i\,\mathbf{f}(\tilde{\mathbf{x}}_i, \tilde{t}_i)\,\Delta t, \tag{8.71}$$

where $\tilde{t}_i = t_k + c_i\,\Delta t$ and $\tilde{\mathbf{x}}_i = \hat{\mathbf{x}}(t_k) + \sum_{j=1}^{s} A_{i,j}\,\mathbf{f}(\tilde{\mathbf{x}}_j, \tilde{t}_j)\,\Delta t$.

As implied by the preceding algorithm, the ordinary Runge–Kutta methods are commonly expressed in terms of a table called the *Butcher tableau*:

$$
\begin{array}{c|cccc}
c_1 & A_{1,1} & & & \\
c_2 & A_{2,1} & A_{2,2} & & \\
\vdots & \vdots & & \ddots & \\
c_s & A_{s,1} & A_{s,2} & \cdots & A_{s,s} \\
\hline
 & \alpha_1 & \alpha_2 & \cdots & \alpha_s
\end{array}
\tag{8.72}
$$

An explicit Runge–Kutta method is said to be consistent if $\sum_{j=1}^{i-1} A_{i,j} = c_i$, for $i = 2, 3, \ldots, s$.

We present the Butcher tableau for two common Runge–Kutta methods. The first method is the forward Euler method (see Algorithm 2.6) and the second the classical fourth-order Runge–Kutta method (see Algorithm 2.8):

Example 8.8 (Forward Euler). *The forward Euler scheme in Algorithm 2.6 has the Butcher tableau:*

$$
\begin{array}{c|c}
0 & 0 \\
\hline
 & 1
\end{array},
\tag{8.73}
$$

which gives the recursion $\hat{\mathbf{x}}(t_{k+1}) = \hat{\mathbf{x}}(t_k) + \mathbf{f}(\hat{\mathbf{x}}(t_k), t_k)\,\Delta t.$

Example 8.9 (Fourth-order Runge–Kutta method). *The classical RK4 method in Algorithm 2.8 has the following Butcher tableau:*

$$
\begin{array}{c|cccc}
0 & & & & \\
\frac{1}{2} & \frac{1}{2} & & & \\
\frac{1}{2} & 0 & \frac{1}{2} & & \\
1 & 0 & 0 & 1 & \\
\hline
 & \frac{1}{6} & \frac{1}{3} & \frac{1}{3} & \frac{1}{6}
\end{array}.
\tag{8.74}
$$

The preceding Runge–Kutta schemes are all so-called explicit schemes. Explicit schemes can suffer from numerical instability, when the solution includes rapidly varying terms. Such problems are called *stiff* equations. Stiff equations require explicit schemes to use small step sizes in order to not diverge from the solution path.

A better-suited family of solvers for stiff problems are the so-called *implicit* Runge–Kutta methods, which provide additional stability to the iterative solution. For implicit methods, the Buther tableau is no longer lower-triangular, but the tableau can be full. The consequence of using a full table is that at every step, a system of algebraic equations has to be solved. This increases the computational cost considerably. The advantage of implicit

Runge–Kutta methods over explicit ones is their greater stability, especially when applied to stiff equations.

The simplest example of an implicit method is the *backward Euler* scheme.

Example 8.10 (Backward Euler). *The implicit backward Euler scheme has the Butcher tableau:*

$$\begin{array}{c|c} 1 & 1 \\ \hline & 1 \end{array}, \tag{8.75}$$

which gives the recursion $\hat{\mathbf{x}}(t_{k+1}) = \hat{\mathbf{x}}(t_k) + \mathbf{f}(\hat{\mathbf{x}}(t_{k+1}), t_k + \Delta t) \Delta t.$

There are a lot of further topics to consider in Runge–Kutta methods, such as stability analysis and adaptive step size methods. We, however, will not discuss these issues here.

Example 8.11 (Comparison of ODE solvers). *We study the two-dimensional nonlinear ordinary differential equation system:*

$$\begin{aligned} \dot{x}_1 &= x_1 - x_2 - x_1^3, \\ \dot{x}_2 &= x_1 + x_2 - x_2^3. \end{aligned} \tag{8.76}$$

This system has only one fixed point, $\mathbf{x} = (0,0)$ *(an unstable spiral), but it also has a limit cycle. We aim to test various Runge–Kutta methods by simulating trajectories of this problem.*

We use a time-span of $[0, 10]$, *with a step size of* $\Delta t = 2^{-4}$. *The methods demonstrated are the forward (explicit) Euler method, the Heun method, the backward (implicit) Euler method, and the classical fourth-order Runge–Kutta method. Figure 8.2 shows six trajectories for each method, starting from respective quadrants. Ideally, the results should be symmetrical, but especially the forward Euler results do not match the rest.*

8.5 Strong Stochastic Runge–Kutta Methods

A practical disadvantage of the Itô–Taylor approximations considered earlier in this chapter is that the derivatives of various orders of the drift and diffusion functions must be determined and evaluated at each step. However, there are discrete time approximations that avoid the use of derivatives. They are in general referred to as *stochastic Runge–Kutta methods*. Stochastic versions of the Runge–Kutta methods are not as simple as the deterministic methods.

As has been discussed earlier, the *Euler–Maruyama* scheme can easily

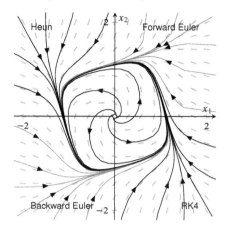

Figure 8.2 Demonstration of four Runge–Kutta schemes for the same problem with a step size of $\Delta t = 2^{-4}$ (see Example 8.11).

be constructed by discretizing the time interval and formulating the solver SDE as a recursive algorithm. Thus, the Euler–Maruyama scheme can be seen as the simplest stochastic Runge–Kutta method – similar to how we interpreted the Euler method as a simple ordinary Runge–Kutta scheme.

In practice, a higher-order stochastic Runge–Kutta method can be derived, for example, by replacing the closed-form derivatives in the Milstein method (Algorithm 8.2 or 8.3) with suitable finite differences (see Kloeden et al., 1994; Kloeden and Platen, 1999). If we heuristically replace the partial differential in Algorithm 8.3 with a finite difference, we can write a method for scalar time-invariant models:

$$\hat{x}(t_{k+1}) = \hat{x}(t_k) + f(\hat{x}(t_k))\,\Delta t + L(\hat{x}(t_k))\,\Delta\beta_k$$
$$+ \frac{1}{2\sqrt{\Delta t}}\left[L(\tilde{x}) - L(\hat{x}(t_k))\right]\left[(\Delta\beta_k)^2 - q\,\Delta t\right] \quad (8.77)$$

with supporting value $\tilde{x} = \hat{x}(t_k) + L(\hat{x}(t_k))\sqrt{\Delta t}$. This method is of strong order 1.0.

However, we still *cannot get rid of the iterated Itô integral* occurring in the Milstein method. An important thing to note is that stochastic versions of Runge–Kutta methods cannot be derived as simple extensions of the deterministic Runge–Kutta methods – see Burrage et al. (2006), which is a response to the article by Wilkie (2004).

To provide a more widely applicable perspective on the methods, we follow a similar derivation as we did for the ordinary Runge–Kutta methods.

Recall the following multidimensional SDE formulation

$$d\mathbf{x} = \mathbf{f}(\mathbf{x}(t), t) \, dt + \mathbf{L}(\mathbf{x}(t), t) \, d\boldsymbol{\beta}, \qquad \mathbf{x}(t_0) \sim p(\mathbf{x}(t_0)), \qquad (8.78)$$

where $\mathbf{x}(t) \in \mathbb{R}^D$ is the state, $\beta(t) \in \mathbb{R}^S$ is a Brownian motion with diffusion matrix $\mathbf{Q}$, and $\mathbf{f}(\mathbf{x}, t)$ and $\mathbf{L}(\mathbf{x}, t)$ are some given functions. In integral form, the equation can be expressed as

$$\mathbf{x}(t) = \mathbf{x}(t_0) + \int_{t_0}^{t} \mathbf{f}(\mathbf{x}(\tau), \tau) \, d\tau + \int_{t_0}^{t} \mathbf{L}(\mathbf{x}(\tau), \tau) \, d\boldsymbol{\beta}(\tau). \qquad (8.79)$$

As we saw in the previous chapter, applying the Itô formula to the terms $\mathbf{f}(\mathbf{x}(t), t)$ and $\mathbf{L}(\mathbf{x}(t), t)$ and collecting the terms gives an Itô–Taylor series expansion of the solution:

$$\begin{aligned}
\mathbf{x}(t) = \mathbf{x}(t_0) &+ \mathbf{f}(\mathbf{x}(t_0), t_0)(t - t_0) + \mathbf{L}(\mathbf{x}(t_0), t_0)(\boldsymbol{\beta}(t) - \boldsymbol{\beta}(t_0)) \\
&+ \int_{t_0}^{t} \int_{t_0}^{\tau} \mathcal{L}_t \mathbf{f}(\mathbf{x}(\tau), \tau) \, d\tau \, d\tau \\
&+ \sum_i \int_{t_0}^{t} \int_{t_0}^{\tau} \mathcal{L}_{\beta,i} \mathbf{f}(\mathbf{x}(\tau), \tau) \, d\beta_i(\tau) \, d\tau \\
&+ \int_{t_0}^{t} \int_{t_0}^{\tau} \mathcal{L}_t \mathbf{L}(\mathbf{x}(\tau), \tau) \, d\tau \, d\boldsymbol{\beta}(\tau) \\
&+ \sum_i \int_{t_0}^{t} \int_{t_0}^{\tau} \mathcal{L}_{\beta,i} \mathbf{L}(\mathbf{x}(\tau), \tau) \, d\beta_i(\tau) \, d\boldsymbol{\beta}(\tau). \qquad (8.80)
\end{aligned}$$

Similarly, as we did in the previous section, we can consider truncated series expansions of various degrees for each of these terms. Collecting the terms can give a similar kind of formulation in terms of tabulated values as we did for the ordinary RK schemes. The extra terms involving the iterated and cross-term Itô integrals complicate the formulation.

Rößler (2010) considers a general class of multidimensional strong order 1.0 stochastic Runge–Kutta schemes, where iterated integrals are avoided in the scheme and they only appear in the supporting values. A more general formulation is given in the next section, where weak order methods are considered. The general multidimensional schemes by Rößler can be summarized as the following algorithm.

Algorithm 8.12 (Stochastic Runge–Kutta method of strong order 1.0). *Start from* $\hat{\mathbf{x}}(t_0) \sim p(\mathbf{x}(t_0))$ *and divide the integration interval* $[t_0, t]$ *into*

M steps of length Δt. The integration method is characterized by its extended Butcher tableau:

$$
\begin{array}{c|c|c}
\mathbf{c}^{(0)} & \mathbf{A}^{(0)} & \mathbf{B}^{(0)} \\
\hline
\mathbf{c}^{(1)} & \mathbf{A}^{(1)} & \mathbf{B}^{(1)} \\
\hline
& \boldsymbol{\alpha}^{\mathsf{T}} & \left[\boldsymbol{\gamma}^{(1)}\right]^{\mathsf{T}} \quad \left[\boldsymbol{\gamma}^{(2)}\right]^{\mathsf{T}}
\end{array}
\tag{8.81}
$$

On each step k, approximate the solution trajectory as follows:

$$
\hat{\mathbf{x}}(t_{k+1}) = \hat{\mathbf{x}}(t_k) + \sum_i \alpha_i\, \mathbf{f}\,(\tilde{\mathbf{x}}_i^{(0)}, t_k + c_i^{(0)}\Delta t)\,\Delta t
$$
$$
+ \sum_{i,n}(\gamma_i^{(1)}\Delta\beta_k^{(n)} + \gamma_i^{(2)}\sqrt{\Delta t})\,\mathbf{L}_n(\tilde{\mathbf{x}}_i^{(n)}, t_k + c_i^{(1)}\Delta t)
\tag{8.82}
$$

with the supporting values

$$
\tilde{\mathbf{x}}_i^{(0)} = \hat{\mathbf{x}}(t_k) + \sum_j A_{i,j}^{(0)}\, \mathbf{f}\,(\tilde{\mathbf{x}}_j^{(0)}, t_k + c_j^{(0)}\Delta t)\,\Delta t
$$
$$
+ \sum_{j,l} B_{i,j}^{(0)}\, \mathbf{L}_l(\tilde{\mathbf{x}}_j^{(l)}, t_k + c_j^{(1)}\Delta t)\,\Delta\beta_k^{(l)},
\tag{8.83}
$$

$$
\tilde{\mathbf{x}}_i^{(n)} = \hat{\mathbf{x}}(t_k) + \sum_j A_{i,j}^{(1)}\, \mathbf{f}\,(\tilde{\mathbf{x}}_j^{(0)}, t_k + c_j^{(0)}\Delta t)\,\Delta t
$$
$$
+ \sum_{j,l} B_{i,j}^{(1)}\, \mathbf{L}_l(\tilde{\mathbf{x}}_j^{(l)}, t_k + c_j^{(1)}\Delta t)\,\frac{\Delta\chi_k^{(l,n)}}{\sqrt{\Delta t}},
\tag{8.84}
$$

for $i = 1, 2, \ldots, s$ and $n = 1, 2, \ldots, S$.

The increments in Algorithm 8.12 are given by the Itô integrals:

$$
\Delta\beta_k^{(i)} = \int_{t_k}^{t_{k+1}} \mathrm{d}\beta_i(\tau) \quad \text{and}
\tag{8.85}
$$

$$
\Delta\chi_k^{(i,j)} = \int_{t_k}^{t_{k+1}} \int_{t_k}^{\tau_2} \mathrm{d}\beta_i(\tau_1)\,\mathrm{d}\beta_j(\tau_2),
\tag{8.86}
$$

for $i, j = 1, 2, \ldots, S$. The increments $\Delta\beta_k^{(i)}$ are normally distributed random variables such that jointly $\Delta\boldsymbol{\beta}_k \sim \mathrm{N}(\mathbf{0}, \mathbf{Q}\,\Delta t)$. The iterated stochastic Itô integrals $\Delta\chi_k^{(i,j)}$ are trickier. For these methods, when $i = j$ the multiple Itô integrals can be rewritten as

$$
\Delta\chi_k^{(i,i)} = \frac{1}{2}\left([\Delta\beta_k^{(i)}]^2 - Q_{i,i}\,\Delta t\right),
\tag{8.87}
$$

which follows from the results given in Equation (8.49). This also generalizes to higher orders. Exact simulation of the integrals $\Delta \chi_k^{(i,j)}$, when $i \neq j$, is not possible, but can be approximated. See Wiktorsson (2001) for an approximative scheme, and Gilsing and Shardlow (2007) for implementation details.

Example 8.13 (Euler–Maruyama Butcher tableau). *The Euler–Maruyama method has the extended Butcher tableau:*

$$
\begin{array}{cc|c}
0 & 0 & 0 \\
\hline
0 & 0 & 0 \\
\hline
 & 1 & 1 & 0
\end{array}
\tag{8.88}
$$

and as we recall from earlier, it is of strong order 0.5.

Coming up with useful and valid stochastic Runge–Kutta schemes is a delicate process, which we will not consider here. Instead, we go through a rather efficient and general scheme proposed by Rößler (2010), which can be formulated as the following algorithm.

Algorithm 8.14 (Strong order 1.0 stochastic Runge–Kutta due to Rößler). *Consider a stochastic Runge–Kutta method with the following extended Butcher tableau:*

$$
\begin{array}{c|cc|cc|ccc}
0 & & & & & \\
1 & 1 & & 0 & & \\
0 & 0 & 0 & 0 & 0 & \\
\hline
0 & & & & & \\
1 & 1 & & 1 & & \\
1 & 1 & 0 & -1 & 0 & \\
\hline
 & \frac{1}{2} & \frac{1}{2} & 0 & 1 & 0 & 0 & 0 & \frac{1}{2} & -\frac{1}{2}
\end{array}
\tag{8.89}
$$

which corresponds to the following iterative scheme

$$
\hat{\mathbf{x}}(t_{k+1}) = \hat{\mathbf{x}}(t_k) + \frac{1}{2}\{\mathbf{f}(\hat{\mathbf{x}}(t_k), t_k) + \mathbf{f}(\tilde{\mathbf{x}}_2^{(0)}, t_k + \Delta t)\}\,\Delta t
$$

$$
+ \sum_n \{\Delta \beta_k^{(n)}\,\mathbf{L}_n(\hat{\mathbf{x}}(t_k), t_k)
$$

$$
+ \frac{1}{2}\sqrt{\Delta t}\,\left(\mathbf{L}_n(\tilde{\mathbf{x}}_2^{(n)}, t_k + \Delta t) - \mathbf{L}_n(\tilde{\mathbf{x}}_3^{(n)}, t_k + \Delta t)\right)\}
\tag{8.90}
$$

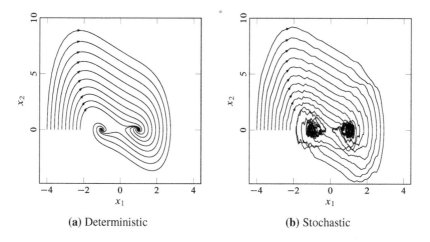

(a) Deterministic   **(b)** Stochastic

Figure 8.3 Trajectories simulated from the Duffing van der Pol
oscillator model in Example 8.15. The deterministic solutions
($q = 0$) converge to either of the two steady states. The
realizations of the noise are identical for each of the stochastic
($q = 0.5^2$) trajectories.

with the supporting values $\tilde{\mathbf{x}}_2^{(0)} = \hat{\mathbf{x}}(t_k) + \mathbf{f}(\hat{\mathbf{x}}(t_k), t_k)\,\Delta t$, *and*

$$\tilde{\mathbf{x}}_2^{(n)} = \hat{\mathbf{x}}(t_k) + \mathbf{f}(\hat{\mathbf{x}}(t_k), t_k)\,\Delta t + \sum_l \mathbf{L}_l(\hat{\mathbf{x}}(t_k), t_k)\,\frac{\Delta \chi_k^{(l,n)}}{\sqrt{\Delta t}}, \qquad (8.91)$$

$$\tilde{\mathbf{x}}_3^{(n)} = \hat{\mathbf{x}}(t_k) + \mathbf{f}(\hat{\mathbf{x}}(t_k), t_k)\,\Delta t - \sum_l \mathbf{L}_l(\hat{\mathbf{x}}(t_k), t_k)\,\frac{\Delta \chi_k^{(l,n)}}{\sqrt{\Delta t}}. \qquad (8.92)$$

Example 8.15 (Duffing van der Pol oscillator)**.** *Consider a simplified version of a Duffing van der Pol oscillator*

$$\ddot{x} + \dot{x} - (\alpha - x^2)\, x = x\, w(t), \quad \alpha \geq 0, \qquad (8.93)$$

*driven by multiplicative white noise $w(t)$ with spectral density q. The cor-
responding two-dimensional Itô stochastic differential equation with the
state $\mathbf{x}(t) = (x, \dot{x})$ is*

$$\begin{pmatrix} dx_1 \\ dx_2 \end{pmatrix} = \begin{pmatrix} x_2 \\ x_1\,(\alpha - x_1^2) - x_2 \end{pmatrix} dt + \begin{pmatrix} 0 \\ x_1 \end{pmatrix} d\beta, \qquad (8.94)$$

*where $\beta(t)$ is a one-dimensional Brownian motion. The deterministic ver-
sion (when $q = 0$) has the steady states $\mathbf{x} = (0,0)$ and $\mathbf{x} = (\pm\sqrt{\alpha}, 0)$,*

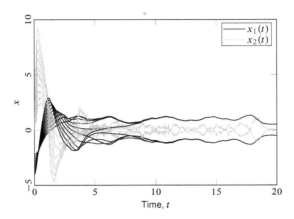

Figure 8.4 Evolution of the trajectories in the stochastic Duffing van der Pol oscillator model in Figure 8.3.

the first of which is also a degenerate stationary state of the stochastic differential equation.

Let $\alpha = 1$. First we study the deterministic solution with no diffusion ($q = 0$). Figure 8.3 shows 10 trajectories, each with different initial values $x_1(0)$. We use a step size of $\Delta t = 2^{-5}$ and a time span of $[0, 20]$. We then replicate the result, but using the SDE model with $q = 0.5^2$ and identical realizations of noise in each trajectory. We use the strong order 1.0 method in Algorithm 8.14 for simulating the trajectories. Figure 8.4 shows the evolution of the trajectories.

Higher-order methods can be formulated by considering more terms in the Itô–Taylor expansion. This, however, might not be very practical, as the number of required function evaluations grows, as does the complexity of the scheme. However, for models with some special structure this might still be feasible. Examples of such cases are models with commutative noise, additive noise models, where $\mathbf{L}(\mathbf{x}, t) \equiv \mathbf{L}(t)$, or diagonal noise models.

A number of stochastic Runge–Kutta methods have also been presented by Kloeden et al. (1994) and Kloeden and Platen (1999) as well as by Rößler (2006). If the noise is additive, then it is possible to derive a Runge–Kutta counterpart of the method in Algorithm 8.4, which uses finite difference approximations instead of the closed-form derivatives (Kloeden and Platen, 1999). When $\mathbf{L}$ is constant, we get the following algorithm.

Algorithm 8.16 (Strong order 1.5 SRK method for constant diffusion).

Draw $\hat{\mathbf{x}}(t_0) \sim p(\mathbf{x}(t_0))$ *and divide the time interval* $[t_0, t]$ *into* M *steps of length* Δt. *At each step* k, *do the following:*

1. *Draw random variables* $\Delta \boldsymbol{\zeta}_k$ *and* $\Delta \boldsymbol{\beta}_k$ *from the joint distribution*

$$\begin{pmatrix} \Delta \boldsymbol{\zeta}_k \\ \Delta \boldsymbol{\beta}_k \end{pmatrix} \sim \mathrm{N}\left(\begin{pmatrix} \mathbf{0} \\ \mathbf{0} \end{pmatrix}, \begin{pmatrix} \mathbf{Q} \, \Delta t^3/3 & \mathbf{Q} \, \Delta t^2/2 \\ \mathbf{Q} \, \Delta t^2/2 & \mathbf{Q} \, \Delta t \end{pmatrix} \right). \tag{8.95}$$

2. *Compute*

$$\hat{\mathbf{x}}(t_{k+1}) = \hat{\mathbf{x}}(t_k) + \mathbf{f}(\hat{\mathbf{x}}(t_k), t_k) \, \Delta t + \mathbf{L} \, \Delta \boldsymbol{\beta}_k$$
$$+ \frac{1}{4} \sum_j \left[\mathbf{f}(\tilde{\mathbf{x}}_j^+, t_{k+1}) - 2 \mathbf{f}(\hat{\mathbf{x}}(t_k), t_k) + \mathbf{f}(\tilde{\mathbf{x}}_j^-, t_{k+1}) \right] \Delta t$$
$$+ \frac{1}{2\sqrt{\Delta t}} \sum_j \left[\mathbf{f}(\tilde{\mathbf{x}}_j^+, t_{k+1}) - \mathbf{f}(\tilde{\mathbf{x}}_j^-, t_{k+1}) \right] \Delta \zeta_{k,j}, \tag{8.96}$$

where the supporting values are given by

$$\tilde{\mathbf{x}}_j^\pm = \hat{\mathbf{x}}(t_k) + \frac{1}{S} \mathbf{f}(\hat{\mathbf{x}}(t_k), t_k) \, \Delta t \pm \mathbf{L}_j \, \sqrt{\Delta t}, \tag{8.97}$$

for $j = 1, 2, \ldots, S$.

8.6 Weak Stochastic Runge–Kutta Methods

Earlier in this chapter, we saw that it is possible to form weak approximations to SDEs, where the interest is not in the solution trajectories, but the distributions of them. It is often computationally convenient to replace the weak Itô–Taylor approximations by Runge–Kutta style approximations that avoid the use of derivatives of the drift and diffusion coefficients.

As an example of such a weak scheme, we consider the following scalar weak order 2.0 Runge–Kutta scheme for time-invariant SDEs due to Platen (see Kloeden and Platen, 1999). The iteration takes the following form:

$$\hat{x}(t_{k+1}) = \hat{x}(t_k) + \frac{1}{2}\left[f(\hat{x}(t_k)) + f(\tilde{x}) \right] \Delta t$$
$$+ \frac{1}{4}\left[L(\tilde{x}^+) + 2 L(\hat{x}(t_k)) + L(\tilde{x}^-) \right] \Delta \hat{\beta}_k$$
$$+ \frac{1}{4\sqrt{\Delta t}}\left[L(\tilde{x}^+) - L(\tilde{x}^-) \right]\left[(\Delta \hat{\beta}_k)^2 - q \, \Delta t \right] \tag{8.98}$$

with supporting values $\tilde{x} = \hat{x}(t_k) + f(\hat{x}(t_k)) \, \Delta t + L(\hat{x}(t_k)) \, \Delta \hat{\beta}_k$ and $\tilde{x}^\pm = \hat{x}(t_k) + f(\hat{x}(t_k)) \, \Delta t \pm L(\hat{x}(t_k)) \, \sqrt{\Delta t}$.

Rößler (2009) considers a general class of multidimensional weak order

2.0 stochastic Runge–Kutta schemes. For simplicity, in the following we will restrict our interest to models driven by standard Brownian motion only. For models with nonidentity diffusion matrices $\mathbf{Q}$, a straightforward approach is to include this scaling into the dispersion matrix. The general multidimensional schemes by Rößler are as follows.

Algorithm 8.17 (Stochastic Runge–Kutta methods of weak order 2.0). *Start from $\hat{\mathbf{x}}(t_0) \sim p(\mathbf{x}(t_0))$ and divide the integration interval $[t_0, t]$ into M steps $t_0 < t_1 < t_2 < \ldots < t_M = t$ such that $\Delta t = t_{k+1} - t_k$. The integration method is characterized by the following extended Butcher tableau:*

$$
\begin{array}{c|c|c}
\mathbf{c}^{(0)} & \mathbf{A}^{(0)} & \mathbf{B}^{(0)} \\
\hline
\mathbf{c}^{(1)} & \mathbf{A}^{(1)} & \mathbf{B}^{(1)} \\
\hline
\mathbf{c}^{(2)} & \mathbf{A}^{(2)} & \mathbf{B}^{(2)} \\
\hline
& \boldsymbol{\alpha}^\mathsf{T} & \left[\boldsymbol{\gamma}^{(1)}\right]^\mathsf{T} \quad \left[\boldsymbol{\gamma}^{(2)}\right]^\mathsf{T} \\
& & \left[\boldsymbol{\gamma}^{(3)}\right]^\mathsf{T} \quad \left[\boldsymbol{\gamma}^{(4)}\right]^\mathsf{T}
\end{array}
\tag{8.99}
$$

On each step k, approximate the solution by the following:

$$
\begin{aligned}
\hat{\mathbf{x}}(t_{k+1}) = \hat{\mathbf{x}}(t_k) &+ \sum_i \alpha_i\, \mathbf{f}\!\left(\tilde{\mathbf{x}}_i^{(0)}, t_k + c_i^{(0)}\Delta t\right) \Delta t \\
&+ \sum_{i,n} \gamma_i^{(1)}\, \mathbf{L}_n\!\left(\tilde{\mathbf{x}}_i^{(n)}, t_k + c_i^{(1)}\Delta t\right) \Delta \hat{\beta}_k^{(n)} \\
&+ \sum_{i,n} \gamma_i^{(2)}\, \mathbf{L}_n\!\left(\tilde{\mathbf{x}}_i^{(n)}, t_k + c_i^{(1)}\Delta t\right) \frac{\Delta \hat{\chi}_k^{(n,n)}}{\sqrt{\Delta t}} \\
&+ \sum_{i,n} \gamma_i^{(3)}\, \mathbf{L}_n\!\left(\bar{\mathbf{x}}_i^{(n)}, t_k + c_i^{(2)}\Delta t\right) \Delta \hat{\beta}_k^{(n)} \\
&+ \sum_{i,n} \gamma_i^{(4)}\, \mathbf{L}_n\!\left(\bar{\mathbf{x}}_i^{(n)}, t_k + c_i^{(2)}\Delta t\right) \sqrt{\Delta t},
\end{aligned}
\tag{8.100}
$$

with supporting values

$$
\begin{aligned}
\tilde{\mathbf{x}}_i^{(0)} = \hat{\mathbf{x}}(t_k) &+ \sum_j A_{i,j}^{(0)}\, \mathbf{f}\!\left(\tilde{\mathbf{x}}_j^{(0)}, t_k + c_j^{(0)}\Delta t\right) \Delta t \\
&+ \sum_{j,l} B_{i,j}^{(0)}\, \mathbf{L}_l\!\left(\tilde{\mathbf{x}}_j^{(l)}, t_k + c_j^{(1)}\Delta t\right) \Delta \hat{\beta}_k^{(l)},
\end{aligned}
\tag{8.101}
$$

$$\tilde{\mathbf{x}}_i^{(n)} = \hat{\mathbf{x}}(t_k) + \sum_j A_{i,j}^{(1)} \, \mathbf{f}(\tilde{\mathbf{x}}_j^{(0)}, t_k + c_j^{(0)} \Delta t) \, \Delta t$$

$$+ \sum_{j,l} B_{i,j}^{(1)} \, \mathbf{L}_l(\tilde{\mathbf{x}}_j^{(l)}, t_k + c_j^{(1)} \Delta t) \, \Delta \hat{\chi}_k^{(l,n)}, \qquad (8.102)$$

$$\tilde{\mathbf{x}}_i^{(n)} = \hat{\mathbf{x}}(t_k) + \sum_j A_{i,j}^{(2)} \, \mathbf{f}(\tilde{\mathbf{x}}_j^{(0)}, t_k + c_j^{(0)} \Delta t) \, \Delta t$$

$$+ \sum_{\substack{j,l \\ l \neq n}} B_{i,j}^{(2)} \, \mathbf{L}_l(\tilde{\mathbf{x}}_j^{(l)}, t_k + c_j^{(1)} \Delta t) \, \frac{\Delta \hat{\chi}_k^{(l,n)}}{\sqrt{\Delta t}}, \qquad (8.103)$$

for $i = 1, 2, \ldots, S$ *and* $n = 1, 2, \ldots, S$.

The increments in Algorithm 8.17 are given by the double Itô integrals (exactly as in the case of the strong stochastic Runge–Kutta schemes), but in the weak schemes we can use the following approximations (see, e.g., Kloeden and Platen, 1999; Rößler, 2009):

$$\Delta \hat{\chi}_k^{(i,j)} = \begin{cases} \frac{1}{2}\big(\Delta \hat{\beta}_k^{(i)} \, \Delta \hat{\beta}_k^{(j)} - \sqrt{\Delta t} \, \hat{\zeta}_k^{(i)}\big), & \text{if } i < j, \\[2mm] \frac{1}{2}\big(\Delta \hat{\beta}_k^{(i)} \, \Delta \hat{\beta}_k^{(j)} + \sqrt{\Delta t} \, \hat{\zeta}_k^{(j)}\big), & \text{if } i > j, \\[2mm] \frac{1}{2}\big([\Delta \hat{\beta}_k^{(i)}]^2 - \Delta t\big), & \text{if } i = j, \end{cases} \qquad (8.104)$$

for $i, j = 1, 2, \ldots, S$. Here only $2S - 1$ independent random variables are needed, and we do not anymore run into problems with the cross-term integrals as we did in the strong stochastic Runge–Kutta schemes. For example, we can choose $\Delta \hat{\beta}_k^{(i)}$ such that they are independent three-point distributed random variables

$$\mathrm{P}\big(\Delta \hat{\beta}_k^{(i)} = \pm\sqrt{3 \, \Delta t}\big) = \frac{1}{6} \quad \text{and} \quad \mathrm{P}\big(\Delta \hat{\beta}_k^{(i)} = 0\big) = \frac{2}{3}, \qquad (8.105)$$

and the supporting variables $\hat{\zeta}_k^{(i)}$ such that they are independent two-point distributed random variables

$$\mathrm{P}\big(\hat{\zeta}_k^{(i)} = \pm\sqrt{\Delta t}\big) = \frac{1}{2}. \qquad (8.106)$$

Rößler (2009) proposes, for example, the following multidimensional weak order 2.0 stochastic Runge–Kutta scheme, which only requires two evaluations of $\mathbf{f}(\bullet, \bullet)$ and only five evaluations of each $\mathbf{L}_i(\bullet, \bullet)$.

Algorithm 8.18 (Weak order 2.0 stochastic Runge–Kutta due to Rößler).

Consider a stochastic Runge–Kutta method with the following extended Butcher tableau:

$$
\begin{array}{c|ccc|cc|cc}
0 & & & & & & & \\
1 & 1 & & & 1 & & & \\
0 & 0 & 0 & & 0 & 0 & & \\
\hline
0 & & & & & & & \\
1 & 1 & & & 1 & & & \\
1 & 1 & 0 & & -1 & 0 & & \\
\hline
0 & & & & & & & \\
1 & 1 & & & 1 & & & \\
1 & 1 & 0 & & -1 & 0 & & \\
\hline
 & \frac{1}{2} & \frac{1}{2} & 0 & \frac{1}{2} & \frac{1}{4} & \frac{1}{4} & 0 & \frac{1}{2} & -\frac{1}{2} \\
 & & & & -\frac{1}{2} & \frac{1}{4} & \frac{1}{4} & 0 & \frac{1}{2} & -\frac{1}{2}
\end{array}
\qquad (8.107)
$$

which corresponds to the following iterative scheme:

$$
\hat{\mathbf{x}}(t_{k+1}) = \hat{\mathbf{x}}(t_k) + \frac{\Delta t}{2}\left[\mathbf{f}(\hat{\mathbf{x}}(t_k), t_k) + \mathbf{f}(\tilde{\mathbf{x}}_2^{(0)}, t_k + \Delta t)\right]
$$

$$
+ \sum_n \left[\frac{1}{2}\mathbf{L}_n(\hat{\mathbf{x}}(t_k), t_k) + \frac{1}{4}\mathbf{L}_n(\tilde{\mathbf{x}}_2^{(n)}, t_k + \Delta t) + \frac{1}{4}\mathbf{L}_n(\tilde{\mathbf{x}}_3^{(n)}, t_k + \Delta t)\right]\Delta\hat{\beta}_k^{(n)}
$$

$$
+ \sum_n \left[\frac{1}{2}\mathbf{L}_n(\tilde{\mathbf{x}}_2^{(n)}, t_k + \Delta t) - \frac{1}{2}\mathbf{L}_n(\tilde{\mathbf{x}}_3^{(n)}, t_k + \Delta t)\right]\frac{\Delta\hat{\chi}_k^{(n,n)}}{\sqrt{\Delta t}}
$$

$$
+ \sum_n \left[\frac{1}{4}\mathbf{L}_n(\bar{\mathbf{x}}_2^{(n)}, t_k + \Delta t) + \frac{1}{4}\mathbf{L}_n(\bar{\mathbf{x}}_3^{(n)}, t_k + \Delta t) - \frac{1}{2}\mathbf{L}_n(\hat{\mathbf{x}}(t_k), t_k)\right]\Delta\hat{\beta}_k^{(n)}
$$

$$
+ \sum_n \left[\frac{1}{2}\mathbf{L}_n(\bar{\mathbf{x}}_2^{(n)}, t_k + \Delta t) - \frac{1}{2}\mathbf{L}_n(\bar{\mathbf{x}}_3^{(n)}, t_k + \Delta t)\right]\sqrt{\Delta t} \qquad (8.108)
$$

with supporting values (note that $\tilde{\mathbf{x}}_1^{(0)} = \tilde{\mathbf{x}}_1^{(n)} = \bar{\mathbf{x}}_1^{(n)} = \hat{\mathbf{x}}(t_k)$)

$$
\tilde{\mathbf{x}}_2^{(0)} = \hat{\mathbf{x}}(t_k) + \mathbf{f}(\hat{\mathbf{x}}(t_k), t_k)\,\Delta t + \sum_l \mathbf{L}_l(\hat{\mathbf{x}}(t_k), t_k)\,\Delta\hat{\beta}_k^{(l)}, \qquad (8.109)
$$

$$
\tilde{\mathbf{x}}_2^{(n)} = \hat{\mathbf{x}}(t_k) + \mathbf{f}(\hat{\mathbf{x}}(t_k), t_k)\,\Delta t + \sum_l \mathbf{L}_l(\hat{\mathbf{x}}(t_k), t_k)\,\Delta\hat{\chi}_k^{(l,n)}, \qquad (8.110)
$$

$$
\tilde{\mathbf{x}}_3^{(n)} = \hat{\mathbf{x}}(t_k) + \mathbf{f}(\hat{\mathbf{x}}(t_k), t_k)\,\Delta t - \sum_l \mathbf{L}_l(\hat{\mathbf{x}}(t_k), t_k)\,\Delta\hat{\chi}_k^{(l,n)}, \qquad (8.111)
$$

$$
\bar{\mathbf{x}}_2^{(n)} = \hat{\mathbf{x}}(t_k) + \mathbf{f}(\hat{\mathbf{x}}(t_k), t_k)\,\Delta t + \sum_{\substack{l \\ l \neq n}} \mathbf{L}_l(\hat{\mathbf{x}}(t_k), t_k)\,\frac{\Delta\hat{\chi}_k^{(l,n)}}{\sqrt{\Delta t}}, \qquad (8.112)
$$

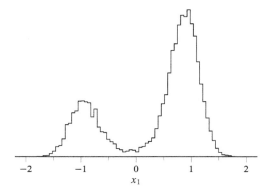

Figure 8.5 A histogram of the state of the Duffing van der Pol oscillator values $x_1(t)$ at $t = 20$ with 10,000 samples simulated by the weak order 2.0 stochastic Runge–Kutta scheme (see Example 8.19).

$$\bar{\mathbf{x}}_3^{(n)} = \hat{\mathbf{x}}(t_k) + \mathbf{f}(\hat{\mathbf{x}}(t_k), t_k)\,\Delta t - \sum_{\substack{l \\ l \neq n}} \mathbf{L}_l(\hat{\mathbf{x}}(t_k), t_k)\,\frac{\Delta\hat{\chi}_k^{(l,n)}}{\sqrt{\Delta t}}. \tag{8.113}$$

Example 8.19 (Weak approximation of the Duffing van der Pol problem). *In Example 8.15, we considered a van der Pol oscillator with two steady states for the zero-diffusion model. Now we are interested in characterizing the solution at $t = 20$ for the initial condition of $\mathbf{x}(0) = (-3, 0)$. We use the stochastic Runge–Kutta method in Algorithm 8.18 that is of weak order 2.0. We consider a time span $[0, 20]$ and a discretization interval $\Delta t = 2^{-4}$. With a Δt this large, the Euler–Maruyama method does not provide plausible results. Figure 8.5 shows the histogram of the values $x_1(20)$.*

8.7 Stochastic Verlet Algorithm

The stochastic Verlet algorithm (see Burrage et al., 2007) is a scheme for simulating numerical solutions to second-order stochastic differential equations. It is best known for its use in integrating Newton's equations of motion, primarily in molecular dynamics, computer graphics, and other engineering applications.

Consider a second-order SDE of the following form:

$$\ddot{x} = f(x) - \eta\,b^2(x)\,\dot{x} + b(x)\,w(t), \tag{8.114}$$

where η is a damping parameter, $w(t)$ a white noise process with spectral density q, and $b(x)$ is a given function. If $b(x)$ is not constant, the model has multiplicative noise. We can rewrite this equation as a pair of first-order equations:

$$\begin{aligned}
\mathrm{d}x(t) &= \mathrm{d}v(t)\,\mathrm{d}t, \\
\mathrm{d}v(t) &= -\eta\,b^2(x(t))\,v(t)\,\mathrm{d}t + f(x(t))\,\mathrm{d}t + b(x(t))\,\mathrm{d}\beta(t).
\end{aligned} \tag{8.115}$$

The Verlet methods are based on the idea of considering the central difference between discretized time steps t_{k-1} and t_{k+1} obtained by the summation of the corresponding two Itô–Taylor expansions. We consider a variant of the Verlet algorithm family known as the *leapfrog* method, which includes a supporting half-step. Under leapfrog methods, the velocity and position are updated successively rather than together. In its simplest form, the leapfrog algorithm is the following.

Algorithm 8.20 (Leapfrog Verlet). *Start from $\hat{x}(t_0)$ and $\hat{v}(t_0)$ drawn from the initial distribution, and divide the time interval $[t_0, t]$ into M steps of length Δt. At each step k, do the following:*

1. *Draw random variable $\Delta\beta_k$ from the distribution*

$$\Delta\beta_k \sim \mathrm{N}(0, q\,\Delta t). \tag{8.116}$$

2. *Compute the half-step supporting value:*

$$\tilde{x} = \hat{x}(t_k) + \frac{1}{2}v(t_k)\,\Delta t. \tag{8.117}$$

3. *Update the velocity and position*

$$\hat{v}(t_{k+1}) = \hat{v}(t_k) - \eta\,b^2(\tilde{x})\,\hat{v}(t_k)\,\Delta t + f(\tilde{x})\,\Delta t + b(\tilde{x})\,\Delta\beta_k, \tag{8.118}$$

$$\hat{x}(t_{k+1}) = \tilde{x} + \frac{1}{2}\,\hat{v}(t_{k+1})\,\Delta t. \tag{8.119}$$

The method becomes the canonical ordinary leapfrog scheme (see, e.g., Allen and Tildesley, 1991) for models with $b(x) = 0$.

The strength of the leapfrog Verlet algorithm is that it maintains the independence of position and velocity. Furthermore, it is time-reversible meaning that its behavior remains the same both in integrating forward and backward. The algorithm does not, however, require any more function evaluations than the Euler–Maruyama method, which makes it appealing for this type of second-order SDEs.

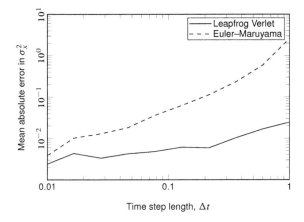

Time step length, Δt

Figure 8.6 Comparison of the mean absolute error given by the leapfrog Verlet and Euler–Maruyama methods in estimating $P_x(t)$ for the spring model (see Example 8.21).

Example 8.21 (Leapfrog solution to the spring model). *Recall the stochastic spring model from Example 3.10. The model is an LTI SDE and we can solve it in closed form as was shown earlier. The model can be rewritten in terms of the second-order formulation in Equation (8.114) by choosing $b(x) = 1$, $\eta = \gamma$, and $f(x) = -v^2\,x$.*

Figure 8.6 shows the mean absolute error of the estimated variance of $P_x(t)$ as a function the time step length Δt. All other parameters were fixed to values of $\gamma = 1$, $v = 1$, and $q = 1$. The leapfrog Verlet algorithm is compared against the Euler–Maruyama method, which is of similar computational complexity, but performs worse for this type of second-order SDEs.

8.8 Exact Algorithm

The exact algorithm (Beskos and Roberts, 2005; Beskos et al., 2006a,b, 2008; Iacus, 2008) is an algorithm that can be used for simulating exact samples (without a discretization error) from SDEs of the form

$$d\mathbf{x} = \mathbf{f}(\mathbf{x})\,dt + d\boldsymbol{\beta}, \tag{8.120}$$

where $\mathbf{f}$ can be expressed as a gradient of a scalar function such that $\mathbf{f}(\mathbf{x}) = \nabla\psi(\mathbf{x})$ and $\boldsymbol{\beta}$ is a vector of standard Brownian motions. By using the Lamperti transform from Section 7.1 (while recalling its limitations) the method can be generalized to other types of SDEs.

First recall that rejection sampling is a method that can be used for sampling from some probability distribution q_1 by using samples from another probability distribution q_2. The method requires that $q_1(\mathbf{x}) \leq \varepsilon q_2(\mathbf{x})$ for some $\varepsilon > 0$. Rejection sampling works as follows:

1. Sample $\mathbf{x} \sim q_2(\mathbf{x})$.

2. Sample uniform random variable $u \sim U(0, 1)$.

3. If $u < \varepsilon \frac{q_1(\mathbf{x})}{q_2(\mathbf{x})}$, accept $\mathbf{x}$, otherwise go back to step 1.

However, it turns out that we can also construct a rejection sampling method alternatively as follows, which is more useful for the exact algorithm:

1. Sample $(\mathbf{x}, I)$ such that $\mathbf{x} \sim q_2(\mathbf{x})$ and $p(I = 1 \mid \mathbf{x}) = \varepsilon \frac{q_1(\mathbf{x})}{q_2(\mathbf{x})}$.

2. If $I = 1$, accept $\mathbf{x}$, otherwise go back to step 1.

In the case of SDEs, we need to replace the ratio of likelihoods $q_1(\mathbf{x})/q_2(\mathbf{x})$ with the Radon–Nikodym derivative. Let us now consider sampling from an SDE by using a Brownian motion as the sampled distribution. Let $P_{\mathbf{x}}$ and $P_{\boldsymbol{\beta}}$ be the measures of the process $\mathbf{x}$ and of the Brownian motion. Then according to the Girsanov theorem, we have the likelihood ratio

$$\frac{dP_{\mathbf{x}}}{dP_{\boldsymbol{\beta}}} = \exp\left\{ \int_0^T \mathbf{f}^{\mathsf{T}}(\boldsymbol{\beta}) \, d\boldsymbol{\beta} - \frac{1}{2} \int_0^T \|\mathbf{f}(\boldsymbol{\beta})\|^2 \, dt \right\}. \tag{8.121}$$

Assume that there exists a scalar function ψ such that

$$\mathbf{f}(\mathbf{x}) = \nabla \psi(\mathbf{x}). \tag{8.122}$$

Applying the Itô formula to $\psi(\boldsymbol{\beta})$ gives

$$d\psi = \sum_i \frac{\partial \psi(\boldsymbol{\beta})}{\partial \beta_i} \, d\beta_i + \frac{1}{2} \sum_{i,j} \frac{\partial^2 \psi(\boldsymbol{\beta})}{\partial \beta_i \, \partial \beta_j} \delta_{i,j} \, dt$$

$$= \mathbf{f}^{\mathsf{T}}(\boldsymbol{\beta}) \, d\boldsymbol{\beta} + \frac{1}{2} (\nabla \cdot \mathbf{f}(\boldsymbol{\beta})) \, dt, \tag{8.123}$$

that is,

$$\int_0^T \mathbf{f}^\top(\boldsymbol{\beta}) \, d\boldsymbol{\beta} - \frac{1}{2} \int_0^T \|\mathbf{f}(\boldsymbol{\beta})\|^2 \, dt$$

$$= \int_0^T [\mathbf{f}^\top(\boldsymbol{\beta}) \, d\boldsymbol{\beta} + \frac{1}{2} \nabla \cdot \mathbf{f}(\boldsymbol{\beta}) \, dt] - \frac{1}{2} \int_0^T [\|\mathbf{f}(\boldsymbol{\beta})\|^2 + \nabla \cdot \mathbf{f}(\boldsymbol{\beta})] \, dt$$

$$= \int_0^T d\psi(t) - \frac{1}{2} \int_0^T [\|\mathbf{f}(\boldsymbol{\beta})\|^2 + \nabla \cdot \mathbf{f}(\boldsymbol{\beta})] \, dt$$

$$= \psi(\boldsymbol{\beta}(T)) - \psi(\boldsymbol{\beta}(0)) - \frac{1}{2} \int_0^T [\|\mathbf{f}(\boldsymbol{\beta})\|^2 + \nabla \cdot \mathbf{f}(\boldsymbol{\beta})] \, dt. \tag{8.124}$$

Thus

$$\frac{dP_\mathbf{x}}{dP_{\boldsymbol{\beta}}} = \exp\left(\psi(\boldsymbol{\beta}(T)) - \psi(\boldsymbol{\beta}(0)) - \frac{1}{2} \int_0^T [\|\mathbf{f}(\boldsymbol{\beta})\|^2 + (\nabla \cdot \mathbf{f}(\boldsymbol{\beta}))] \, dt\right).$$
$$\tag{8.125}$$

Let us now define a biased Brownian motion $\tilde{\boldsymbol{\beta}}$ such the final point has the distribution with a density

$$h(\tilde{\boldsymbol{\beta}}(T)) \propto \exp\left(\psi(\tilde{\boldsymbol{\beta}}(T)) - \frac{1}{2T} \|\tilde{\boldsymbol{\beta}}(T)\|^2\right). \tag{8.126}$$

It then turns out that

$$\frac{dP_\mathbf{x}}{dP_{\tilde{\boldsymbol{\beta}}}} \propto \exp\left(-\frac{1}{2} \int_0^T [\|\mathbf{f}(\boldsymbol{\beta})\|^2 + (\nabla \cdot \mathbf{f}(\boldsymbol{\beta}))] \, dt\right). \tag{8.127}$$

Provided that $\|\mathbf{f}(\boldsymbol{\beta})\|^2 + \nabla \cdot \mathbf{f}(\boldsymbol{\beta})$ is bounded from below, we can now find a nonnegative function ϕ such that

$$\frac{dP_\mathbf{x}}{dP_{\tilde{\boldsymbol{\beta}}}} \propto \exp\left(-\int_0^T \phi(\boldsymbol{\beta}) \, dt\right) \le 1. \tag{8.128}$$

That function has the form

$$\phi(\boldsymbol{\beta}) = \frac{1}{2}[\|\mathbf{f}(\boldsymbol{\beta})\|^2 + (\nabla \cdot \mathbf{f}(\boldsymbol{\beta}))] - c, \tag{8.129}$$

where c is a suitable constant. Let us also assume that $\phi(\boldsymbol{\beta}) \le M$. Then we get the following theorem.

Theorem 8.22. *Let Φ be a homogeneous Poisson process with unit intensity on $[0, T] \times [0, M]$ and let N be the number of points of the process* *found below the graph $(t, \phi(t))$. Then we have*

$$p(N = 0) = \exp\left(-\int_0^T \phi(t)\,dt\right). \tag{8.130}$$

We can now take the event $N = 0$ to be the event $I = 1$ in our rejection sampling. The key observation is now that we do not need to sample the whole path of $\boldsymbol{\beta}$ to be able to determine if $N = 0$ – we only need to know the values of $\phi(\boldsymbol{\beta}(t))$ at the time points of the sampled Poisson process, and for that, being able sample those points (i.e., a skeleton of the process) is enough.

The basic algorithm now takes the following form (called EA1 in Beskos et al., 2006b).

Algorithm 8.23 (Exact algorithm). *The exact algorithm (EA) consists of the following steps:*

1. *Simulate τ from the Poisson distribution with intensity $\lambda = T\,M$.*
2. *Simulate τ samples (t_i, x_i) uniformly on $[0, T] \times [0, M]$.*
3. *Simulate the skeleton of biased Brownian motion $\tilde{\boldsymbol{\beta}}$ at the time points $\{t_i\}$, such that $\tilde{\boldsymbol{\beta}}(T) \sim h$ and $\tilde{\boldsymbol{\beta}}(0) = 0$.*
4. *Evaluate N, by checking how many times $x_i < \phi(\tilde{\boldsymbol{\beta}}(t_i))$.*
5. *If $N = 0$, go to step 6 and otherwise go to step 1.*
6. *Simulate Brownian bridges between all the pairs of points (t_{i+1}, t_i).*

Various extensions to unbounded ϕ are presented in articles of Beskos and Roberts (2005) and Beskos et al. (2006a,b, 2008).

Example 8.24 (Exact simulation of sine diffusion). *Consider the sine diffusion model*

$$dx = \sin(x)\,dt + d\beta, \quad x(0) = 0, \tag{8.131}$$

where $\beta(t)$ is a standard Brownian motion. We can choose $\psi(x) = 1 - \cos(x)$, which gives

$$\phi(\beta) = \frac{1}{2}\left[\sin^2(\beta) + \cos(\beta) + 1\right]. \tag{8.132}$$

We simulate from this SDE using the Exact algorithm by choosing $T = 1.5$, and plot the histogram of the solutions at time T. The histogram of the sample distribution is shown in Figure 8.7.

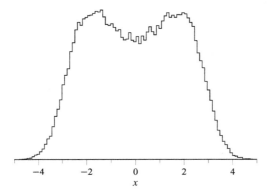

Figure 8.7 Histogram of $N = 100,000$ sample paths of the sine diffusion model at $T = 1.5$ simulated by the Exact algorithm (see Example 8.24).

8.9 Exercises

8.1   Consider the following scalar SDE:

$$dx = -c\, x\, dt + g\, x\, d\beta, \qquad x(0) = x_0, \qquad (8.133)$$

where c, g, and x_0 are positive constants and $\beta(t)$ is a standard Brownian motion.

(a) Check using the Itô formula that the solution to this equation is

$$x(t) = x_0 \exp\left[(-c - g^2/2)\, t + g\, \beta(t)\right].$$

(b) Simulate trajectories from the equation using the Milstein method with parameters $x_0 = 1, c = 1/10, g = 1/10$, and check that the histogram at $t = 1$ looks the same as obtained by sampling from the preceding exact solution.

8.2   Consider the following Beneš SDE:

$$dx = \tanh(x)\, dt + d\beta, \quad x(0) = 0, \qquad (8.134)$$

where $\beta(t)$ is a standard Brownian motion. Recall that it has the probability density

$$p(x, t) = \frac{1}{\sqrt{2\pi t}}\, \cosh(x)\, \exp\left(-\frac{1}{2}t\right) \exp\left(-\frac{1}{2t}x^2\right).$$

Use the following numerical methods for simulating from the model:

(a) Simulate 1,000 trajectories from the SDE with the strong order 1.5 Itô–
Taylor series–based method presented earlier. Compare the histogram to
the exact solution at $t = 5$.

(b) Simulate 1,000 trajectories from the SDE with the weak order 2.0 Itô–
Taylor series–based method presented earlier using both (i) Gaussian
increments and (ii) the three-point distributed random variables. Compare the histograms to the exact distribution at $t = 5$.

(c) What can you say about the simulated trajectories when $t \in [0, 5]$ for
the different methods?

8.3   Consider a simple strong order 1.0 stochastic Runge–Kutta method with the
following extended Butcher tableau:

$$
\begin{array}{c|ccc|ccc|ccc}
0 \\
0 & 0 & & & 0 \\
0 & 0 & 0 & & 0 & 0 \\
\hline
0 \\
0 & 0 & & & 1 \\
0 & 0 & 0 & & -1 & 0 \\
\hline
  & 1 & 0 & 0 & 1 & 0 & 0 & 0 & \frac{1}{2} & -\frac{1}{2}
\end{array}
\tag{8.135}
$$

(a) Write down the iteration equations required for evaluating the method
corresponding to the table in Equation (8.135).

(b) Consider the Duffing van der Pol oscillator model:

$$
\begin{pmatrix} dx_1 \\ dx_2 \end{pmatrix} = \begin{pmatrix} x_2 \\ x_1\,(\alpha - x_1^2) - x_2 \end{pmatrix} dt + \begin{pmatrix} 0 \\ x_1 \end{pmatrix} d\beta,
$$

where $\beta(t)$ is a one-dimensional Brownian motion ($q = 0.5^2$) and $\alpha =
1$. Use the method you just constructed for drawing trajectories starting
from $x_2(0) = 0$ and $x_1(0) = -4, -3.9, \ldots, -2$. Use a time span $[0, 10]$.
Plot the results in the (x_1, x_2) plane.

(c) Experiment with different step sizes $\Delta t = 2^{-k}, k = 0, 2, 4, 6$ and visually compare the trajectories produced by the method implemented in
(b) to the Euler–Maruyama scheme.

8.4   Consider the following two-dimensional SDE:

$$
\begin{pmatrix} dx_1 \\ dx_2 \end{pmatrix} = \begin{pmatrix} \frac{3}{2}x_1 \\ \frac{3}{2}x_2 \end{pmatrix} dt + \begin{pmatrix} \frac{1}{10}x_1 & 0 \\ 0 & \frac{1}{10}x_2 \end{pmatrix} d\beta,
$$

where $\beta(t) = (\beta_1(t), \beta_2(t))$ such that each $\beta_i(t)$ is a standard Brownian
motion. The initial value is $\mathbf{x}(0) = (1/10\,,\,1/10)$.

(a) Implement the Euler–Maruyama scheme for this problem.

(b) Implement the following weak order 2.0 Runge–Kutta method for this problem (following Algorithm 8.17):

$$
\begin{array}{cccc|ccc|ccc}
0 & & & & 1 & & & & & \\
\frac{2}{3} & \frac{2}{3} & & & 0 & 0 & & & & \\
\frac{2}{3} & -\frac{1}{3} & 1 & & 0 & 0 & & & & \\
\hline
0 & & & & 1 & & & & & \\
1 & 1 & & & -1 & 0 & & & & \\
1 & 1 & 0 & & -1 & 0 & & & & \\
\hline
0 & & & & 1 & & & & & \\
0 & 0 & & & 1 & & & & & \\
0 & 0 & 0 & & -1 & 0 & & & & \\
\hline
 & \frac{1}{4} & \frac{1}{2} & \frac{1}{4} & \frac{1}{2} & \frac{1}{4} & \frac{1}{4} & 0 & \frac{1}{2} & -\frac{1}{2} \\
 & & & & -\frac{1}{2} & \frac{1}{4} & \frac{1}{4} & 0 & \frac{1}{2} & -\frac{1}{2}
\end{array}
$$

(c) Simulate 1,000 trajectories from the SDE with Euler–Maruyama and the weak order 2.0 Runge–Kutta method. Use step sizes $\Delta t = 2^{-k}, k = 0, 1, \ldots, 6$. Compare your results to the expected value given by

$$
E[x_i(t)] = \frac{1}{10} \exp\left(\frac{3}{2}t\right)
$$

for $i = 1, 2$, and plot the absolute errors as a function of step size.

8.5 Consider the following SDE ($D = 2, S = 4$) describing stochastic flow on a torus:

$$
dx = L(x)\,d\beta,
$$

where $\beta(t) = (\beta_1(t), \beta_2(t), \beta_3(t), \beta_4(t))$ such that each $\beta_i(t)$ is a standard Brownian motion. The diffusion is given such that the columns in $L(x)$ are (use $\alpha = 1$):

$$
L^1(x) = \begin{pmatrix} \cos\alpha \\ \sin\alpha \end{pmatrix} \sin(x_1), \qquad
L^2(x) = \begin{pmatrix} \cos\alpha \\ \sin\alpha \end{pmatrix} \cos(x_1),
$$

$$
L^3(x) = \begin{pmatrix} -\sin\alpha \\ \cos\alpha \end{pmatrix} \sin(x_2), \qquad
L^4(x) = \begin{pmatrix} -\sin\alpha \\ \cos\alpha \end{pmatrix} \cos(x_2).
$$

(a) Consider a set of initial points $x(0)$ on a uniform 15×15 grid on $[0, 2\pi] \times [0, 2\pi]$. Use the Euler–Maruyama method with the same realization of Brownian motion (reset the random seed) for each trajectory and a step size of $\Delta t = 2^{-4}$. Plot what the distribution looks like at $t = 0.5, 1.0, 2.0, 4.0$ (consider x_i modulo 2π for staying on the torus).

(b) Implement the weak order 2.0 Runge–Kutta scheme presented in Algorithm 8.18, and repeat the preceding experiment.

8.6   Consider the double-well potential model

$$dx = (x - x^3)\, dt + \frac{1}{2}\sqrt{1 + x^2}\, d\beta,$$

where $\beta(t)$ is a standard Brownian motion. Simulate trajectories from the stochatic differential equation to characterize the solution at $x(10)$ starting from $x(0) = 1$:

(a) Use the Euler–Maruyama scheme.
(b) Use the weak order 2.0 stochastic Runge–Kutta scheme in Algorithm 8.18.
(c) Study the effect of step size Δt.

8.7   Consider the second-order SDE model that fits under the family of models suitable for the Verlet integration scheme with $f(x) = x - x^3$, $b(x) = x$, $\eta = 1$, and $q = 1$.

(a) Write the Itô SDE corresponding to the given model.
(b) Simulate 10,000 trajectories on the interval $[0, 10]$ from the model using (i) the Leapfrog Verlet algorithm and (ii) the Euler–Maruyama method. Use a step size of $\Delta t = 1/10$ and the initial value $\mathbf{x} = (1, 0)$. Compare the results.
(c) Study the behavior of the two methods as Δt increases.

8.8   Consider the use of the Exact algorithm for sampling from the Beneš SDE $dx = \tanh(x)\, dt + d\beta$. What happens to the likelihood ratio in this case? Implement the resulting algorithm and compare the result to the exact density.

9

Approximation of Nonlinear SDEs

This chapter is concerned with approximations of nonlinear SDEs aiming for computing statistics of SDEs and for forming approximations to the transition densities. However, many of them can also be used for numerical simulation of SDEs. These approximations will also turn out to be useful in filtering and smoothing theory as well as in SDE parameter estimation, which we consider in the subsequent chapters. In this chapter, we start with Gaussian approximations, which are especially common in filtering and smoothing applications (such as target tracking). These approximations include various linearization, Taylor series, and sigma-point approximations. We then proceed to temporal Taylor series approximations of the moments, Hermite expansions of the transition densities, discretization approximations to the Fokker–Planck–Kolmogorov equation, and finally pathwise series expansions of Brownian motions, which are related to the Wong–Zakai theorem and Wiener chaos expansions.

9.1 Gaussian Assumed Density Approximations

In Section 5.5, we saw that the differential equations for the mean and covariance of the solution to the SDE

$$dx = f(x, t) dt + L(x, t) d\beta \tag{9.1}$$

are

$$\frac{dm}{dt} = E[f(x, t)], \tag{9.2}$$

$$\frac{dP}{dt} = E[f(x, t)(x - m)^T] + E[(x - m) f^T(x, t)]$$
$$+ E[L(x, t) Q L^T(x, t)]. \tag{9.3}$$

165

If we write down the expectation integrals explicitly, these equations can be seen to have the form

$$\frac{d\mathbf{m}}{dt} = \int \mathbf{f}(\mathbf{x}, t)\, p(\mathbf{x}, t)\, d\mathbf{x}, \tag{9.4}$$

$$\frac{d\mathbf{P}}{dt} = \int \mathbf{f}(\mathbf{x}, t)\, (\mathbf{x} - \mathbf{m})^{\mathsf{T}}\, p(\mathbf{x}, t)\, d\mathbf{x}$$

$$+ \int (\mathbf{x} - \mathbf{m})\, \mathbf{f}^{\mathsf{T}}(\mathbf{x}, t)\, p(\mathbf{x}, t)\, d\mathbf{x}$$

$$+ \int \mathbf{L}(\mathbf{x}, t)\, \mathbf{Q}\, \mathbf{L}^{\mathsf{T}}(\mathbf{x}, t)\, p(\mathbf{x}, t)\, d\mathbf{x}. \tag{9.5}$$

Because $p(\mathbf{x}, t)$ is the solution of the Fokker–Planck–Kolmogorov equation (5.11), these equations cannot usually be solved in practice. However, one very useful class of approximations can be obtained by replacing the FPK solution with a Gaussian approximation as follows:

$$p(\mathbf{x}, t) \approx \mathrm{N}(\mathbf{x} \mid \mathbf{m}(t), \mathbf{P}(t)), \tag{9.6}$$

where $\mathbf{m}(t)$ and $\mathbf{P}(t)$ are the mean and covariance of the state, respectively. This approximation is referred to as the *Gaussian assumed density approximation* (Kushner, 1967; Särkkä and Sarmavuori, 2013), because we do the computations under the assumption that the state distribution is indeed Gaussian. It is also related to a *Gaussian process approximation* (Archambeau and Opper, 2011; Ala-Luhtala et al., 2015) of the SDE. The approximation method (as it is presented in Särkkä and Sarmavuori, 2013) can be written as the following algorithm.

Algorithm 9.1 (Gaussian assumed density approximation I). *A Gaussian process approximation to the SDE* (9.1) *can be obtained by integrating the following differential equations from the initial conditions* $\mathbf{m}(t_0) = \mathrm{E}[\mathbf{x}(t_0)]$ *and* $\mathbf{P}(t_0) = \mathrm{Cov}[\mathbf{x}(t_0)]$ *to the target time* t:

$$\frac{d\mathbf{m}}{dt} = \int \mathbf{f}(\mathbf{x}, t)\, \mathrm{N}(\mathbf{x} \mid \mathbf{m}, \mathbf{P})\, d\mathbf{x},$$

$$\frac{d\mathbf{P}}{dt} = \int \mathbf{f}(\mathbf{x}, t)\, (\mathbf{x} - \mathbf{m})^{\mathsf{T}}\, \mathrm{N}(\mathbf{x} \mid \mathbf{m}, \mathbf{P})\, d\mathbf{x}$$

$$+ \int (\mathbf{x} - \mathbf{m})\, \mathbf{f}^{\mathsf{T}}(\mathbf{x}, t)\, \mathrm{N}(\mathbf{x} \mid \mathbf{m}, \mathbf{P})\, d\mathbf{x} \tag{9.7}$$

$$+ \int \mathbf{L}(\mathbf{x}, t)\, \mathbf{Q}\, \mathbf{L}^{\mathsf{T}}(\mathbf{x}, t)\, \mathrm{N}(\mathbf{x} \mid \mathbf{m}, \mathbf{P})\, d\mathbf{x}.$$

If we denote the Gaussian expectation as

$$E_N[\mathbf{g}(\mathbf{x})] = \int \mathbf{g}(\mathbf{x}) \, N(\mathbf{x} \mid \mathbf{m}, \mathbf{P}) \, d\mathbf{x}, \qquad (9.8)$$

then the equations can also be written as

$$
\begin{aligned}
\frac{d\mathbf{m}}{dt} &= E_N[\mathbf{f}(\mathbf{x}, t)], \\
\frac{d\mathbf{P}}{dt} &= E_N[(\mathbf{x} - \mathbf{m}) \, \mathbf{f}^\mathsf{T}(\mathbf{x}, t)] + E_N[\mathbf{f}(\mathbf{x}, t) \, (\mathbf{x} - \mathbf{m})^\mathsf{T}] \\
&\quad + E_N[\mathbf{L}(\mathbf{x}, t) \, \mathbf{Q} \, \mathbf{L}^\mathsf{T}(\mathbf{x}, t)].
\end{aligned}
\qquad (9.9)
$$

If the function $\mathbf{x} \mapsto \mathbf{f}(\mathbf{x}, t)$ is differentiable, the covariance differential equation can be simplified by using the following well-known property of Gaussian random variables (also known as Stein's lemma).

Theorem 9.2. *Let $\mathbf{f}(\mathbf{x}, t)$ be differentiable with respect to $\mathbf{x}$ and let $\mathbf{x} \sim N(\mathbf{m}, \mathbf{P})$. Then the following identity holds (see, e.g., Papoulis, 1984; Särkkä and Sarmavuori, 2013):*

$$
\int \mathbf{f}(\mathbf{x}, t) \, (\mathbf{x} - \mathbf{m})^\mathsf{T} \, N(\mathbf{x} \mid \mathbf{m}, \mathbf{P}) \, d\mathbf{x}
$$

$$
= \left[\int \mathbf{F}_{\mathbf{x}}(\mathbf{x}, t) \, N(\mathbf{x} \mid \mathbf{m}, \mathbf{P}) \, d\mathbf{x} \right] \mathbf{P}, \quad (9.10)
$$

where $\mathbf{F}_{\mathbf{x}}(\mathbf{x}, t)$ is the Jacobian matrix of $\mathbf{f}(\mathbf{x}, t)$ with respect to $\mathbf{x}$.

Using the theorem, the mean and covariance Equations (9.9) can be equivalently written as follows.

Algorithm 9.3 (Gaussian assumed density approximation II). *A Gaussian process approximation to the SDE (9.1) can be obtained by integrating the following differential equations from the initial conditions $\mathbf{m}(t_0) = E[\mathbf{x}(t_0)]$ and $\mathbf{P}(t_0) = \mathrm{Cov}[\mathbf{x}(t_0)]$ to the target time t:*

$$
\begin{aligned}
\frac{d\mathbf{m}}{dt} &= E_N[\mathbf{f}(\mathbf{x}, t)], \\
\frac{d\mathbf{P}}{dt} &= \mathbf{P} \, E_N[\mathbf{F}_{\mathbf{x}}(\mathbf{x}, t)]^\mathsf{T} + E_N[\mathbf{F}_{\mathbf{x}}(\mathbf{x}, t)] \, \mathbf{P} + E_N[\mathbf{L}(\mathbf{x}, t) \, \mathbf{Q} \, \mathbf{L}^\mathsf{T}(\mathbf{x}, t)],
\end{aligned}
$$
$$(9.11)$$

where $E_N[\bullet]$ denotes the expectation with respect to $\mathbf{x} \sim N(\mathbf{m}, \mathbf{P})$.

The approximations presented in this section are formally equivalent to so-called statistical linearization approximations (Gelb, 1974; Socha,

2008), and they are also closely related to the variational approximations of Archambeau and Opper (2011).

Although the preceding algorithms provide a generic Gaussian assumed density approximation framework for SDEs, to implement the methods, one is required to compute the following kind of n-dimensional Gaussian integrals:

$$E_N[\mathbf{g}(\mathbf{x},t)] = \int \mathbf{g}(\mathbf{x},t)\, N(\mathbf{x} \mid \mathbf{m}, \mathbf{P})\, d\mathbf{x}. \tag{9.12}$$

A classical approach that is very common in filtering theory (Jazwinski, 1970; Maybeck, 1982a) is to linearize (via Taylor series) the drift $\mathbf{f}(\mathbf{x},t)$ around the mean $\mathbf{m}$ as

$$\mathbf{f}(\mathbf{x},t) \approx \mathbf{f}(\mathbf{m},t) + \mathbf{F_x}(\mathbf{m},t)\,(\mathbf{x} - \mathbf{m}) \tag{9.13}$$

and to approximate the expectation of the diffusion part as

$$\mathbf{L}(\mathbf{x},t) \approx \mathbf{L}(\mathbf{m},t). \tag{9.14}$$

This leads to the following approximation, which is commonly used in extended Kalman filters (EKF).

Algorithm 9.4 (Linearization approximation of SDE). *A linearization-based approximation to the SDE* (9.1) *can be obtained by integrating the following differential equations from the initial conditions* $\mathbf{m}(t_0) = E[\mathbf{x}(t_0)]$ *and* $\mathbf{P}(t_0) = \mathrm{Cov}[\mathbf{x}(t_0)]$ *to the target time* t:

$$\begin{aligned} \frac{d\mathbf{m}}{dt} &= \mathbf{f}(\mathbf{m},t), \\ \frac{d\mathbf{P}}{dt} &= \mathbf{P}\,\mathbf{F_x^T}(\mathbf{m},t) + \mathbf{F_x}(\mathbf{m},t)\,\mathbf{P} + \mathbf{L}(\mathbf{m},t)\,\mathbf{Q}\,\mathbf{L^T}(\mathbf{m},t). \end{aligned} \tag{9.15}$$

Another general class of approximations is the Gauss–Hermite cubature type of approximations, where the integrals are approximated as weighted sums

$$\int \mathbf{f}(\mathbf{x},t)\, N(\mathbf{x} \mid \mathbf{m}, \mathbf{P})\, d\mathbf{x} \approx \sum_i W^{(i)}\,\mathbf{f}(\mathbf{x}^{(i)},t), \tag{9.16}$$

where $\mathbf{x}^{(i)}$ and $W^{(i)}$ are the sigma points (abscissas) and their accompanying weights, which have been selected using a method-specific deterministic rule. These kinds of rules are nowadays commonly used in the context of filtering theory (cf. Särkkä and Sarmavuori, 2013; Särkkä, 2013). In n-dimensional Gauss–Hermite integration, the unscented transform, and cubature integration, the sigma points are selected as follows:

$$\mathbf{x}^{(i)} = \mathbf{m} + \sqrt{\mathbf{P}}\,\boldsymbol{\xi}_i, \tag{9.17}$$

where the matrix square root is defined by $\mathbf{P} = \sqrt{\mathbf{P}}\sqrt{\mathbf{P}}^{\mathsf{T}}$ (typically Cholesky factorization), and the points $\boldsymbol{\xi}_i$ and the weights $W^{(i)}$ are selected as follows:

Gauss–Hermite integration: This product rule–based method uses a set of m^n vectors $\boldsymbol{\xi}_i$, which have been formed as a Cartesian product of zeros of the Hermite polynomials of order m. The weights $W^{(i)}$ are formed as products of the corresponding one-dimensional Gauss–Hermite integration weights (for details, see Ito and Xiong, 2000; Wu et al., 2006).

Unscented transform: This method uses a zero vector (origin) and $2n$ unit coordinate vectors $\mathbf{e}_i$ as follows (the method can also be generalized a bit):

$$\boldsymbol{\xi}_0 = \mathbf{0},$$

$$\boldsymbol{\xi}_i = \begin{cases} \sqrt{\lambda + n}\,\mathbf{e}_i, & i = 1, 2, \ldots, n, \\ -\sqrt{\lambda + n}\,\mathbf{e}_{i-n}, & i = n+1, n+2, \ldots, 2n, \end{cases} \tag{9.18}$$

and the weights are defined as follows:

$$W^{(0)} = \frac{\lambda}{n + \kappa},$$
$$W^{(i)} = \frac{1}{2(n + \kappa)}, \qquad i = 1, 2, \ldots, 2n, \tag{9.19}$$

where κ and λ are parameters of the method and n is the dimensionality of $\mathbf{x}$ (see, e.g., Julier et al., 1995; Wan and van der Merwe, 2001; Julier and Uhlmann, 2004; Särkkä, 2013). This rule is a special case of more general symmetric integrations rules that can be constructed up to an arbitrary order (McNamee and Stenger, 1967).

Cubature method: This spherical third-degree method uses only $2n$ vectors as follows:

$$\boldsymbol{\xi}_i = \begin{cases} \sqrt{n}\,\mathbf{e}_i, & i = 1, 2, \ldots, n, \\ -\sqrt{n}\,\mathbf{e}_{i-n}, & i = n+1, n+2, \ldots, 2n, \end{cases} \tag{9.20}$$

and the weights are defined as $W^{(i)} = 1/(2n)$, for $i = 1, 2, \ldots, 2n$. This rule is in fact a special case of the preceding rule, but has turned out to be useful in filtering theory (Arasaratnam and Haykin, 2009; Arasaratnam et al., 2010; Särkkä and Solin, 2012; Särkkä, 2013).

The sigma-point methods lead to the following approximations to the mean and covariance differential equations.

Algorithm 9.5 (Sigma-point approximation of SDEs). *A sigma point–based approximation to the SDE* (9.1) *can be obtained by integrating the following differential equations from the initial conditions* $\mathbf{m}(t_0) = \mathrm{E}[\mathbf{x}(t_0)]$ *and* $\mathbf{P}(t_0) = \mathrm{Cov}[\mathbf{x}(t_0)]$ *to the target time* t:

$$\frac{d\mathbf{m}}{dt} = \sum_i W^{(i)} \mathbf{f}(\mathbf{m} + \sqrt{\mathbf{P}}\,\boldsymbol{\xi}_i, t),$$

$$\begin{aligned}
\frac{d\mathbf{P}}{dt} = &\sum_i W^{(i)} \mathbf{f}(\mathbf{m} + \sqrt{\mathbf{P}}\,\boldsymbol{\xi}_i, t)\,\boldsymbol{\xi}_i^\mathsf{T} \sqrt{\mathbf{P}}^\mathsf{T} \\
&+ \sum_i W^{(i)} \sqrt{\mathbf{P}}\,\boldsymbol{\xi}_i\,\mathbf{f}^\mathsf{T}(\mathbf{m} + \sqrt{\mathbf{P}}\,\boldsymbol{\xi}_i, t) \\
&+ \sum_i W^{(i)} \mathbf{L}(\mathbf{m} + \sqrt{\mathbf{P}}\,\boldsymbol{\xi}_i, t)\,\mathbf{Q}\,\mathbf{L}^\mathsf{T}(\mathbf{m} + \sqrt{\mathbf{P}}\,\boldsymbol{\xi}_i, t).
\end{aligned} \tag{9.21}$$

Once the Gaussian integral approximation has been selected, the solutions to the resulting ordinary differential equations can be computed, for example, by the fourth-order Runge–Kutta method or some similar numerical ODE solution method. It would also be possible to approximate the integrals using various other methods from filtering theory (see, e.g., Jazwinski, 1970; Wu et al., 2006; Särkkä and Sarmavuori, 2013).

Example 9.6 (Linearization and Gauss–Hermite approximations). *The linearization-based and sigma-point approximations of the Beneš SDE*

$$dx = \tanh(x)\,dt + d\beta, \quad x(0) = x_0, \tag{9.22}$$

where $\beta(t)$ *is a standard Brownian motion, which we already considered in Example 7.10, look like the following:*

- *The linearization approximation is*

$$\begin{aligned}
\frac{dm}{dt} &= \tanh(m), \\
\frac{dP}{dt} &= 2(1 - \tanh^2(m))\,P + 1.
\end{aligned} \tag{9.23}$$

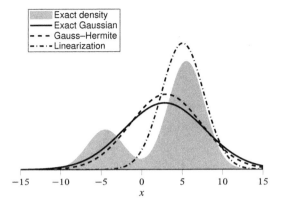

Figure 9.1 Comparison of the Gaussian approximations in
Example 9.6 to the exact density and to the exact Gaussian fit.

- *A generic sigma-point approximation is given as*

$$\frac{dm}{dt} = \sum_i W^{(i)} \tanh(m + \sqrt{P}\,\xi_i),$$

$$\frac{dP}{dt} = 1 + 2 \sum_i W^{(i)} \sqrt{P}\,\xi_i \tanh(m + \sqrt{P}\,\xi_i).$$

(9.24)

*The comparison of these approximations to the exact solutions is shown
in Figure 9.1. The approximations are compared to the exact density and
to the exact Gaussian fit formed by moment-matching to the exact density.
The starting point was $x(0) = 1/2$ and the solution was computed at time
$t = 5$. The sigma-point method was a 10-point Gauss–Hermite quadrature approximation. As can be seen in the figure, the sigma-point method
produces a more accurate match to the moment-matched Gaussian than
the linearization approximation. However, it is apparent that a Gaussian
approximation poorly approximates the true multi-modal distribution.*

Example 9.7 (Gaussian approximation of a nonlinear trigonometric SDE).
*Consider the nonlinear Itô stochastic differential equation model that was
already covered in Example 8.6:*

$$dx = -\left(\frac{1}{10}\right)^2 \sin(x)\,\cos^3(x)\,dt + \frac{1}{10}\cos^2(x)\,d\beta, \quad x(0) = x_0,$$

(9.25)

where $\beta(t)$ is a standard Brownian motion. The SDE has the solution

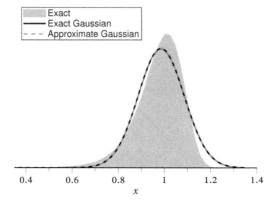

Figure 9.2 An example of a sigma point–based Gaussian approximation to a nonlinear SDE in Example 9.7. The exact solution at $t = 10$ is shown by the patch, and the solid line illustrates the Gaussian fit to it. The approximation is shown by the dashed line.

$x(t) = \arctan(1/10\ \beta(t) + \tan(x_0))$. *In this example, let* $x_0 = 1$. *The goal is to characterize the solution at* $t = 10$ *using a Gaussian approximation* $p(x(t)) \approx \mathrm{N}(x(t) \mid m(t), P(t))$ *of the exact solution.*

In the model, we have the drift $f(x) = -(1/10)^2 \sin(x) \cos^3(x)$ *and dispersion coefficient* $L(x) = 1/10 \cos^2(x)$. *Applying the cubature integration sigma-point scheme* ($\xi = \pm 1, W^{(i)} = 1/2, i = 1, 2$) *gives the following mean and covariance differential equations:*

$$\frac{dm}{dt} = \frac{1}{2} f(m - \sqrt{P}) + \frac{1}{2} f(m + \sqrt{P}), \tag{9.26}$$

$$\frac{dP}{dt} = \sqrt{P}\, f(m - \sqrt{P}) - \sqrt{P}\, f(m + \sqrt{P})$$

$$+ \frac{1}{2}\left[L(m - \sqrt{P})\right]^2 + \frac{1}{2}\left[L(m + \sqrt{P})\right]^2, \tag{9.27}$$

where $m(0) = x_0$, $P(0) = 0$. *We use the fourth-order Runge–Kutta scheme for solving* $m(10)$ *and* $P(10)$ *with a step size of* $\Delta t = 2^{-6}$. *Figure 9.2 illustrates the exact solution of* $x(10)$ *and shows the moment-matched Gaussian fit to it by a solid line. The ODE-based approximative solution* $\mathrm{N}(x(10) \mid m(10), P(10))$ *is shown by a dashed line, and it coincides well with the Gaussian fit.*

It is worth noting that all the Gaussian assumed density approximations, including the linearization and sigma-point methods discussed in this section (Algorithms 9.1–9.5), also provide Gaussian approximations for the transition densities.

Algorithm 9.8 (Gaussian assumed density transition density approximation). *The Gaussian assumed density approximation for transition density $p(\mathbf{x}(t) \mid \mathbf{x}(s))$ where $t > s$ can be constructed by starting from the initial conditions $\mathbf{m}(s) = \mathbf{x}(s)$, $\mathbf{P}(s) = \mathbf{0}$ in any of the Algorithms 9.1–9.5 and solving the equations at time t. If we denote the resulting mean and covariance as $\mathbf{m}(t \mid s)$ and $\mathbf{P}(t \mid s)$, respectively, then the transition density approximation is given as*

$$p(\mathbf{x}(t) \mid \mathbf{x}(s)) \approx \mathrm{N}(\mathbf{x}(t) \mid \mathbf{m}(t \mid s), \mathbf{P}(t \mid s)). \tag{9.28}$$

Because all the Gaussian assumed density approximations produce a local approximation to the transition density, they can also be used as SDE simulation methods. The basic idea is to fix a small Δt and sequentially simulate transitions over Δt intervals by using the transition density approximation. This can be written as the following algorithm.

Algorithm 9.9 (SDE simulation by Gaussian assumed density approximation). *Gaussian assumed density approximation–based numerical simulation of an SDE solution over an interval $[t_0, t]$ can be performed as follows:*

- *Divide the interval to M subintervals of length Δt. These interval lengths can also vary if required.*
- *Starting from $\hat{\mathbf{x}}(t_0) \sim p(\mathbf{x}(t_0))$, sequentially at each of the M subintervals k perform a draw from the approximate transition density as follows:*

$$\hat{\mathbf{x}}(t_{k+1}) \sim \mathrm{N}(\mathbf{m}(t_{k+1} \mid t_k), \mathbf{P}(t_{k+1} \mid t_k)), \tag{9.29}$$

where the transition density approximation of the right is computed with Algorithm 9.8 with $\hat{\mathbf{x}}(t_k)$ as the starting point.

Algorithm 9.9 has the property that unlike, for example, the Euler–Maruyama method, it is exact for linear systems. The convergence properties are still to be investigated, but they could be expected to be similar to the methods of Ozaki (1992, 1993) and Shoji and Ozaki (1998). Numerical comparison of the Gaussian assumed density approximation–based simulation method to local linearization is provided in Example 9.14.

The Gaussian assumed density approximations can be generalized to the exponential family of distributions by considering the projections of SDEs on manifolds of exponential densities (Brigo et al., 1999). The linearization approximations can also be formed along a nominal trajectory instead of the approximated mean trajectory (Maybeck, 1982a). Yet another possibility is to use a sigma-point version of the nominal trajectory methodology (García-Fernández et al., 2015, 2017).

9.2 Linearized Discretizations

One way to approximate the solutions of SDEs using Gaussian distributions is to first use methods such as Itô–Taylor series expansions or stochastic Runge–Kutta methods for forming a discrete-time approximation to the SDE, and then approximate it using Gaussian assumed density approximations. That is, if we originally have an SDE

$$\mathrm{d}\mathbf{x} = \mathbf{f}(\mathbf{x}, t)\, \mathrm{d}t + \mathbf{L}(\mathbf{x}, t)\, \mathrm{d}\boldsymbol{\beta}, \tag{9.30}$$

we then form an approximation as

$$\mathbf{x}(t_{k+1}) = \mathbf{f}_k^{\Delta t}(\mathbf{x}(t_k), \mathbf{e}_k), \tag{9.31}$$

where $\mathbf{f}_k^{\Delta t}$ is defined by the discretization method and $\mathbf{e}_k$ is a noise process related to the discretization. We can then step over $t_1, t_2, \ldots$ such that we always compute the mean and covariance of the next SDE state $\mathbf{x}(t_{k+1})$ assuming that the previous state $\mathbf{x}(t_k)$ was Gaussian with the mean and covariance that we calculated for it.

As we assume Gaussianity, the integrals to be recursively computed for $k = 1, 2, \ldots$ are the following:

$$\mathbf{m}_{k+1} = \iint \mathbf{f}_k^{\Delta t}(\mathbf{x}(t_k), \mathbf{e}_k)\, \mathrm{N}(\mathbf{x}(t_k) \mid \mathbf{m}_k, \mathbf{P}_k)\, p(\mathbf{e}_k)\, \mathrm{d}\mathbf{x}(t_k)\, \mathrm{d}\mathbf{e}_k, \tag{9.32}$$

$$\mathbf{P}_{k+1} = \iint \left[\mathbf{f}_k^{\Delta t}(\mathbf{x}(t_k), \mathbf{e}_k) - \mathbf{m}_{k+1} \right] \left[\mathbf{f}_k^{\Delta t}(\mathbf{x}(t_k), \mathbf{e}_k) - \mathbf{m}_{k+1} \right]^\mathsf{T}$$
$$\times \mathrm{N}(\mathbf{x}(t_k) \mid \mathbf{m}_k, \mathbf{P}_k)\, p(\mathbf{e}_k)\, \mathrm{d}\mathbf{x}(t_k)\, \mathrm{d}\mathbf{e}_k. \tag{9.33}$$

If the noises $\mathbf{e}_k$ are Gaussian as well, then the joint distribution of $(\mathbf{x}(t_k), \mathbf{e}_k)$ is Gaussian and we can use the numerical integration methods outlined in the previous section, because the integrals have the Gaussian form (9.12). Because $\mathbf{e}_k$ is independent of $(\mathbf{x}(t_k), \mathbf{e}_k)$, we can also often integrate the noise out in closed form.

Example 9.10 (Approximative Itô–Taylor method). *In the Itô–Taylor* *method in Algorithm 8.4, we have*

$$\mathbf{f}_k^{\Delta t}(\mathbf{x}(t_k), \mathbf{e}_k) = \hat{\mathbf{x}}(t_k) + \mathbf{f}(\hat{\mathbf{x}}(t_k), t_k) \, \Delta t + \mathbf{L} \, \Delta \boldsymbol{\beta}_k$$
$$+ \mathbf{a}_k(\hat{\mathbf{x}}(t_k)) \frac{\Delta t^2}{2} + \sum_j \mathbf{b}_{k,j}(\hat{\mathbf{x}}(t_k)) \, \Delta \boldsymbol{\zeta}_k, \quad (9.34)$$

where the terms $\mathbf{a}_k(\hat{\mathbf{x}}(t_k))$ *and* $\mathbf{b}_{k,j}(\hat{\mathbf{x}}(t_k))$ *are given by Equation (8.48) and the terms* $\Delta \boldsymbol{\beta}_k, \Delta \boldsymbol{\zeta}_k$ *are jointly Gaussian as defined in Equation (8.46). If we put* $\mathbf{e}_k = (\Delta \boldsymbol{\beta}_k, \Delta \boldsymbol{\zeta}_k)$, *then the mean and covariance recursions can be written in form (9.33), where the integrations can be approximated, for example, using cubature type of approximations discussed in the previous section.*

The preceding kind of approximation has been used in filtering context, for example, by Särkkä and Solin (2012), where the authors also compared it with the approximations described in the previous section.

9.3 Local Linearization Methods of Ozaki and Shoji

The local linearization methods of Ozaki (1992, 1993) and Shoji and Ozaki (1998) (see also Rao, 1999; Iacus, 2008) can be seen as instances of linearization methods, where the linearization is chosen in a specific way. The basic idea of the method of Ozaki (1993) can be described as follows. We aim to approximate the solution of the following scalar SDE:

$$\mathrm{d}x = f(x) \, \mathrm{d}t + \mathrm{d}\beta. \quad (9.35)$$

The idea is to approximate the solution on interval $[t, t + \Delta t]$, conditioned on $x(t)$ as a linear SDE

$$\mathrm{d}x = F \, x \, \mathrm{d}t + \mathrm{d}\beta. \quad (9.36)$$

The reason why this is useful is that the discretization for linear equation is then readily given by Equations (6.24) and (6.25) as

$$A(\Delta t) = \exp(F \, \Delta t),$$
$$\Sigma(\Delta t) = \int_0^{\Delta t} \exp(2F \, (\tau - \Delta t)) \, q \, \mathrm{d}\tau \quad (9.37)$$
$$= \frac{q}{2F} [\exp(2F \, \Delta t) - 1],$$

giving the Gaussian approximation

$$p(x(t + \Delta t) \mid x(t)) \approx \mathrm{N}(x(t + \Delta t) \mid A(\Delta t) \, x(t), \Sigma(\Delta t)). \quad (9.38)$$

To derive the method, let us now consider the noise-free equation

$$\frac{d\xi(t)}{dt} = f(\xi), \quad \xi(t) = x(t). \tag{9.39}$$

Differentiating once gives

$$\frac{d^2\xi}{dt^2} = f'(\xi)\frac{d\xi}{dt}, \tag{9.40}$$

where f' is the derivative of f. Let us now assume that the derivative is constant with the value of $f'(x(t))$. Then, by solving the differential equation (9.40), we get

$$\frac{d\xi}{dt}(t + \Delta t) = \exp(f'(x(t))\,\Delta t)\,\frac{d\xi(t)}{dt} = \exp(f'(x(t))\,\Delta t)\,f(x(t)). \tag{9.41}$$

Integrating from 0 to Δt then gives

$$\begin{aligned}
\xi(t + \Delta t) &= x(t) + \int_0^{\Delta t} \exp(f'(x(t))\,\tau)\,f(x(t))\,d\tau \\
&= x(t) + \frac{1}{f'(x(t))}[\exp(f'(x(t))\,\Delta t) - 1]\,f(x(t)) \\
&= \left(1 + \frac{f(x(t))}{x(t)\,f'(x(t))}[\exp(f'(x(t))\,\Delta t) - 1]\right) x(t). \tag{9.42}
\end{aligned}$$

We then finally wish to determine F such that $\exp(f'(x(t))\,\Delta t)$ matches the coefficient of $x(t)$ in the preceding, that is,

$$\exp(F\,\Delta t) = 1 + \frac{f(x(t))}{x(t)\,f'(x(t))}[\exp(f'(x(t))\,\Delta t) - 1], \tag{9.43}$$

which gives

$$F = \frac{1}{\Delta t}\log\left(1 + \frac{f(x(t))}{x(t)\,f'(x(t))}[\exp(f'(x(t))\,\Delta t) - 1]\right). \tag{9.44}$$

This method can also be extended to multivariate SDEs, but it becomes more complicated and less useful.

In the scalar case, we get the following algorithm.

Algorithm 9.11 (Local linearization of Ozaki). *Given a scalar SDE of the form*

$$dx = f(x)\,dt + d\beta, \tag{9.45}$$

we approximate its solution on interval $[t, t + \Delta t]$ by a linear time-invariant SDE

$$dx = F x \, dt + d\beta, \tag{9.46}$$

where

$$F = \frac{1}{\Delta t} \log \left(1 + \frac{f(x(t))}{x(t) \, f'(x(t))} [\exp(f'(x(t)) \, \Delta t) - 1] \right). \tag{9.47}$$

The method of Shoji and Ozaki (1998) is a modification to the local linearization method of Ozaki (1992, 1993), which extends more easily to multivariate case. We aim to approximate the solution of a scalar SDE

$$dx = f(x, t) \, dt + d\beta \tag{9.48}$$

with a linear SDE

$$dx = G x \, dt + a t \, dt + b \, dt + d\beta. \tag{9.49}$$

Recall that the Itô formula for f gives

$$df(x, t) = \frac{\partial f(x, t)}{\partial t} dt + \frac{\partial f(x, t)}{\partial x} dx + \frac{1}{2} \frac{\partial^2 f(x, t)}{\partial x^2} dx^2$$

$$= \left[\frac{\partial f(x, t)}{\partial t} + \frac{q}{2} \frac{\partial^2 f(x, t)}{\partial x^2} \right] dt + \frac{\partial f(x, t)}{dx} dx. \tag{9.50}$$

Let us now assume that $\partial f / \partial x$, $\partial^2 f / \partial x^2$, and $\partial f / \partial t$ are constant. Then on the interval $[t, u]$, this gives

$$f(x(u), u) - f(x(t), t) = \left[\frac{\partial f(x, t)}{\partial t} + \frac{q}{2} \frac{\partial^2 f(x, t)}{\partial x^2} \right] (u - t)$$

$$+ \frac{\partial f(x, t)}{\partial x} (x(u) - x(t)). \tag{9.51}$$

We get the approximation

$$f(x(u), u) = G(t) x(u) + a(t) u + b(t), \tag{9.52}$$

where the terms are given by

$$G(t) = \frac{\partial f(x, t)}{\partial x},$$

$$a(t) = \frac{\partial f(x, t)}{\partial t} + \frac{q}{2} \frac{\partial^2 f(x, t)}{\partial x^2},$$

$$b(t) = f(x(t), t) - \frac{\partial f(x, t)}{\partial x} x(t) - \left[\frac{\partial f(x, t)}{\partial t} + \frac{q}{2} \frac{\partial^2 f(x, t)}{\partial x^2} \right] t.$$

$$\tag{9.53}$$

Here we can now pick up the coefficients for the approximation. In algorithmic form the method is the following.

Algorithm 9.12 (Local linearization of Shoji and Ozaki). *Given a scalar SDE of the form*

$$\mathrm{d}x = f(x, t)\,\mathrm{d}t + \mathrm{d}\beta, \tag{9.54}$$

we approximate it on $[t, t + \Delta t]$ is as a linear SDE

$$\mathrm{d}x = G\,x\,\mathrm{d}t + a\,t\,\mathrm{d}t + b\,\mathrm{d}t + \mathrm{d}\beta, \tag{9.55}$$

where

$$G = \frac{\partial f(x(t), t)}{\partial x},$$

$$a = \frac{\partial f(x(t), t)}{\partial t} + \frac{q}{2}\frac{\partial^2 f(x(t), t)}{\partial x^2},$$

$$b = f(x(t), t) - \frac{\partial f(x(t), t)}{\partial x}\,x(t) - \left[\frac{\partial f(x(t), t)}{\partial t} + \frac{q}{2}\frac{\partial^2 f(x(t), t)}{\partial x^2}\right]t.$$

$$\tag{9.56}$$

Due to the appearance of $x(t)$ in the linearization coefficients, local linearization methods are not suitable for forming Gaussian (process) approximations to SDEs in the same sense as Gaussian assumed density approximations are. However, they were originally developed for numerical simulation of SDEs as well as for one-step approximation of SDEs for parameter estimation. Numerical simulation with local linearization can be done analogously to Algorithm 9.9 – at each small subinterval of length Δt, we draw from the local linearization approximation instead of the true transition density. This can be written in the following algorithmic form.

Algorithm 9.13 (SDE simulation by local linearization). *Local linearization methods can be used for simulating trajectories from an SDE as follows:*

1. Divide the simulation interval $[t_0, t]$ into M subintervals of length Δt.
2. At each subinterval, sequentially draw a new sample from the linear SDE corresponding to the local linearization approximation.

The following example provides a comparison of the local linearization methods with the Gaussian assumed density approximations outlined in Section 9.1.

Example 9.14 (Local linearization vs. Gaussian approximations). *For example, consider the Beneš SDE*

$$dx = \tanh(x)\, dt + d\beta, \quad x(0) = x_0. \tag{9.57}$$

The first local linearization in Algorithm 9.11 gives

$$dx = F\, x\, dt + d\beta, \tag{9.58}$$

where

$$F = \frac{1}{\Delta t} \log\left(1 + \frac{\tanh(x_0)}{x_0\,[1 - \tanh^2(x_0)]}\right.$$

$$\left. \times\, [\exp([1 - \tanh^2(x_0)]\,\Delta t) - 1]\right). \tag{9.59}$$

The coefficients for the second linearization are

$$G = 1 - \tanh^2(x_0),$$
$$a = q\, \tanh(x_0)\, (\tanh^2(x_0) - 1), \tag{9.60}$$
$$b = \tanh(x_0) - x_0\, (\tanh^2(x_0) - 1).$$

Also recall that the corresponding (classical) linearization and sigma-point approximations are given by Equations (9.23) and (9.24) with initial conditions $m(0) = x_0$ and $P(0) = 0$.

Figure 9.3 shows simulation results computed with the local linearization methods, the sigma point–based simulation method (cf. Example 9.6), and the strong Itô–Taylor method of order 1.5. In order to test the differences between the methods, we computed the histograms of the simulation results at time $t = 5$ with initial condition $x(0) = 1/2$ and by using a relatively large $\Delta t = 1$ in each method. The sigma-point method was a Gauss–Hermite–based method with 10 points. As can be seen, in this case the sigma-point method produces a more accurate histogram than the local linearization methods.

9.4 Taylor Series Expansions of Moment Equations

One way to approximate the moments of an SDEs on an interval $[t, t + \Delta t]$ is to use the Taylor series expansion of the function (Kessler, 1997; Aït-Sahalia, 2002, 2008)

$$\Delta t \mapsto E[\phi(\mathbf{x}(t + \Delta t)) \mid \mathbf{x}(t)]. \tag{9.61}$$

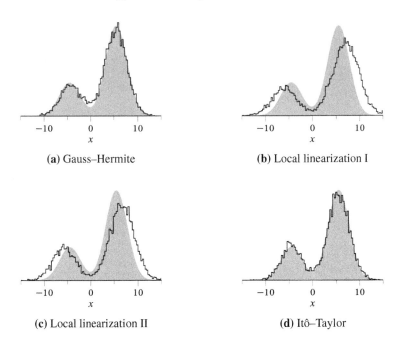

(a) Gauss–Hermite

(b) Local linearization I

(c) Local linearization II

(d) Itô–Taylor

Figure 9.3 Comparison of numerical simulation of SDEs with local linearization methods in Example 9.14, the Gauss–Hermite–based simulation method, and the strong Itô–Taylor method of order 1.5. The solid density represents the ground-truth reference.

The Taylor series expansion centered at $\Delta t = 0$, is now given as

$$\mathrm{E}[\phi(\mathbf{x}(t + \Delta t))] = \sum_{k=0}^{\infty} \frac{1}{k!} \frac{\mathrm{d}^k \, \mathrm{E}[\phi(\mathbf{x}(t))]}{\mathrm{d}t^k} \Delta t^k, \qquad (9.62)$$

for which we need the time derivatives of $\mathrm{E}[\phi(\mathbf{x}(t))]$. Fortunately, we know that the time derivative of this kind of expectation is given by the formula (5.13), which says that

$$\frac{\mathrm{d} \, \mathrm{E}[\phi(\mathbf{x}(t))]}{\mathrm{d}t} = \mathrm{E}[\mathcal{A}\phi(\mathbf{x}(t))], \qquad (9.63)$$

where $\mathcal{A}$ is the generator of the diffusion. Applying Equation (9.63) again to the time derivative of $\mathcal{A}\phi$ then gives

$$\frac{\mathrm{d} \, \mathrm{E}[\mathcal{A}\phi(\mathbf{x}(t))]}{\mathrm{d}t} = \mathrm{E}[\mathcal{A}^2\phi(\mathbf{x}(t))], \qquad (9.64)$$

and by using $E[\mathcal{A}\phi(\mathbf{x}(t))] = d\,E[\phi(\mathbf{x}(t))]/dt$ on the left-hand side, we get

$$\frac{d^2\,E[\phi(\mathbf{x}(t))]}{dt^2} = E[\mathcal{A}^2\phi(\mathbf{x}(t))]. \tag{9.65}$$

Applying this further to $\mathcal{A}^2\phi(\mathbf{x}(t))$, $\mathcal{A}^3\phi(\mathbf{x}(t))$, ... then gives the general formula

$$\frac{d^n\,E[\phi(\mathbf{x}(t))]}{dt^n} = E[\mathcal{A}^n\phi(\mathbf{x}(t))]. \tag{9.66}$$

By using this, we can form the Taylor series expansion of $E[\phi]$ as

$$E[\phi(\mathbf{x}(t + \Delta t))] = \sum_{k=0}^{\infty} \frac{1}{k!} \frac{d^k\,E[\phi(\mathbf{x}(t))]}{dt^k} \Delta t^k$$

$$= \sum_{k=0}^{\infty} \frac{1}{k!} E[\mathcal{A}^k\phi(\mathbf{x}(t))] \Delta t^k. \tag{9.67}$$

However, if we now condition on $\mathbf{x}(t)$, the expectations in the series disappear and we get the following algorithm.

Algorithm 9.15 (Taylor series expansion of moments). *Given an SDE*

$$d\mathbf{x} = \mathbf{f}(\mathbf{x}, t)\,dt + \mathbf{L}(\mathbf{x}, t)\,d\boldsymbol{\beta}, \tag{9.68}$$

the conditional expectation of a function $\phi(\mathbf{x})$ on interval $[t, t + \Delta t]$ can be approximated by truncating the Taylor series expansion

$$E[\phi(\mathbf{x}(t + \Delta t)) \mid \mathbf{x}(t)] = \sum_{k=0}^{\infty} \frac{1}{k!} \mathcal{A}^k \phi(\mathbf{x}(t)) \Delta t^k, \tag{9.69}$$

where $\mathcal{A}$ is the generator of the diffusion in Definition 5.2.

The Taylor series approximation can be used to form Gaussian approximations in the same way as we did in Section 9.2. The Taylor series provides asymptotically exact expressions for the mean and covariance, and hence they can be used to replace the discretization-based approximations that we used in Equation (9.33). These kinds of methods have been proposed, for example, in Kessler (1997), Rao (1999), and Iacus (2008).

The moment computation is illustrated in the following example, where the series expansion turns out to be finite and hence the moments can be computed exactly from the Taylor series.

Example 9.16 (Moments of the Beneš SDE). *Let*

$$dx = \tanh(x)\,dt + d\beta, \tag{9.70}$$

where $\beta(t)$ is a standard Brownian motion. In this case, we have

$$\mathcal{A} = \tanh(x) \frac{\partial}{\partial x} + \frac{1}{2} \frac{\partial^2}{\partial x^2}. \tag{9.71}$$

Applying this to function $\phi(x) = x$, then gives

$$\begin{aligned}
\mathcal{A}^0 \phi(x) &= x, \\
\mathcal{A}^1 \phi(x) &= \tanh(x), \\
\mathcal{A}^n \phi(x) &= 0, \quad \text{for } n \geq 2,
\end{aligned} \tag{9.72}$$

and to function $\phi(x) = x^2$:

$$\begin{aligned}
\mathcal{A}^0 \phi(x) &= x^2, \\
\mathcal{A}^1 \phi(x) &= 2x \tanh(x) + 1, \\
\mathcal{A}^2 \phi(x) &= 2, \\
\mathcal{A}^n \phi(x) &= 0, \quad \text{for } n \geq 3.
\end{aligned} \tag{9.73}$$

For function $\phi(x) = x^3$, we get:

$$\begin{aligned}
\mathcal{A}^0 \phi(x) &= x^3, \\
\mathcal{A}^1 \phi(x) &= 3x + 3x^2 \tanh(x), \\
\mathcal{A}^2 \phi(x) &= 6x + 6 \tanh(x), \\
\mathcal{A}^3 \phi(x) &= 6 \tanh(x), \\
\mathcal{A}^n \phi(x) &= 0, \quad \text{for } n \geq 4.
\end{aligned} \tag{9.74}$$

Thus we get that Taylor series expansions for the first three noncentral moments of the SDE on interval $[t, t + \Delta t]$, conditioned on $x(t)$, are

$$\begin{aligned}
\mathrm{E}[x(t + \Delta t) \mid x(t)] &= x(t) + \tanh(x(t)) \, \Delta t, \\
\mathrm{E}[x^2(t + \Delta t) \mid x(t)] &= x^2(t) + 2x(t) \tanh(x(t)) \, \Delta t + \Delta t + \Delta t^2, \\
\mathrm{E}[x^3(t + \Delta t) \mid x(t)] &= x^3(t) + 3x(t) \, \Delta t + 3x^2(t) \tanh(x(t)) \, \Delta t \\
&\quad + 3x(t) \, \Delta t^2 + 3 \tanh(x(t)) \, \Delta t^2 + \tanh(x(t)) \, \Delta t^3,
\end{aligned} \tag{9.75}$$

which are indeed exact. If we used the first two moments in the preceding to form a Gaussian approximation as in Example 9.6, we would recover the exact Gaussian fit shown in Figure 9.1. However, this approximation could be improved by using higher-order moments to form a non-Gaussian density approximation.

Similarly to Gaussian assumed density approximations in Section 9.1 and local linearization methods in Section 9.3, the local transition density

approximation implied by the Taylor series expansion can be used to construct SDE simulation methods (cf. Algorithms 9.9 and 9.13).

9.5 Hermite Expansions of Transition Densities

Provided that the transition density of the SDE is close to a normal distribution, we can use a Fourier–Hermite series to approximate it. This idea is
presented by Aït-Sahalia (2002, 2008) and can also be found, for example,
in Rao (1999) and Iacus (2008).

Consider the following SDE:

$$dx = f(x) dt + L d\beta, \tag{9.76}$$

where β is a vector of Brownian motions with diffusion matrix Q. We
also need to assume that $L Q L^T$ is invertible. In the articles of Aït-Sahalia
(2002, 2008), it was assumed that the equation had already been transformed into unit diffusion by the Lamperti transform, but here we will directly apply the method to the preceding SDE with nonunit diffusion.

We now aim to form an approximation to the transition density

$$p_x(x(t + \Delta t) \mid x(t)), \tag{9.77}$$

where $\Delta t > 0$. Let us now define a "pseudonormalized" increment

$$z = [L Q L^T \Delta t]^{-1/2} [x(t + \Delta t) - x(t)], \tag{9.78}$$

which should be quite close to a $N(0, I)$ random variable. If we denote
the probability density of z given $x(t)$ as $q(z \mid x(t))$, then the transition
density of the original process can be rewritten as

$$
\begin{aligned}
&p(x(t + \Delta t) \mid x(t)) \\
&= |L Q L^T \Delta t|^{-1/2} q([L Q L^T \Delta t]^{-1/2} [x(t + \Delta t) - x(t)] \mid x(t)).
\end{aligned}
\tag{9.79}
$$

Let $H_\alpha(z)$ be the multivariate probabilists' Hermite polynomials with
multi-indexing $\alpha = (\alpha_1, \alpha_2, \ldots, \alpha_n)$, which can be defined as follows:

$$H_\alpha(z) = H_{\alpha_1}(z_1) \times H_{\alpha_2}(z_1) \times \cdots \times H_{\alpha_n}(z_n), \tag{9.80}$$

where H_j are univariate Hermite polynomials defined as

$$H_j(z) = (-1)^j \exp(z^2/2) \frac{d^j}{dz^j} \exp(-z^2/2). \tag{9.81}$$

We can now form the Jth order Fourier–Hermite series expansion of q/N as follows:

$$\frac{q(\mathbf{z} \mid \mathbf{x}(t))}{N(\mathbf{z} \mid \mathbf{0}, \mathbf{I})} \approx \sum_{|\alpha| \leq J} \frac{c_\alpha}{\alpha!} H_\alpha(\mathbf{z}), \qquad (9.82)$$

where

$$c_\alpha = \int H_\alpha(\mathbf{z})\, q(\mathbf{z} \mid \mathbf{x}(t))\, d\mathbf{z}$$
$$= E\left[H_\alpha([\mathbf{L}\,\mathbf{Q}\,\mathbf{L}^{\mathsf{T}}\,\Delta t]^{-1/2}\,[\mathbf{x}(t + \Delta t) - \mathbf{x}(t)]) \mid \mathbf{x}(t) \right]. \qquad (9.83)$$

This series will converge provided that the tails of $q(\mathbf{z} \mid \mathbf{x}(t))$ are "thin" enough (see Aït-Sahalia, 2002, 2008, for mathematical details). We can then recover the transition density approximation from Equation (9.79) as

$$p(\mathbf{x}(t + \Delta t) \mid \mathbf{x}(t))$$
$$= \frac{N(\mathbf{z} \mid \mathbf{0}, \mathbf{I})}{|\mathbf{L}\,\mathbf{Q}\,\mathbf{L}^{\mathsf{T}}\,\Delta t|^{1/2}}\, q([\mathbf{L}\,\mathbf{Q}\,\mathbf{L}^{\mathsf{T}}\,\Delta t]^{-1/2}\,[\mathbf{x}(t + \Delta t) - \mathbf{x}(t)] \mid \mathbf{x}(t))$$
$$\approx \frac{N(\mathbf{z} \mid \mathbf{0}, \mathbf{I})}{|\mathbf{L}\,\mathbf{Q}\,\mathbf{L}^{\mathsf{T}}\,\Delta t|^{1/2}} \sum_{|\alpha| \leq J} \frac{c_\alpha}{\alpha!} H_\alpha([\mathbf{L}\,\mathbf{Q}\,\mathbf{L}^{\mathsf{T}}\,\Delta t]^{-1/2}\,[\mathbf{x}(t + \Delta t) - \mathbf{x}(t)]),$$

$$(9.84)$$

In order to implement the method, we also need a way to compute the expectations appearing in Equation (9.83). For that purpose Aït-Sahalia (2002, 2008) suggests to use the moment Taylor series expansion that was discussed in Section 9.4. By selecting the orders of the Taylor series and Hermite series suitably, we can control the approximation accuracy. The resulting expansion is "closed form" in the sense that it is possible to tabulate expansions such that we only need to plug in the derivatives of the drift. Such expansions with Taylor series of order three and Hermite polynomials of order six, for scalar SDEs, have been tabulated in Aït-Sahalia (2002).

The algorithm is thus the following.

Algorithm 9.17 (Hermite expansion of transition density). *The Hermite expansion approximation of the transition density of an SDE*

$$d\mathbf{x} = \mathbf{f}(\mathbf{x})\, dt + \mathbf{L}\, d\boldsymbol{\beta} \qquad (9.85)$$

can be formed as follows:

1. *Compute the coefficients c_α in Equation (9.83) by using a truncated moment Taylor series expansion introduced in Section 9.4.*

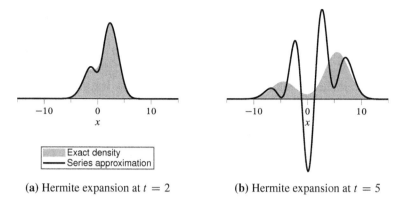

(a) Hermite expansion at $t = 2$ (b) Hermite expansion at $t = 5$

Figure 9.4 Illustration of the Hermite expansion approximation in Example 9.18 for the Beneš SDE.

2. *Approximate the transition density of the SDE using a truncated Hermite expansion in Equation* (9.84).

The expansion is illustrated in the following example.

Example 9.18 (Hermite expansion of Beneš SDE). *We used the coefficients tabulated in Aït-Sahalia (2002) to approximate the solution to the Beneš SDE*

$$\mathrm{d}x = \tanh(x)\,\mathrm{d}t + \mathrm{d}\beta, \quad x(0) = x_0. \tag{9.86}$$

Figure 9.4 shows the approximation at $t = 2$ and $t = 5$ with $x_0 = 1/2$. The approximation at $t = 2$ is still quite accurate, but the approximation at $t = 5$ is already quite inaccurate.

9.6 Discretization of FPK

We can also use partial differential equation (PDE) approximations to obtain approximate solutions to the Fokker–Planck–Kolmogorov PDE or its transition density. The Hermite-expansion and Gaussian approximations can indeed be seen as methods in this class, but here we consider the finite differences and basis function type of (Galerkin) approaches that can be seen as methods to discretize the FPK, that is, to approximate it as a finite-dimensional equation. More details on PDE methodology can be found, for example, in the books of Kreyszig (1993) and Brenner and Scott (2002).

Recall that the Fokker–Planck–Kolmorogov equation has the form

$$\frac{\partial p(\mathbf{x}, t)}{\partial t} = \mathcal{A}^* p(\mathbf{x}, t), \tag{9.87}$$

where $\mathcal{A}^*$ is the operator defined in Equation (5.36). It turns out that we can approximate this equation as a finite-dimensional system, which is just a homogeneous linear system that we can easily numerically solve. How to obtain this finite-dimensional system is, though, the key question, and here we take a look at two methods to do that: the finite-differences and basis function approximations.

We already saw in Exercise 5.3 that we can use finite differences to approximate the PDE. For that purpose, we need to discretize the state space to a finite grid $\{\mathbf{x}_i : i = 1, 2, \ldots, N\}$ and then approximate the derivatives as finite differences. A similar finite-differences approximation was also used to generate the reference result in Figure 7.3. In the one-dimensional case, the approximation could be given as

$$\begin{aligned}
\frac{\partial p(x, t)}{\partial x} &\approx \frac{p(x + h, t) - p(x - h, t)}{2h}, \\
\frac{\partial^2 p(x, t)}{\partial x^2} &\approx \frac{p(x + h, t) - 2p(x, t) + p(x - h, t)}{h^2},
\end{aligned} \tag{9.88}$$

and analogously in the multivariate case. If we collect the grid point evaluations $p(\mathbf{x}_i, t)$ into a vector $\mathbf{p}(t)$, this kind of approximation corresponds to certain matrix multiplication of $\mathbf{p}(t)$. Thus the whole FPK can be rewritten as a linear system

$$\frac{d\mathbf{p}}{dt} = \mathbf{F}_{\mathrm{fd}}(t) \, \mathbf{p}, \tag{9.89}$$

where $\mathbf{F}_{\mathrm{fd}}(t)$ is the finite-difference approximation matrix for the operator $\mathcal{A}^*$. An advantage of finite-difference approximations is that the matrix $\mathbf{F}_{\mathrm{fd}}(t)$ tends to be very sparse and hence sparse matrix routines can be used to make the computations faster.

Given the grid evaluations of the initial conditions $p(\mathbf{x}_i, t_0)$ collected into a vector $\mathbf{p}(t_0)$, this equation can now be solved as

$$\mathbf{p}(t) = \mathbf{\Psi}_{\mathrm{fd}}(t, t_0) \, \mathbf{p}(t_0), \tag{9.90}$$

where $\mathbf{\Psi}_{\mathrm{fd}}$ is the transition matrix corresponding to $\mathbf{F}_{\mathrm{fd}}(t)$. In a time-invariant case, we thus simply have

$$\mathbf{p}(t) = \exp((t - t_0) \, \mathbf{F}_{\mathrm{fd}}) \, \mathbf{p}(t_0). \tag{9.91}$$

However, in practice we often use methods like backward Euler or Runge–Kutta for the temporal integration, because the computation of the transition matrix or the matrix exponential can be computationally heavy or even intractable. In particular, the matrix exponential tends to be a dense matrix even when $\mathbf{F}_{fd}$ is a sparse matrix, and hence the speedup provided by sparse routines is lost. It is also possible to jointly discretize the time direction to get a single equation for the full space–time solution.

Another useful approach for the FPK solution is to fix a set of basis functions $\{\phi_i(\mathbf{x}) : i = 1, 2, \ldots\}$ and then approximate the PDE solution as

$$p(\mathbf{x}, t) \approx \sum_{i=1}^{N} p_i(t) \, \phi_i(\mathbf{x}), \qquad (9.92)$$

where the time evolution of the coefficients $p_i(t)$ is to be determined – note that the coefficients are not usually the same as with the finite-difference approximation, although the finite-difference approximation is a special case of this kind of expansions.

It now turns out that we can approximate the operator $\mathcal{A}^*$ on the basis $\{\phi_i\}$ with a matrix $\mathbf{F}_{bf}(t)$ that operates on the coefficients. The operation $\mathcal{A}^* p$ then corresponds to multiplication of the coefficients in the expansion (9.92) with this matrix:

$$\mathcal{A}^* p \leftrightarrow \mathbf{F}_{bf}(t) \, \mathbf{p}, \qquad (9.93)$$

where $\mathbf{p} = (p_1, p_2, \ldots)$ is a vector formed from the series expansion coefficients. Because the basis functions ϕ_i are independent of time, we also have the correspondence

$$\frac{\partial p}{\partial t} \leftrightarrow \frac{d\mathbf{p}}{dt}. \qquad (9.94)$$

Thus we can express the FPK PDE on the basis $\{\phi_i\}$ as follows:

$$\frac{d\mathbf{p}}{dt} = \mathbf{F}_{bf}(t) \, \mathbf{p}. \qquad (9.95)$$

Note that this equation has the same form as Equation (9.89), and we can use either matrix exponential (in the time-independent case) or the transition matrix corresponding to $\mathbf{F}_{bf}(t)$ to express the solution as

$$\mathbf{p}(t) = \mathbf{\Psi}_{bf}(t, t_0) \, \mathbf{p}(t_0), \qquad (9.96)$$

where $\mathbf{p}(t_0)$ is the vector of projection coefficients of $p(\mathbf{x}, t_0)$.

How can we obtain the matrix $\mathbf{F}_{bf}(t)$ then? Let us put

$$g(\mathbf{x}) = \sum_{i=1}^{N} g_i\, \phi_i(\mathbf{x}) \quad \text{and} \quad f(\mathbf{x}) = \sum_{i=1}^{N} f_i\, \phi_i(\mathbf{x}), \tag{9.97}$$

and consider the transformation $g = \mathcal{A}^* f$, which now looks like this:

$$\sum_{i=1}^{N} g_i\, \phi_i(\mathbf{x}) = \sum_{i=1}^{N} f_i\, \mathcal{A}^*\phi_i(\mathbf{x}). \tag{9.98}$$

Assume that we know $\{f_i\}$ and wish to determine $\{g_i\}$, and denote the corresponding vectors as $\mathbf{f}$ and $\mathbf{g}$. We now have a few options to determine the required matrix:

Point collocation: One way is to select N points $\{\mathbf{x}_j\}$ in the state space and insist that the equation must hold on each of them:

$$\sum_{i=1}^{N} g_i\, \phi_i(\mathbf{x}_1) = \sum_{i=1}^{N} f_i\, \mathcal{A}^*\phi_i(\mathbf{x}_1),$$

$$\vdots \tag{9.99}$$

$$\sum_{i=1}^{N} g_i\, \phi_i(\mathbf{x}_N) = \sum_{i=1}^{N} f_i\, \mathcal{A}^*\phi_i(\mathbf{x}_N).$$

If we define matrices $\mathbf{\Phi}$ and $\mathbf{\Gamma}$ such that $\Phi_{ij} = \phi_j(\mathbf{x}_i)$, $\Gamma_{ij} = \mathcal{A}^*\phi_j(\mathbf{x}_i)$, then this equation has the form

$$\mathbf{\Phi}\,\mathbf{g} = \mathbf{\Gamma}\,\mathbf{f}, \quad \text{which gives} \quad \mathbf{g} = \mathbf{\Phi}^{-1}\mathbf{\Gamma}\,\mathbf{f} \tag{9.100}$$

and thus the matrix is determined via $\mathbf{F}_{bf}(t) = \mathbf{\Phi}^{-1}\mathbf{\Gamma}$.

Ritz–Galerkin method: We can also fix an inner product $\langle \bullet, \bullet \rangle$ and choose another set of N functions $\{\varphi_i\}$. We can now take inner products of Equations (9.98) with each of these functions to give

$$\sum_{i=1}^{N} g_i\, \langle \phi_i, \varphi_1 \rangle = \sum_{i=1}^{N} f_i\, \langle \mathcal{A}^*\phi_i, \varphi_1 \rangle,$$

$$\vdots \tag{9.101}$$

$$\sum_{i=1}^{N} g_i\, \langle \phi_i, \varphi_N \rangle = \sum_{i=1}^{N} f_i\, \langle \mathcal{A}^*\phi_i, \varphi_N \rangle.$$

Defining matrices $\mathbf{\Phi}$ and $\mathbf{\Gamma}$ via $\Phi_{ij} = \langle \phi_j, \varphi_i \rangle$ and $\Gamma_{ij} = \langle \mathcal{A}^* \phi_j, \varphi_i \rangle$, this again leads to the matrix

$$\mathbf{F}_{\mathrm{bf}}(t) = \mathbf{\Phi}^{-1} \, \mathbf{\Gamma}. \qquad (9.102)$$

Other basis function methods: We can also get several other methods as special cases of the Ritz–Galerkin method:

1. A very common choice of basis functions is to put $\varphi_i = \phi_i$ and select the basis functions to be orthonormal $\langle \phi_i, \phi_j \rangle = \delta_{ij}$. This class of methods is often called spectral methods. An example of such basis is the canonical Fourier basis consisting of sines and cosines. Due to the orthonormality, the matrix $\mathbf{F}_{\mathrm{bf}}(t)$ directly consists of the elements $\langle \mathcal{A}^* \phi_j, \phi_i \rangle$.

2. Using basis functions or "elements" that have a finite support leads to finite-element methods (FEM), which are very popular in engineering applications. However, in order to allow for simpler (typically piecewise linear) basis functions, we need to integrate $\langle \mathcal{A}^* \phi_j, \varphi_i \rangle$ by parts to have only single differentiation on both ϕ_j and φ_i, which complicates the equations a bit. For details, see Brenner and Scott (2002). The advantage of FEM is that it is suitable for sparse matrix routines, as the finite support of the elements induces sparseness in the matrices.

3. The point collocation corresponds to the selection $\varphi_i(\mathbf{x}) = \delta(\mathbf{x} - \mathbf{x}_i)$ with an L_2 inner product.

We can now formulate the following generic algorithm.

Algorithm 9.19 (Discretization approximation of FPK). *The following is a generic algorithm for discretization approximation of FPK targeted for solving $p(\mathbf{x}, t)$ given $p(\mathbf{x}, t_0)$ where $t > t_0$:*

1. *Construct a finite-dimensional vector $\mathbf{p}(t_0) = \big(p_1(t_0), \ldots, p_N(t_0) \big)$ such that we can represent the initial $p(\mathbf{x}, t_0)$ approximately as a series expansion*

$$p(\mathbf{x}, t_0) \approx \sum_{i=1}^{N} p_i(t_0) \, \phi_i(\mathbf{x}). \qquad (9.103)$$

The basis functions can be piecewise constant (as in finite differences), Fourier basis functions, or any of the other Galerkin or finite-element type of basis functions discussed previously.

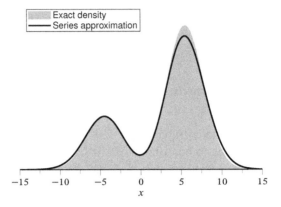

Figure 9.5 Illustration of the series expansion of FPK equation for the Beneš SDE in Example 9.20.

2. *Construct a discretization matrix* $\mathbf{F}(t)$ *by projecting or otherwise approximating the operator on the basis using the previously described methods.*

3. *Solve the linear system*

$$\frac{d\mathbf{p}}{dt} = \mathbf{F}(t)\,\mathbf{p}, \quad \mathbf{p}(t_0) = given, \tag{9.104}$$

at time t using backward Euler or Runge–Kutta or by computing the matrix exponential in the case of a time-invariant system (and if the nonsparseness is not an issue).

4. *Approximate the probability density at t as*

$$p(\mathbf{x}, t) \approx \sum_{i=1}^{N} p_i(t)\,\phi_i(\mathbf{x}). \tag{9.105}$$

If we are interested in the transition density, then the initial coefficients $\mathbf{p}(t_0)$ *should be selected to approximate the Dirac delta function.*

An example of FPK approximation using an orthonormal basis is presented in the following.

Example 9.20 (Discretized FPK for the Beneš SDE). *Consider the model*

$$dx = \tanh(x)\,dt + d\beta, \tag{9.106}$$

which gives the FPK

$$\frac{\partial p(x,t)}{\partial t} = -\frac{\partial}{\partial x}\left[\tanh(x)\, p(x,t)\right] + \frac{1}{2}\frac{\partial^2 p(x,t)}{\partial x^2}. \tag{9.107}$$

Let us select the basis

$$\phi_i(x) = \frac{1}{\sqrt{L}}\sin(\lambda_i\,(x+L)), \tag{9.108}$$

where $\lambda_i = \frac{i\pi}{2L}$, which is also the eigenbasis of the operator $\partial^2/\partial x^2$ with Dirichlet boundary conditions at $x = -L$ and $x = L$. If we define

$$\langle f, g \rangle = \int_{-L}^{L} f(x)\, g(x)\, dx, \tag{9.109}$$

then $\langle \phi_i, \phi_j \rangle = \delta_{ij}$ and the matrix elements are given by

$$
\begin{aligned}
[F_{\mathrm{bf}}]_{ij} &= \int_{-L}^{L}\left(-\frac{\partial}{\partial x}\left[\tanh(x)\,\phi_j(x)\right] + \frac{1}{2}\frac{\partial^2\phi_j(x)}{\partial x^2}\right)\phi_i(x)\, dx \\
&= \int_{-L}^{L}\left([\tanh^2(x)-1]\,\phi_j(x) - \tanh(x)\frac{\partial\phi_j(x)}{\partial x}\right)\phi_i(x)\, dx - \frac{1}{2}\lambda_i^2\,\delta_{ij} \\
&= \frac{1}{L}\int_{-L}^{L}\Bigg([\tanh^2(x)-1]\,\sin(\lambda_j\,(x+L)) \\
&\qquad\qquad - \tanh(x)\,\lambda_j\,\cos(\lambda_j\,(x+L))\Bigg)\sin(\lambda_i\,(x+L))\, dx \\
&\quad - \frac{1}{2}\lambda_i^2\,\delta_{ij},
\end{aligned}
\tag{9.110}
$$

where we have used $\langle \partial^2\phi_j/\partial x^2, \phi_i \rangle = -\lambda_i^2\delta_{ij}$ and the remaining integral can be evaluated with numerical integration.

The series expansion approximation with 20 terms is shown in Figure 9.5. The linear system resulting from the discretization was solved using a matrix exponential. As can be seen in the figure, the approximation is already quite close to the exact density. However, there is one disadvantage: the approximation becomes negative at some points and hence the interpretation of it directly as a probability density approximation can be hard.

9.7 Simulated Likelihood Methods

The basic idea in simulated likelihood methods (Pedersen, 1995; Brandt and Santa-Clara, 2002; Iacus, 2008) is to use a simple simulation to approximate the transition density $p(\mathbf{x}(t) \mid \mathbf{x}(t_0))$. We first divide the interval $[t_0, t]$ into M steps of length $(t - t_0)/M$. Then we use an Euler–Maruyama approximation on each subinterval. Instead of running Euler–Maruyama over all the intervals, we stop at step $m = M - 1$. We then compute the expectation of the Euler–Maruyama induced transition density approximation over that last subinterval. That is, we replace the last step in the Euler–Maruyama approximation with the transition density implied by the Euler approximation. This simple trick allows us to obtain a continuous approximation to the transition density which can be shown to converge to the true density (Pedersen, 1995).

The algorithm is the following.

Algorithm 9.21 (Simulated likelihood method). *The transition density approximation in the simulated likelihood method is formed as follows:*

1. *Divide the interval $[t_0, t]$ to M subintervals of length Δt.*

2. *Run Euler–Maruyama over $M - 1$ steps, that is, until the second-to-last step $t_{M-1} = t - \Delta t$.*

3. *Perform the preceding simulation N times, which results in the samples $\hat{\mathbf{x}}^{(n)}(t_{M-1})$ for the second-to-last step, where $n = 1, \ldots, N$.*

4. *The transition density is then approximated as*

$$p(\mathbf{x}(t) \mid \mathbf{x}(t_0))$$
$$\approx \frac{1}{N} \sum_{n=1}^{N} \mathrm{N}(\mathbf{x}(t) \mid \hat{\mathbf{x}}^{(n)}(t_{M-1}) + \mathbf{f}(\hat{\mathbf{x}}^{(n)}(t_{M-1}), t_{M-1}) \Delta t,$$
$$\mathbf{L}(\hat{\mathbf{x}}^{(n)}(t_{M-1}), t_{M-1}) \mathbf{Q} \mathbf{L}^{\mathsf{T}}(\hat{\mathbf{x}}^{(n)}(t_{M-1}), t_{M-1}) \Delta t). \quad (9.111)$$

Instead of the applying Euler–Maruyama approximation, we can also use any of the other SDE discretization methods from Chapter 8 that allow for closed-form transition density approximation. The approach can also be seen as a kernel density estimator, and we can replace the transition density–based kernel with some other kernel while still getting convergence of the approximation.

9.8 Pathwise Series Expansions and the Wong–Zakai Theorem

As discussed in Section 7.2, if we fix the time interval $[0, T]$, then on that interval standard Brownian motion has a series expansion of the form (see, e.g., Luo, 2006, and Section 7.2)

$$\beta(t) = \sum_{n=1}^{\infty} z_n \int_0^t \varphi_n(\tau) \, d\tau, \qquad (9.112)$$

where $z_n \sim N(0, 1)$ for $n = 1, 2, \ldots$ are independent Gaussian random variables and $\{\varphi_n(t)\}$ are a suitable set of basis functions. One possible choice is the set of functions corresponding to the Karhunen–Loeve expansion given in Equation (7.29). The series expansion can be interpreted as the following representation for the differential of standard Brownian motion:

$$d\beta(t) = \sum_{n=1}^{\infty} z_n \varphi_n(t) \, dt. \qquad (9.113)$$

We can now consider approximating the following equation by substituting a finite number N of terms from the preceding sum for the term $d\beta(t)$ in the scalar SDE

$$dx = f(x, t) \, dt + L(x, t) \, d\beta. \qquad (9.114)$$

In the limit $N \to \infty$, we could then expect to get the exact solution. However, it has been shown by Wong and Zakai (1965) that this approximation actually converges to the *Stratonovich* SDE

$$dx = f(x, t) \, dt + L(x, t) \circ d\beta. \qquad (9.115)$$

That is, we can approximate the preceding Stratonovich SDE with an equation of the form

$$dx = f(x, t) \, dt + L(x, t) \sum_{n=1}^{N} z_n \varphi_n(t) \, dt, \qquad (9.116)$$

which actually is just an ordinary differential equation

$$\frac{dx}{dt} = f(x, t) + L(x, t) \sum_{n=1}^{N} z_n \varphi_n(t), \qquad (9.117)$$

and the solution converges to the exact solution, when $N \to \infty$. The solution of an Itô SDE can be approximated by first converting it into the corresponding Stratonovich equation and then approximating the resulting equation.

Now an obvious extension is to consider a multivariate version of this approximation. Because any multivariate Brownian motion can be formed as a linear combination of independent standard Brownian motions, it is possible to form analogous multivariate approximations. Unfortunately, in the multivariate case the approximation does not generally converge to the Stratonovich solution. There exists basis functions for which this is true (e.g., Haar wavelets), but the convergence is not generally guaranteed. Anyway, we get the following algorithm (cf. Lyons et al., 2012, 2014).

Algorithm 9.22 (Pathwise series expansion of SDEs). *An approximation to a Stratonovich SDE*

$$d\mathbf{x} = \mathbf{f}(\mathbf{x}, t)\, dt + \mathbf{L}(\mathbf{x}, t) \circ d\boldsymbol{\beta}, \qquad (9.118)$$

via a series expansion of the Brownian motion can be formed as follows:

1. *Approximate the vector of Brownian motions with a series expansion of the form*

$$\boldsymbol{\beta}(t) = \sqrt{\mathbf{Q}} \sum_{n=1}^{N} \mathbf{z}_n \int_0^t \varphi_n(\tau)\, d\tau, \qquad (9.119)$$

where $\mathbf{z}_n \sim \mathrm{N}(\mathbf{0}, \mathbf{I})$ are independent Gaussian random variables. For example, we can select $\varphi_n(t)$ to be the Fourier cosine basis in (7.29).

2. *The corresponding approximation to the SDE is then given by the ordinary differential equation*

$$\frac{d\mathbf{x}}{dt} = \mathbf{f}(\mathbf{x}, t) + \mathbf{L}(\mathbf{x}, t) \sqrt{\mathbf{Q}} \sum_{n=1}^{N} \mathbf{z}_n\, \varphi_n(t). \qquad (9.120)$$

However, the convergence of the preceding algorithm with $N \to \infty$ needs to be carefully investigated case-by-case (Lyons et al., 2014). Also recall that in the additive noise case the Itô and Stratonovich SDEs are equivalent, and hence in that case, if the approximation converges, it converges to the corresponding Itô SDE.

The series expansion approximation can be used both for forming parametric approximations to SDE solutions as well as for simulating trajectories from SDEs.

Another type of series expansion is the so-called *Wiener chaos expansion* (see, e.g., Cameron and Martin, 1947; Luo, 2006). Assume that we indeed are able to solve the Equation (9.120) with any given countably infinite number of values $\{z_1, z_2, \ldots\}$. Then we can see the solution as a

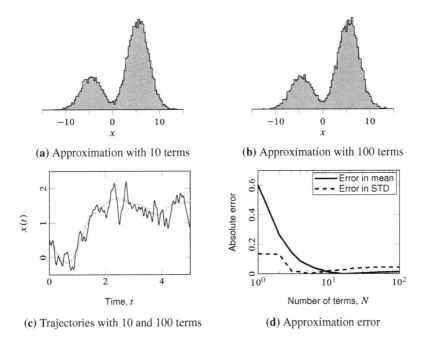

(a) Approximation with 10 terms   (b) Approximation with 100 terms

(c) Trajectories with 10 and 100 terms   (d) Approximation error

Figure 9.6 The subfigures (a) and (b) show the probability density approximations resulting from the series expansions in the Beneš SDE in Example 9.23 with $N = 10$ and $N = 100$ terms. The subfigure (c) shows the corresponding trajectory approximations and (d) illustrates the accuracy of the approximation by showing the errors in the empirical approximations of the mean and standard deviation (STD).

function (or functional) of the form

$$x(t) = U(t; z_1, z_2, \ldots). \qquad (9.121)$$

The Wiener chaos expansion is the multivariate Fourier–Hermite series for the right-hand side. That is, it is a polynomial expansion of a generic functional of Brownian motion in terms of Gaussian random variables. Hence the expansion is also called *polynomial chaos*.

Example 9.23 (Pathwise series expansion of Beneš SDE). *The series expansion approximation of the Beneš SDE*

$$dx = \tanh(x)\, dt + d\beta, \quad x(0) = x_0, \qquad (9.122)$$

with $x_0 = 1/2$ at time $t = 5$, is shown in Figure 9.6. The series expansion *was the cosine series given in Equation (7.29) with $T = 5$, which thus corresponds to the approximation*

$$\frac{dx}{dt} = \tanh(x) + \sum_{n=1}^{N} z_n \left(\frac{2}{T}\right)^{1/2} \cos\left(\frac{(2n-1)\pi}{2T} t\right), \qquad (9.123)$$

where $z_n \sim N(0, 1)$. As can be seen from the figure, the distribution of the SDE is already well approximated with $N = 10$ basis functions, although the trajectories themselves are much smoother than actual Brownian paths.

9.9 Exercises

9.1   Gaussian approximation of SDEs:

   (a) Form a Gaussian assumed density approximation to the SDE in Equation (8.134) in the time interval $t \in [0, 5]$ and compare it to the exact solution. Compute the Gaussian integrals numerically on a uniform grid.

   (b) Form a Gaussian assumed density approximation to Equation (8.133) and numerically compare it to the histogram obtained in Exercise 8.1.

9.2   Derive a similar series of moments as in Example 9.16 for the sine diffusion model

$$dx = \sin(x)\, dt + d\beta, \qquad (9.124)$$

where $\beta(t)$ is a standard Brownian motion, and compare it to moments computed from the Euler–Maruyama method.

9.3   Derive the local linearization for the model given in Equation (9.124) and numerically compare it to a cubature-based sigma-point approximation.

9.4   Use a finite-differences approximation to the FPK equation for the model given in Equation (9.124):

   (a) Select the same basis as in Example 9.20 and write out the evolution equation.

   (b) Solve it numerically and sketch the evolution of the probability density with a suitably chosen initial condition. Compare the result to the Euler–Maruyama method.

9.5   Implement a pathwise series approximation to the model (9.124) using the same basis as in Example 9.23.

10

Filtering and Smoothing Theory

In this chapter, we consider a solution to the following problem. Assume that we have a pair of processes $(\mathbf{x}(t), \mathbf{y}(t))$ such that $\mathbf{y}(t)$ is observed and $\mathbf{x}(t)$ is hidden. Now the question is as follows: Given that we have observed $\mathbf{y}(t)$ on some in interval, what can we say (in statistical sense) about the hidden process $\mathbf{x}(t)$? This problem is a classical one and has been throughly treated – from different points of view – for example, in the books of Stratonovich (1968), Jazwinski (1970), Maybeck (1979 and 1982a), Särkkä (2006 and 2013), and Crisan and Rozovskiĭ (2011).

The most fruitful statistical approach to the problem is the so-called probabilistic or Bayesian approach, which is used in the aforementioned references and which we also follow here. The main question in a statistical sense is that of the conditional probability distribution of the hidden process $\mathbf{x}(t)$ given the observed process $\mathbf{y}(t)$. This distribution contains all statistical information that can be known about the processes given the model and the observations. It is important to understand that from the modeling point of view, randomness does not "exist", but it is merely used as a representation of uncertainty.

Although in the context of this book it is most natural to model both the processes $\mathbf{x}(t)$ and $\mathbf{y}(t)$ as continuous-time processes defined as solutions to SDEs, this is not always the case in the real world. In the real world, the process $\mathbf{x}(t)$ is indeed often a continuous-time process, but the observations are not; due to the sampling process, we only get to see samples $\mathbf{y}_k$, which are connected to discrete-time samples $\mathbf{x}(t_k)$ of the hidden process. The time points $t_1, t_2, \dots$ are sometimes uniformly spaced, that is, sometimes $\Delta t_k = t_{k+1} - t_k$ is constant, but more often in multisensor applications they are not due to asynchrony of the sensors. This asynchronous case is where the SDE-based models have the most advantages; we can evaluate them at arbitrary time points and thus form arbitrary discretizations and sampling schemes without changing the underlying continuous time (SDE) model. This property is sometimes called "discretization invariance" and is extremely useful in multisensor systems.

10.1 Statistical Inference on SDEs

Let us now assume that we have a physical system whose dynamics are governed by the differential equation

$$\frac{d\mathbf{x}(t)}{dt} = \mathbf{f}(\mathbf{x}(t), t). \tag{10.1}$$

We then model the uncertainty in the physical system by including a noise process $\mathbf{w}(t)$ to the system, which leads to a white noise–driven differential equation:

$$\frac{d\mathbf{x}(t)}{dt} = \mathbf{f}(\mathbf{x}(t), t) + \mathbf{L}(\mathbf{x}, t)\mathbf{w}(t), \tag{10.2}$$

which we now, based on the learnings from the previous chapters, rather reinterpret as an Itô stochastic differential equation

$$d\mathbf{x}(t) = \mathbf{f}(\mathbf{x}(t), t)\,dt + \mathbf{L}(\mathbf{x}, t)\,d\boldsymbol{\beta}(t). \tag{10.3}$$

When it comes to the measurements, we can model our ideal measurements as functions of the state

$$\mathbf{y}(t) = \mathbf{h}(\mathbf{x}(t), t), \tag{10.4}$$

where $\mathbf{h}(\bullet, \bullet)$ models the sensor that we have. If we directly measure the state, then $\mathbf{h}$ is an identity function, but for example, in the case of distance measurement, the function can be a nonlinear function of the position coordinates in the state $\mathbf{x}(t)$. To account for the uncertainties and inaccuracies in the sensor measurements, we can assume that they are corrupted by some white noise $\boldsymbol{\varepsilon}(t)$:

$$\mathbf{y}(t) = \mathbf{h}(\mathbf{x}(t), t) + \boldsymbol{\varepsilon}(t). \tag{10.5}$$

In order to interpret this as an SDE, we need to define a process $\mathbf{z}(t)$ such that formally $\mathbf{y}(t) = d\mathbf{z}(t)/dt$, which enables us to write the measurement model as a stochastic differential equation

$$d\mathbf{z}(t) = \mathbf{h}(\mathbf{x}(t), t)\,dt + d\boldsymbol{\eta}(t), \tag{10.6}$$

with $\boldsymbol{\varepsilon}(t) = d\boldsymbol{\eta}(t)/dt$. The resulting model is called a continuous-time state-space model, which we can now define as follows.

Definition 10.1 (Continuous-time state-space model). *A continuous time state-space model is a model of the form*

$$
\begin{aligned}
d\mathbf{x} &= \mathbf{f}(\mathbf{x},t)\,dt + \mathbf{L}(\mathbf{x},t)\,d\boldsymbol{\beta}, \\
d\mathbf{z} &= \mathbf{h}(\mathbf{x},t)\,dt + d\boldsymbol{\eta},
\end{aligned}
\tag{10.7}
$$

where the first equation is the dynamic model and the second the measurement model. In the equation, $\mathbf{x}(t) \in \mathbb{R}^D$ is the state process, $\mathbf{z}(t) \in \mathbb{R}^M$ is the (integrated) measurement process, $\mathbf{f}$ is the drift function, $\mathbf{h}$ is the measurement model function, $\mathbf{L}(\mathbf{x},t)$ is the dispersion matrix, and $\boldsymbol{\beta}(t) \in \mathbb{R}^S$ and $\boldsymbol{\eta}(t) \in \mathbb{R}^M$ are independent Brownian motions with diffusion matrices $\mathbf{Q}$ and $\mathbf{R}$, respectively.

This model can also be interpreted as a white noise–driven model

$$
\begin{aligned}
\frac{d\mathbf{x}}{dt} &= \mathbf{f}(\mathbf{x},t) + \mathbf{L}(\mathbf{x},t)\,\mathbf{w}, \\
\mathbf{y} &= \mathbf{h}(\mathbf{x},t) + \boldsymbol{\varepsilon},
\end{aligned}
\tag{10.8}
$$

where we formally have $\mathbf{w}(t) = d\boldsymbol{\beta}(t)/dt$, $\mathbf{y}(t) = d\mathbf{z}(t)/dt$, and $\boldsymbol{\varepsilon}(t) = d\boldsymbol{\eta}(t)/dt$.

As discussed previously, in the real world we often obtain the measurements at discrete time instants rather than continuously as in the preceding model. In that case, we can model our measurements as

$$
\mathbf{y}_k = \mathbf{h}_k(\mathbf{x}(t_k)) + \mathbf{r}_k,
\tag{10.9}
$$

where $t_k \in \{t_1, t_2, \ldots\}$ is some sequence of measurement times and $\mathbf{r}_k$ is a discrete-time sequence of independent Gaussian noises $N(\mathbf{0}, \mathbf{R}_k)$. This is the most important class of models that we encounter in practical applications.

However, the model essentially says that

$$
p(\mathbf{y}_k \mid \mathbf{x}(t_k)) = N(\mathbf{y}_k \mid \mathbf{h}_k(\mathbf{x}(t_k)), \mathbf{R}_k),
\tag{10.10}
$$

which has the limitation that it assumes that the noise is Gaussian. We can get rid of this limitation by letting our measurement model have a more general form

$$
\mathbf{y}_k \sim p(\mathbf{y}_k \mid \mathbf{x}(t_k)),
\tag{10.11}
$$

which means that the measurement model is constructed by defining the conditional distribution of measurements $p(\mathbf{y}_k \mid \mathbf{x}(t_k))$. This distribution can also be non-Gaussian. The resulting model is the continuous-discrete state-space model.

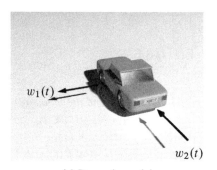

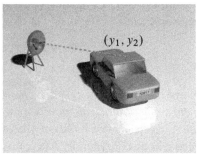

(a) Dynamic model   (b) Measurement model

Figure 10.1 Illustration of (a) a dynamic and (b) a measurement model of a car from Example 10.3. In the dynamic model, the unknown forces $w_1(t)$ and $w_2(t)$ are modeled as white noise processes. The measurements (y_1, y_2) are modeled as noise corrupted observations of the car's position.

Definition 10.2 (Continuous-discrete state-space model). *A general continuous-discrete state-space model can be formulated as*

$$\mathrm{d}\mathbf{x} = \mathbf{f}(\mathbf{x}, t)\,\mathrm{d}t + \mathbf{L}(\mathbf{x}, t)\,\mathrm{d}\boldsymbol{\beta},$$
$$\mathbf{y}_k \sim p(\mathbf{y}_k \mid \mathbf{x}(t_k)), \tag{10.12}$$

where $\mathbf{x}(t) \in \mathbb{R}^D$ is the state, and $\mathbf{y}_k \in \mathbb{R}^M$ is the measurement obtained at time instant t_k. $\mathbf{f}$ is the drift function, $\mathbf{L}$ is the dispersion matrix, and $\boldsymbol{\beta}(t) \in \mathbb{R}^S$ is Brownian motion with diffusion matrix $\mathbf{Q}$. Furthermore, $p(\mathbf{y}_k \mid \mathbf{x}(t_k))$ is the measurement model, which defines the distribution (or likelihood) of the measurement $\mathbf{y}_k$ given the state $\mathbf{x}(t_k)$. The measurements are assumed to be conditionally independent of the state.

In practice, we often construct the measurement model $p(\mathbf{y}_k \mid \mathbf{x}(t_k))$ as a noise-corrupted measurement of the form

$$\mathbf{y}_k = \mathbf{h}_k(\mathbf{x}(t_k)) + \mathbf{r}_k, \tag{10.13}$$

where $\mathbf{r}_k \sim \mathrm{N}(\mathbf{0}, \mathbf{R}_k)$ is a Gaussian measurement noise. This thus implies that our measurement model is given as

$$p(\mathbf{y}_k \mid \mathbf{x}(t_k)) = \mathrm{N}(\mathbf{y}_k \mid \mathbf{h}_k(\mathbf{x}(t_k)), \mathbf{R}_k). \tag{10.14}$$

Example 10.3 describes a typical statistical inference problem.

Example 10.3 (Car tracking model). *Recall that in Example 3.4 we modeled the dynamics of a car via the white noise–force Newton's law*

$$\frac{d^2 x_1}{dt^2} = w_1(t) \quad and \quad \frac{d^2 x_2}{dt^2} = w_2(t)$$

(see Figure 10.1a), which then resulted in an SDE model of the following form (in white noise interpretation):

$$\frac{d\mathbf{x}}{dt} = \mathbf{F} \mathbf{x} + \mathbf{L} \mathbf{w}.$$

Let us now assume that we use a radar to obtain noisy measurements (y_1, y_2) *of the car's position, which thus can be modeled as*

$$y_1(t) = x_1(t) + \varepsilon_1(t),$$
$$y_2(t) = x_2(t) + \varepsilon_2(t),$$

where $\varepsilon_1(t)$ *and* $\varepsilon_2(t)$ *are white noise processes (see Figure 10.1b). The measurement model can now be written as*

$$\mathbf{y}(t) = \mathbf{H} \mathbf{x}(t) + \boldsymbol{\varepsilon}(t), \qquad \mathbf{H} = \begin{pmatrix} 1 & 0 & 0 & 0 \\ 0 & 1 & 0 & 0 \end{pmatrix}, \qquad (10.15)$$

where $\boldsymbol{\varepsilon} = (\varepsilon_1, \varepsilon_2)$.

If we now interpret both the dynamic model and measurement model as proper SDEs, the resulting model can be written as

$$d\mathbf{x} = \mathbf{F} \mathbf{x} \, dt + \mathbf{L} \, d\boldsymbol{\beta},$$
$$d\mathbf{z} = \mathbf{H} \mathbf{x} \, dt + d\boldsymbol{\eta}, \qquad (10.16)$$

where formally $\mathbf{y} = d\mathbf{z}/dt$ *and* $\boldsymbol{\varepsilon} = d\boldsymbol{\eta}/dt$.

The corresponding continuous-discrete model is obtained by assuming that we measure

$$y_{1,k} = x_1(t_k) + r_{1,k},$$
$$y_{2,k} = x_2(t_k) + r_{2,k}, \qquad (10.17)$$

with $\mathbf{r}_k = (r_{1,k}, r_{2,k}) \sim \mathrm{N}(\mathbf{0}, \mathbf{R}_k)$. *The resulting model then has the form*

$$d\mathbf{x} = \mathbf{F} \mathbf{x} \, dt + \mathbf{L} \, d\boldsymbol{\beta},$$
$$\mathbf{y}_k = \mathbf{H} \mathbf{x}(t_k) + \mathbf{r}_k. \qquad (10.18)$$

The models in Equations (10.16) and (10.18) are canonical examples of state-space models, where our interest is to infer the state (the position and velocity of the car) from the noisy measurements $\mathbf{z}(t)$ or $\mathbf{y}_k$. In a statistical sense, this corresponds to computation of the conditional distribution or the posterior distribution of the state $\mathbf{x}(t)$ given the measurements. Given the conditional distribution, we can, for example, compute the conditional mean of the state, which is also its minimum mean squared estimate, as well as its covariance, which measures the accuracy of the estimate.

We can now consider different kinds of statistical estimation problems on the model (for the discrete-time analogues, see Särkkä, 2013):

Batch estimation problem: Reconstruct the whole state trajectory, say, $\mathcal{X}_{t_T} = \{\mathbf{x}(\tau) \mid 0 \leq \tau \leq t_T\}$, from the measurements $\mathcal{Z}_{t_T} = \{\mathbf{z}(\tau) \mid 0 \leq \tau \leq t_T\}$ or in the continuous-discrete case from $\mathbf{y}_{1:T} = \{\mathbf{y}_1, \mathbf{y}_2, \ldots, \mathbf{y}_T\}$. The solution involves computing the posterior distribution $p(\mathcal{X}_{t_T} \mid \mathcal{Z}_{t_T})$ or $p(\mathcal{X}_{t_T} \mid \mathbf{y}_{1:T})$, which turns out to be intractable in general. Therefore, it is convenient to consider the so-called filtering and smoothing problems that follow. In the previous car example, the batch estimate corresponds to determining the joint conditional distribution of the whole state trajectory given all the measurements.

Filtering problem: Determine the state $\mathbf{x}(t)$ at time t given the history of measurements obtained so far, that is, $\mathcal{Z}_t = \{\mathbf{z}(\tau) \mid 0 \leq \tau \leq t\}$ or in continuous-discrete time given $\mathbf{y}_{1:k}$ such that $t_k \leq t$. The corresponding posterior distribution is called the filtering distribution, and it is the distribution $p(\mathbf{x}(t) \mid \mathcal{Z}_t)$ or $p(\mathbf{x}(t) \mid \mathbf{y}_{1:k})$. Strictly speaking, when $t > t_k$, the latter distribution is a predicted or prediction distribution, but we often call it simply filtering distribution as well. In the aforementioned car example, this corresponds to determining the conditional distribution of the car position and velocity given the history of measurements obtained so far. These kinds of estimates are useful in all kinds of target tracking applications where we are interested in the current state of the system given the measurements that we have already seen (without looking into the future).

Smoothing problem: Determine the states $\mathbf{x}(\tau)$ for each $\tau \in [0, t_T]$ given the history of measurements $\mathcal{Z}_{t_T} = \{\mathbf{z}(\tau) \mid 0 \leq \tau \leq t_T\}$ or in continuous-discrete time given $\mathbf{y}_{1:T}$. The posterior distributions to be computed are $p(\mathbf{x}(\tau) \mid \mathcal{Z}_{t_T})$ or $p(\mathbf{x}(\tau) \mid \mathbf{y}_{1:T})$ for all $\tau \in [0, t_T]$. The difference to the batch estimate is that we only compute the distributions for each τ separately and do not attempt to reconstruct the joint distribution over all times. In the car tracking example, this corresponds to determining the conditional distributions of the history of the states at each time during the measurement interval. These kinds of smoothing solutions are useful in refining the state estimates of tracking problems by also using information "from the future".

10.2 Batch Trajectory Estimates

Let us now take a brief look at batch estimation problem and see why it is computationally inconvenient. It is easiest to start by considering the continuous-discrete problem in Definition 10.2. If we denote the transition density of the SDE in Equation (10.12) as $p(\mathbf{x}(t_{k+1}) \mid \mathbf{x}(t_k))$, then we can form the distribution of the states at the measurement times $t_1, t_2, \ldots, t_T$ and the initial state t_0 as follows:

$$p(\mathbf{x}(t_0), \ldots, \mathbf{x}(t_T)) = p(\mathbf{x}_0) \prod_{k=0}^{T-1} p(\mathbf{x}(t_{k+1}) \mid \mathbf{x}(t_k)). \tag{10.19}$$

Similarly, we could form any other finite-dimensional distribution of the process, but let us keep with the measurement times for notational simplicity. Due to conditional independence of the measurements, we can write

$$p(\mathbf{y}_1, \ldots, \mathbf{y}_T \mid \mathbf{x}(t_0), \ldots, \mathbf{x}(t_T)) = \prod_{k=1}^{T} p(\mathbf{y}_k \mid \mathbf{x}(t_k)). \tag{10.20}$$

In principle, for a given T we could simply compute the posterior distribution of the states by Bayes' rule:

$$
\begin{aligned}
&p(\mathbf{x}(t_0), \ldots, \mathbf{x}(t_T) \mid \mathbf{y}_1, \ldots, \mathbf{y}_T) \\
&= \frac{p(\mathbf{y}_1, \ldots, \mathbf{y}_T \mid \mathbf{x}(t_0), \ldots, \mathbf{x}(t_T)) \, p(\mathbf{x}(t_0), \ldots, \mathbf{x}(t_T))}{p(\mathbf{y}_1, \ldots, \mathbf{y}_T)} \\
&\propto p(\mathbf{y}_1, \ldots, \mathbf{y}_T \mid \mathbf{x}(t_0), \ldots, \mathbf{x}(t_T)) \, p(\mathbf{x}(t_0), \ldots, \mathbf{x}(t_T)). \tag{10.21}
\end{aligned}
$$

This problem is now equivalent to a discrete-time batch state estimation problem whose shortcomings are discussed, for example, in Särkkä (2013). The challenge with this formulation is that the computational complexity scales at least cubically with the number of measurements, and another problem is that there is no way to include more measurements to the posterior distribution without recomputing the whole state posterior distribution.

Thus even with a finite number of time points, as in the preceding discrete-time formulation, the explicit batch estimation problem quickly becomes computationally heavy and even intractable. In the current case, the problem is even worse – we have the whole state trajectory $\mathcal{X}_{t_T}$ to estimate, not only some of its finite-dimensional distributions.

Another problem in the batch formulation is that the joint distribution $p(\mathcal{X}_{t_T})$ is not as simply defined as in the finite-dimensional case. The joint distribution is actually a measure in the same sense as the Wiener measure that we discussed in Section 7.2. In principle, Equation (10.19) does indeed define the measure provided that we let the time points be arbitrary, but this definition is not computationally very helpful.

It is, however, possible to compute maximum a posteriori (MAP) estimates of the whole state trajectory by using the Onsager–Machlup functional (Ikeda and Watanabe, 1981; Dutra et al., 2014), which has the form

$$J[\phi] = -\frac{1}{2} \int_0^T \left[\left(\frac{d\phi}{dt} - \mathbf{f}(\phi, t) \right)^{\mathsf{T}} \mathbf{W}^{-1} \left(\frac{d\phi}{dt} - \mathbf{f}(\phi, t) \right) \right.$$
$$\left. + \nabla \cdot \mathbf{f}(\phi, t) \right] dt, \quad (10.22)$$

where $\mathbf{W} = \mathbf{L} \mathbf{Q} \mathbf{L}^{\mathsf{T}}$ is assumed to be invertible and independent of the state and time. The functional can be extended to noninvertible and time-dependent cases, but here we use the restricted version of a notational convenience.

The interpretation of this functional is that it is the limit of log-probabilities of ε-tubes around ϕ. That is, if we define

$$B_\phi^\varepsilon = \left\{ \sup_{t \in [0,T]} \|\mathbf{x}(t) - \phi(t)\| \le \varepsilon \right\}, \quad (10.23)$$

then we have

$$\lim_{\varepsilon \downarrow 0} \frac{P(B_\phi^\varepsilon)}{P(B_\varphi^\varepsilon)} = \frac{\exp(J[\phi]) \, p(\phi(0))}{\exp(J[\varphi]) \, p(\varphi(0))}, \quad (10.24)$$

where $p(\varphi(0))$ denotes the distribution $p(\mathbf{x}(0))$ evaluated at $\mathbf{x}(0) = \varphi(0)$.

It turns out (Dutra et al., 2014) that we can combine the measurement model to the functional and compute the MAP path for a continuous-discrete model as shown in Algorithm 10.4.

Algorithm 10.4 (Continuous-discrete MAP path estimate). *The continuous-discrete MAP path can be computed by maximizing the* *functional*

$$H[\phi] = J[\phi] + \log p(\phi(0)) + \sum_{k=1}^{K} \log p(\mathbf{y}_k \mid \phi(t_k))$$

$$= -\frac{1}{2} \int_0^T \left[\left(\frac{d\phi}{dt} - \mathbf{f}(\phi, t) \right)^\mathsf{T} \mathbf{W}^{-1} \left(\frac{d\phi}{dt} - \mathbf{f}(\phi, t) \right) + \nabla \cdot \mathbf{f}(\phi, t) \right] dt$$

$$+ \log p(\phi(0)) + \sum_{k=1}^{K} \log p(\mathbf{y}_k \mid \phi(t_k)). \quad (10.25)$$

The Onsager–Machlup functional is also closely related to negative energy functional, which is given as

$$J_e[\phi] = -\frac{1}{2} \int_0^T \left[\left(\frac{d\phi}{dt} - \mathbf{f}(\phi, t) \right)^\mathsf{T} \mathbf{W}^{-1} \left(\frac{d\phi}{dt} - \mathbf{f}(\phi, t) \right) \right] dt. \quad (10.26)$$

The difference is thus the lack of the divergence term $\nabla \cdot \mathbf{f}(\phi, t)$. In classical state-estimation literature (Jazwinski, 1970), this seems to be a more common formulation, although it corresponds to the MAP estimate of the noise instead of the MAP estimate of the path (Dutra et al., 2014). The minimum energy path can indeed be solved by replacing $J[\phi]$ in the preceding algorithm with $J_e[\phi]$.

The continuous-time measurement model version of the MAP path estimator has been discussed, for example, in Aihara and Bagchi (1999) and Zeitouni (1989). Continuous-time minimum energy functionals and probability density functionals are discussed in Jazwinski (1970). However, it appears that a continuous-time version of the MAP estimate takes the form shown in Algorithm 10.5.

Algorithm 10.5 (Continuous-time MAP path estimate). *The continuous-time MAP path can be computed by maximizing the following functional:*

$$
\begin{aligned}
H[\boldsymbol{\phi}] &= J[\boldsymbol{\phi}] - \frac{1}{2} \int_0^T \mathbf{h}^\mathsf{T}(\boldsymbol{\phi}(t), t) \, \mathbf{R}^{-1} \, \mathbf{h}(\boldsymbol{\phi}(t), t) \, dt \\
&\quad + \int_0^T \mathbf{h}^\mathsf{T}(\boldsymbol{\phi}(t), t) \, \mathbf{R}^{-1} \, d\mathbf{z} \\
&= -\frac{1}{2} \int_0^T \left[\left(\frac{d\boldsymbol{\phi}}{dt} - \mathbf{f}(\boldsymbol{\phi}, t) \right)^\mathsf{T} \mathbf{W}^{-1} \left(\frac{d\boldsymbol{\phi}}{dt} - \mathbf{f}(\boldsymbol{\phi}, t) \right) + \nabla \cdot \mathbf{f}(\boldsymbol{\phi}, t) \right] dt \\
&\quad - \frac{1}{2} \int_0^T \mathbf{h}^\mathsf{T}(\boldsymbol{\phi}(t), t) \, \mathbf{R}^{-1} \, \mathbf{h}(\boldsymbol{\phi}(t), t) \, dt + \int_0^T \mathbf{h}^\mathsf{T}(\boldsymbol{\phi}(t), t) \, \mathbf{R}^{-1} \, d\mathbf{z}.
\end{aligned}
\tag{10.27}
$$

10.3 Kushner–Stratonovich and Zakai Equations

In this section, we present the classical continuous-time filtering equations that provide the formal (Bayesian) solution to the filtering problem. Recall that in Section 5.3 we concluded that the Fokker–Planck–Kolmogorov equation (5.11) can be compactly written as follows:

$$
\frac{\partial p}{\partial t} = \mathcal{A}^* p,
\tag{10.28}
$$

where $\mathcal{A}^*$ is the operator defined in Equation (5.36).

The continuous-time optimal filtering equation, which computes $p(\mathbf{x}(t) \mid \mathcal{Z}_t)$, is called the Kushner–Stratonovich (KS) equation (Kushner, 1964; Bucy, 1965) and can be derived as the continuous-time limits of the so-called Bayesian filtering equations (see, e.g., Särkkä, 2013). A Stratonovich calculus version of the equation was studied by Stratonovich already in the late 1950's (cf. Stratonovich, 1968).

Algorithm 10.6 (Kushner–Stratonovich equation). *The stochastic partial differential equation for the filtering density $p(\mathbf{x}, t \mid \mathcal{Z}_t) \triangleq p(\mathbf{x}(t) \mid \mathcal{Z}_t)$ is*

$$
\begin{aligned}
dp(\mathbf{x}, t \mid \mathcal{Z}_t) &= \mathcal{A}^* \, p(\mathbf{x}, t \mid \mathcal{Z}_t) \, dt \\
&\quad + (\mathbf{h}(\mathbf{x}, t) - \mathrm{E}[\mathbf{h}(\mathbf{x}, t) \mid \mathcal{Z}_t])^\mathsf{T} \, \mathbf{R}^{-1} \\
&\quad \times (d\mathbf{z} - \mathrm{E}[\mathbf{h}(\mathbf{x}, t) \mid \mathcal{Z}_t] \, dt) \, p(\mathbf{x}, t \mid \mathcal{Z}_t), \quad (10.29)
\end{aligned}
$$

where $\mathrm{d}p(\mathbf{x}, t \mid \mathcal{Z}_t) = p(\mathbf{x}, t + \mathrm{d}t \mid \mathcal{Z}_{t+\mathrm{d}t}) - p(\mathbf{x}, t \mid \mathcal{Z}_t)$ *and*

$$E[\mathbf{h}(\mathbf{x}, t) \mid \mathcal{Z}_t] = \int \mathbf{h}(\mathbf{x}, t) \, p(\mathbf{x}, t \mid \mathcal{Z}_t) \, \mathrm{d}\mathbf{x}. \tag{10.30}$$

This equation is only formal in the sense that as such it is quite impossible to work with. However, it is possible derive moment equations from it, as well as form approximations to the solutions. What makes the equation difficult is that it is a stochastic partial differential equation – recall that the operator $\mathcal{A}^*$ contains partial derivatives. Furthermore, the equation is nonlinear, as could be seen by expanding the expectation integrals in the equation (recall that they are integrals over $p(\mathbf{x}, t \mid \mathcal{Z}_t)$). The stochasticity is generated by the observation process $\mathbf{z}(t)$.

The nonlinearity in the KS equation can be eliminated by deriving an equation for an unnormalized filtering distribution instead of the normalized one. This leads to the so-called Zakai equation (Zakai, 1969).

Algorithm 10.7 (Zakai equation). *Let* $q(\mathbf{x}, t \mid \mathcal{Z}_t) \triangleq q(\mathbf{x}(t) \mid \mathcal{Z}_t)$ *be the solution to Zakai's stochastic partial differential equation*

$$\mathrm{d}q(\mathbf{x}, t \mid \mathcal{Z}_t) = \mathcal{A}^* \, q(\mathbf{x}, t \mid \mathcal{Z}_t) \, \mathrm{d}t + \mathbf{h}^\mathsf{T}(\mathbf{x}, t) \, \mathbf{R}^{-1} \, \mathrm{d}\mathbf{z} \, q(\mathbf{x}, t \mid \mathcal{Z}_t), \tag{10.31}$$

where $\mathrm{d}q(\mathbf{x}, t \mid \mathcal{Z}_t) = q(\mathbf{x}, t + \mathrm{d}t \mid \mathcal{Z}_{t+\mathrm{d}t}) - q(\mathbf{x}, t \mid \mathcal{Z}_t)$ *and* $\mathcal{A}^*$ *is the Fokker–Planck–Kolmogorov operator defined in Equation (5.36). Then we have the following:*

$$p(\mathbf{x}(t) \mid \mathcal{Z}_t) = \frac{q(\mathbf{x}(t) \mid \mathcal{Z}_t)}{\int q(\mathbf{x}(t) \mid \mathcal{Z}_t) \, \mathrm{d}\mathbf{x}(t)}. \tag{10.32}$$

Closed-form solutions to the KS and Zakai equations can only be found in isolated special cases. In Section 10.4, we consider the solution of the linear filtering problem, which is called the Kalman–Bucy filter. However, there are some non-Gaussian closed-form solutions as well, and one of them is given in the following example.

Example 10.8 (Beneš filter). *The Beneš filter (Beneš, 1981) is the exact solution to the Kushner–Stratonovich and Zakai equations for the following model:*

$$\mathrm{d}x = \tanh(x) \, \mathrm{d}t + \mathrm{d}\beta, \tag{10.33}$$

$$\mathrm{d}z = x \, \mathrm{d}t + \mathrm{d}\eta, \tag{10.34}$$

where $\beta(t)$ *and* $\eta(t)$ *are standard Brownian motions,* $x(0) = 0$, *and*

$z(0) = 0$. *The equations for the sufficient statistics $\mu(t)$ and $P(t)$ of the posterior distribution are*

$$d\mu(t) = P(t)\,dz - P(t)\,\mu(t)\,dt,$$
$$\frac{dP(t)}{dt} = 1 - P^2(t), \tag{10.35}$$

with $\mu(0) = 0$ and $P(0) = 0$. The posterior distribution is then of the form

$$p(x(t) \mid \mathcal{Z}_t) \propto \cosh(x(t))\,\exp\left\{-\frac{(x(t) - \mu(t))^2}{2P(t)}\right\}. \tag{10.36}$$

10.4 Linear and Extended Kalman–Bucy Filtering

The car model in Example 10.3 was actually a linear Gaussian filtering problem, which refers to a problem where the functions $\mathbf{f}$ and $\mathbf{h}$ are linear in $\mathbf{x}$. In that case, the filtering solution is Gaussian, and we can solve the filtering equations in closed form. The general filtering solution to all continuous-time linear Gaussian problems can be computed in closed form as well, and it is given in the following.

The *Kalman–Bucy filter* (Kalman and Bucy, 1961) is the exact solution to the continuous-time linear Gaussian filtering problem

$$d\mathbf{x} = \mathbf{F}(t)\,\mathbf{x}\,dt + \mathbf{L}(t)\,d\boldsymbol{\beta},$$
$$d\mathbf{z} = \mathbf{H}(t)\,\mathbf{x}\,dt + d\boldsymbol{\eta}, \tag{10.37}$$

where $\mathbf{x}(t) \in \mathbb{R}^D$ is the state process, $\mathbf{z}(t) \in \mathbb{R}^M$ is the (integrated) measurement process; $\mathbf{F}(t)$ is the dynamic model matrix; $\mathbf{H}(t)$ is the measurement model matrix; $\mathbf{L}(t)$ is an arbitrary time-varying matrix, independent of $\mathbf{x}(t)$ and $\mathbf{y}(t)$; and $\boldsymbol{\beta}(t) \in \mathbb{R}^S$ and $\boldsymbol{\eta}(t) \in \mathbb{R}^M$ are independent Brownian motions with diffusion matrices $\mathbf{Q}$ and $\mathbf{R}$, respectively. The solution is given as follows.

Algorithm 10.9 (Kalman–Bucy filter). *The Bayesian filter, which computes the posterior distribution $p(\mathbf{x}(t) \mid \mathcal{Z}_t) = \mathrm{N}(\mathbf{x}(t) \mid \mathbf{m}(t), \mathbf{P}(t))$ for the system (10.37), is as follows:*

$$\mathbf{K} = \mathbf{P}\,\mathbf{H}^\mathsf{T}(t)\,\mathbf{R}^{-1},$$
$$d\mathbf{m} = \mathbf{F}(t)\,\mathbf{m}\,dt + \mathbf{K}\,[d\mathbf{z} - \mathbf{H}(t)\,\mathbf{m}\,dt], \tag{10.38}$$
$$\frac{d\mathbf{P}}{dt} = \mathbf{F}(t)\,\mathbf{P} + \mathbf{P}\,\mathbf{F}^\mathsf{T}(t) + \mathbf{L}(t)\,\mathbf{Q}\,\mathbf{L}^\mathsf{T}(t) - \mathbf{K}\,\mathbf{R}\,\mathbf{K}^\mathsf{T}.$$

An example of a linear Gaussian model and the corresponding Kalman–Bucy filter is given in Example 10.10.

Example 10.10 (Kalman–Bucy filter for the Ornstein–Uhlenbeck model).
Let us consider the following continuous-time state estimation problem:

$$\begin{aligned} dx &= -\lambda\, x\, dt + d\beta, \\ dz &= x\, dt + d\eta, \end{aligned} \tag{10.39}$$

where $\lambda > 0$, and $\beta(t)$ and $\eta(t)$ are independent Brownian motions with diffusion coefficients $q > 0$ and $r > 0$, respectively. The model thus corresponds to an Ornstein–Uhlenbeck process, whose measurements are corrupted with Gaussian noise.

The Kalman–Bucy filter for computing the mean $m(t)$ and variance $P(t)$ such that $p(x(t) \mid Z_t) = N(x(t) \mid m(t), P(t))$ is the following:

$$\begin{aligned} dm &= -\lambda\, m\, dt + \frac{P}{r}\,(dz - m\, dt), \\ \frac{dP}{dt} &= -2\lambda\, P + q - \frac{P^2}{r}. \end{aligned} \tag{10.40}$$

The Kalman–Bucy filter also provides a useful basis for approximating (intractable) solutions to nonlinear filtering problems. There exists various approximation methods to cope with nonlinear models, for example, based on Monte Carlo approximations, series expansions of processes and densities, Gaussian (process) approximations, and many others (see, e.g., Crisan and Rozovskiĭ, 2011). Still, the most common approach is to use Gaussian approximations outlined in the beginning of Chapter 9 in order to transform the nonlinear filtering problem into an approximate Kalman–Bucy filtering problem. The classical filtering theory is very much based on this idea, and a typical approach is to use Taylor series expansions of the drift function (Jazwinski, 1970). The use of Gaussian sigma-point types of approximations in this context has been studied in Särkkä (2007) and Särkkä and Sarmavuori (2013). In this section, we only outline Gaussian approximation–based approximate filtering; for other methods, the reader is referred to Crisan and Rozovskiĭ (2011).

The extended Kalman–Bucy filter (see, e.g., Gelb, 1974) is perhaps the most common and the simplest possible extension of the Kalman–Bucy filter to nonlinear models of the form (10.7). It can be derived by using first-order Taylor series expansions on the functions **f** and **h** around the current mean estimate.

Algorithm 10.11 (Extended Kalman–Bucy filter). *The equations of the* *extended Kalman–Bucy filter (EKBF) are as follows:*

$$\mathbf{K} = \mathbf{P}\,\mathbf{H}_{\mathbf{x}}^{\mathsf{T}}(\mathbf{m}, t)\,\mathbf{R}^{-1},$$
$$\mathrm{d}\mathbf{m} = \mathbf{f}(\mathbf{m}, t)\,\mathrm{d}t + \mathbf{K}\,[\mathrm{d}\mathbf{z} - \mathbf{h}(\mathbf{m}, t)\,\mathrm{d}t],$$
$$\frac{\mathrm{d}\mathbf{P}}{\mathrm{d}t} = \mathbf{F}_{\mathbf{x}}(\mathbf{m}, t)\,\mathbf{P} + \mathbf{P}\,\mathbf{F}_{\mathbf{x}}^{\mathsf{T}}(\mathbf{m}, t) \qquad (10.41)$$
$$+\,\mathbf{L}(\mathbf{m}, t)\,\mathbf{Q}\,\mathbf{L}^{\mathsf{T}}(\mathbf{m}, t) - \mathbf{K}\,\mathbf{R}\,\mathbf{K}^{\mathsf{T}},$$

where $\mathbf{F}_{\mathbf{x}}$ is the Jacobian matrix of $\mathbf{f}$ with elements $[\mathbf{F}_{\mathbf{x}}]_{ij} = \partial f_i / \partial x_j$, and $\mathbf{H}_{\mathbf{x}}$ is the Jacobian matrix of $\mathbf{h}$ with elements $[\mathbf{H}_{\mathbf{x}}]_{ij} = \partial h_i / \partial x_j$.

Example 10.12 describes an EKBF.

Example 10.12 (Extended Kalman–Bucy filter). *The extended Kalman–Bucy filter for the Beneš filtering problem in Example 10.8 is the following:*

$$\mathrm{d}m = \tanh(m)\,\mathrm{d}t + P\,(\mathrm{d}z - m\,\mathrm{d}t),$$
$$\frac{\mathrm{d}P}{\mathrm{d}t} = 2\,(1 - \tanh^2(m))\,P + 1 - P^2. \qquad (10.42)$$

It is now easy to see that in Algorithm 10.11 we have actually employed the linearization approximation from Algorithm 9.4. Taking a step backward lets us now use Algorithm 9.1 to formulate the following general Gaussian approximation to the nonlinear filtering problem (see, e.g., Särkkä and Sarmavuori, 2013).

Algorithm 10.13 (Continuous-time Gaussian assumed density filter). *The equations of the continuous-time Gaussian assumed density filter are as follows:*

$$\mathbf{K} = \mathrm{E}_{\mathrm{N}}[(\mathbf{x} - \mathbf{m})\,\mathbf{h}^{\mathsf{T}}(\mathbf{x}(t), t)]\,\mathbf{R}^{-1}, \qquad (10.43)$$
$$\mathrm{d}\mathbf{m} = \mathrm{E}_{\mathrm{N}}[\mathbf{f}(\mathbf{x}, t)]\,\mathrm{d}t + \mathbf{K}\,(\mathrm{d}\mathbf{z} - \mathrm{E}_{\mathrm{N}}[\mathbf{h}(\mathbf{x}, t)]\,\mathrm{d}t), \qquad (10.44)$$
$$\frac{\mathrm{d}\mathbf{P}}{\mathrm{d}t} = \mathrm{E}_{\mathrm{N}}[(\mathbf{x} - \mathbf{m})\,\mathbf{f}^{\mathsf{T}}(\mathbf{x}, t)] + \mathrm{E}_{\mathrm{N}}[\mathbf{f}(\mathbf{x}, t)\,(\mathbf{x} - \mathbf{m})^{\mathsf{T}}]$$
$$+\,\mathrm{E}_{\mathrm{N}}[\mathbf{L}(\mathbf{x}, t)\,\mathbf{Q}\,\mathbf{L}^{\mathsf{T}}(\mathbf{x}, t)] - \mathbf{K}\,\mathbf{R}\,\mathbf{K}^{\mathsf{T}}, \qquad (10.45)$$

where the expectations are taken with respect to $\mathbf{x}(t) \sim \mathrm{N}(\mathbf{m}(t), \mathbf{P}(t))$.

Various sigma-point approximations to the continuous-time filtering problem can now be generated by replacing the preceding Gaussian expectations with sigma-point approximations analogously to Algorithm 9.5. The generic form resulting from approximating Equations (10.45) is described in Algorithm 10.14.

Algorithm 10.14 (Continuous-time sigma-point filter). *The equations of a generic continuous-time sigma-point filter are as follows:*

$$\mathbf{K} = \sum_i W^{(i)} \sqrt{\mathbf{P}}\, \boldsymbol{\xi}_i\, \mathbf{h}^\mathsf{T}(\mathbf{m} + \sqrt{\mathbf{P}}\, \boldsymbol{\xi}_i, t)\, \mathbf{R}^{-1}, \tag{10.46}$$

$$\mathrm{d}\mathbf{m} = \sum_i W^{(i)} \mathbf{f}(\mathbf{m} + \sqrt{\mathbf{P}}\, \boldsymbol{\xi}_i, t)\, \mathrm{d}t$$

$$+ \mathbf{K}\left(\mathrm{d}\mathbf{z} - \sum_i W^{(i)} \mathbf{h}(\mathbf{m} + \sqrt{\mathbf{P}}\, \boldsymbol{\xi}_i, t)\, \mathrm{d}t \right), \tag{10.47}$$

$$\frac{\mathrm{d}\mathbf{P}}{\mathrm{d}t} = \sum_i W^{(i)} \mathbf{f}(\mathbf{m} + \sqrt{\mathbf{P}}\, \boldsymbol{\xi}_i, t)\, \boldsymbol{\xi}_i^\mathsf{T}\, \sqrt{\mathbf{P}}^\mathsf{T}$$

$$+ \sum_i W^{(i)} \sqrt{\mathbf{P}}\, \boldsymbol{\xi}_i\, \mathbf{f}^\mathsf{T}(\mathbf{m} + \sqrt{\mathbf{P}}\, \boldsymbol{\xi}_i, t)$$

$$+ \sum_i W^{(i)} \mathbf{L}(\mathbf{m} + \sqrt{\mathbf{P}}\, \boldsymbol{\xi}_i, t)\, \mathbf{Q}\, \mathbf{L}^\mathsf{T}(\mathbf{m} + \sqrt{\mathbf{P}}\, \boldsymbol{\xi}_i, t)$$

$$- \mathbf{K}\, \mathbf{R}\, \mathbf{K}^\mathsf{T}. \tag{10.48}$$

For details on selection of sigma points $\boldsymbol{\xi}_i$ and weights $W^{(i)}$, see Section 9.1. For example, by selecting the unscented transform sigma points and weights we get the unscented Kalman–Bucy filter (Särkkä, 2007).

10.5 Continuous-Discrete Bayesian Filtering Equations

In applications involving digital computers and computer controlled sensors, we do not usually obtain measurements in continuous time, but we are only able to get samples from the underlying process at discrete instants of time. For these kinds of models, the continuous-discrete formulation of the problem is more appropriate. A general continuous-discrete filtering problem was formulated in Definition 10.2.

The filtering problem is now to determine the distributions

$$p(\mathbf{x}(t_k) \mid \mathbf{y}_{1:k}), \tag{10.49}$$

which are thus the posterior distributions of the states at the measurements times t_k given the measurements obtained so far. A bit more generally, we might be interested in determining the probability densities

$$p(\mathbf{x}(t) \mid \mathbf{y}_{1:k}), \qquad t \in [t_k, t_{k+1}), \tag{10.50}$$

which also give the distributions of the state between the last and the next measurement. Although these distributions are often called predicted distributions (e.g., Särkkä, 2013), as we already discussed earlier in this chapter, we often call them filtering distributions as well to keep the terminology consistent with the continuous-time case.

A conceptually simple way of dealing with the continuous-discrete-time filtering problem is to solve the transition densities $p(\mathbf{x}(t_{k+1}) \mid \mathbf{x}(t_k))$ from the Fokker–Planck–Kolmogorov forward partial differential equation (see Section 5.4, Theorem 5.10). The filtering problem then takes the form

$$\begin{aligned}
\mathbf{x}(t_{k+1}) &\sim p(\mathbf{x}(t_{k+1}) \mid \mathbf{x}(t_k)), \\
\mathbf{y}_k &\sim p(\mathbf{y}_k \mid \mathbf{x}(t_k)),
\end{aligned} \tag{10.51}$$

which is a canonical discrete-time filtering problem (Särkkä, 2013) – provided that we introduce the notation $\mathbf{x}_k \triangleq \mathbf{x}(t_k)$.

The filtering distributions at times $t_1, t_2, \ldots$ can now be computed by starting from a prior distribution $p(\mathbf{x}(t_0))$ and by using the following Bayesian filter recursions (see Särkkä, 2013).

Algorithm 10.15 (Bayesian filter). *The Bayesian filter recursion starts from the prior distribution $p(\mathbf{x}(t_0))$. Then for each $k = 1, 2, \ldots$ we do the following:*

- Prediction step: *The predictive distribution of the state $\mathbf{x}(t_k)$ at the time t_k, given the dynamic model, can be computed by the Chapman–Kolmogorov equation:*

$$p(\mathbf{x}(t_k) \mid \mathbf{y}_{1:k-1}) = \\ \int p(\mathbf{x}(t_k) \mid \mathbf{x}(t_{k-1})) \, p(\mathbf{x}(t_{k-1}) \mid \mathbf{y}_{1:k-1}) \, d\mathbf{x}(t_{k-1}). \tag{10.52}$$

- Update step: *Given the measurement $\mathbf{y}_k$ at time t_k, the posterior distribution of the state $\mathbf{x}(t_k)$ can be computed by Bayes' rule*

$$p(\mathbf{x}(t_k) \mid \mathbf{y}_{1:k}) = \frac{1}{Z_k} p(\mathbf{y}_k \mid \mathbf{x}(t_k)) \, p(\mathbf{x}(t_k) \mid \mathbf{y}_{1:k-1}), \tag{10.53}$$

where the normalization constant Z_k *is given by the following:*

$$Z_k = \int p(\mathbf{y}_k \mid \mathbf{x}(t_k)) \, p(\mathbf{x}(t_k) \mid \mathbf{y}_{1:k-1}) \, d\mathbf{x}(t_k). \tag{10.54}$$

Note that given $p(\mathbf{x}(t_k) \mid \mathbf{y}_{1:k})$, we can then reapply the Chapman–Kolmogorov equation to get the predicted (or filtering) distribution

$p(\mathbf{x}(t) \mid \mathbf{y}_{1:k})$ for $t > t_k$ as follows:

$$p(\mathbf{x}(t) \mid \mathbf{y}_{1:k}) = \int p(\mathbf{x}(t) \mid \mathbf{x}(t_k))\, p(\mathbf{x}(t_k) \mid \mathbf{y}_{1:k})\, d\mathbf{x}(t_k). \quad (10.55)$$

It is also useful to notice that we can cope with the case of a missing measurement just by skipping the update step.

The equations in Algorithm 10.15 are exactly the discrete-time Bayesian filtering equations, and thus all the methods in Särkkä (2013) are directly applicable. However, to use these equations we need to compute or approximate the transition density of the SDE. This can be done using the various numerical methods introduced in Chapter 9.

Instead of first forming the equivalent discrete-time system, it is also possible to derive a Bayesian filter directly for the continuous-discrete filtering problem (10.12). The result is the following algorithm.

Algorithm 10.16 (Continuous-discrete Bayesian filter). *The continuous-discrete Bayesian filter recursion starts from the prior distribution* $p(\mathbf{x}(t_0))$. *Then for each* $k = 1, 2, \ldots$ *we do the following:*

- Prediction step: *Solve the predicted probability density at time of the measurement* $p(\mathbf{x}(t_k) \mid \mathbf{y}_{1:k-1})$ *by integrating the Fokker–Planck–Kolmogorov equation (see Theorem 5.4) from the filtering density at the previous measurement time step* t_{k-1} *to the current time* t_k:

$$\frac{\partial p(\mathbf{x}, t)}{\partial t} = -\sum_i \frac{\partial}{\partial x_i}[f_i(\mathbf{x}, t)\, p(\mathbf{x}, t)]$$

$$+ \frac{1}{2}\sum_{i,j} \frac{\partial^2}{\partial x_i\, \partial x_j}\left\{[\mathbf{L}(\mathbf{x}, t)\, \mathbf{Q}\, \mathbf{L}^\mathsf{T}(\mathbf{x}, t)]_{ij}\, p(\mathbf{x}, t)\right\}, \quad (10.56)$$

where we have denoted $p(\mathbf{x}, t) \triangleq p(\mathbf{x}(t) \mid \mathbf{y}_{1:k-1})$ *and the initial condition is* $p(\mathbf{x}, t_{k-1}) \triangleq p(\mathbf{x}(t_{k-1}) \mid \mathbf{y}_{1:k-1})$.
- Update step: *Use Bayes' rule (10.53) for calculating the conditional density* $p(\mathbf{x}(t_k) \mid \mathbf{y}_{1:k})$, *given the new measurement* $\mathbf{y}_k$.

Again, if we are interested in solving $p(\mathbf{x}(t) \mid \mathbf{y}_{1:k})$ for $t > t_k$ given $p(\mathbf{x}(t_k) \mid \mathbf{y}_{1:k})$, we can reapply the prediction step for integrating the solution from t_k up to t.

As in the continuous-time case, the Bayesian filtering equations have closed-form solutions only in isolated special cases. Before going to the linear Gaussian special case, let us consider the following nonlinear filtering problem, which is a continuous-discrete version of the Beneš filter in Example 10.8.

Example 10.17 (Beneš–Daum filter). *By Equation (7.62) in Example 7.10, the transition density of the Beneš SDE*

$$\mathrm{d}x = \tanh(x)\,\mathrm{d}t + \mathrm{d}\beta \tag{10.57}$$

is

$$p(x(t_k) \mid x(t_{k-1})) = \frac{1}{\sqrt{2\pi\Delta t_{k-1}}} \frac{\cosh(x(t_k))}{\cosh(x(t_{k-1}))} \exp\left(-\frac{1}{2}\Delta t_{k-1}\right)$$
$$\times \exp\left(-\frac{1}{2\Delta t_{k-1}}(x(t_k) - x(t_{k-1}))^2\right), \tag{10.58}$$

where $\Delta t_{k-1} = t_k - t_{k-1}$. This is also the solution to the associated FPK equation. Assume that the measurements are obtained from the model

$$p(y_k \mid x(t_k)) = \mathrm{N}(y_k \mid x(t_k), \sigma^2). \tag{10.59}$$

If we assume that the filtering solution at time t_{k-1} is of the form

$$p(x(t_{k-1}) \mid y_{1:k-1}) \propto \cosh(x(t_{k-1})) \exp\left(-\frac{1}{2P_{k-1}}(x(t_{k-1}) - m_{k-1})^2\right), \tag{10.60}$$

for some known m_{k-1}, P_{k-1}, then the Chapman–Kolmogorov equation gives

$$p(x(t_k) \mid y_{1:k-1}) \propto \cosh(x(t_k)) \exp\left(-\frac{1}{2P_k^-}(x(t_k) - m_k^-)^2\right), \tag{10.61}$$

where

$$\begin{aligned} m_k^- &= m_{k-1}, \\ P_k^- &= P_{k-1} + \Delta t_{k-1}. \end{aligned} \tag{10.62}$$

The Bayes' rule gives

$$p(x(t_k) \mid y_{1:k}) \propto \cosh(x(t_k)) \exp\left(-\frac{1}{2P_k}(x(t_k) - m_k)^2\right), \tag{10.63}$$

where the equations for the parameters (sufficient statistics) can be written

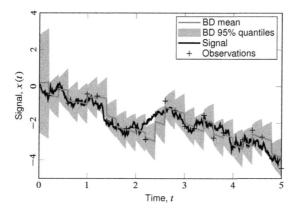

Figure 10.2 Illustration of the result of the Beneš–Daum filter in Example 10.17. The evolution of the predicted distribution between the observation is shown as well. The characteristic discontinuous nature of a continuous discrete filter can be seen here quite well – the uncertainty in predictions grows until the next observation and conditioning of observation with Bayes' rule cause the filtering distribution to jump.

as follows:

$$m_k = m_k^- + \left(\frac{P_k^-}{P_k^- + \sigma^2}\right)[y_k - m_k^-],$$

$$P_k = P_k^- - \frac{(P_k^-)^2}{P_k^- + \sigma^2}.$$

(10.64)

This result can be easily obtained by completing the squares in the exponent. These prediction and update equations of the Beneš–Daum filter (Daum, 1984) are functionally exactly the same as the equations for a discretely observed Brownian motion, but now the probability density is given by the Equation (10.63). Particularly, the conditional mean and variance are given as

$$\mathrm{E}[x(t_k) \mid y_{1:k}] = m_k + P_k \tanh(m_k),$$

$$\mathrm{Var}[x(t_k) \mid y_{1:k}] = P_k + (1 - \tanh^2(m_k)) P_k^2.$$

(10.65)

Figure 10.2 shows an example of the filtering result. The predicted distributions between the measurements have been obtained by replacing Δt_k in Equation (10.62) with a value $\Delta t_\star$ running over the range $[0, \Delta t_k]$.

10.6 Kalman Filtering

As in the continuous-time case the solution to the continuous-discrete linear Gaussian filtering problem is solvable in closed form. A linear Gaussian special case of the continuous-discrete state-space model in Definition 10.2 has the general form

$$\begin{aligned} d\mathbf{x} &= \mathbf{F}(t)\,\mathbf{x}\,dt + \mathbf{L}(t)\,d\boldsymbol{\beta}, \\ \mathbf{y}_k &= \mathbf{H}_k\,\mathbf{x}(t_k) + \mathbf{r}_k, \end{aligned} \tag{10.66}$$

where $\mathbf{r}_k \sim \mathrm{N}(\mathbf{0}, \mathbf{R}_k)$ and $\boldsymbol{\beta}$ is a Brownian motion with diffusion matrix $\mathbf{Q}$. This model is of the type that we already encountered in Example 10.3. From Sections 6.1 and 6.2, we now deduce that the corresponding discrete-time model has the form

$$\begin{aligned} \mathbf{x}(t_{k+1}) &= \mathbf{A}_k\,\mathbf{x}(t_k) + \mathbf{q}_k, \\ \mathbf{y}_k &= \mathbf{H}_k\,\mathbf{x}(t_k) + \mathbf{r}_k, \end{aligned} \tag{10.67}$$

where $\mathbf{q}_k \sim \mathrm{N}(\mathbf{0}, \boldsymbol{\Sigma}_k)$ with $\mathbf{A}_k$ and $\boldsymbol{\Sigma}_k$ given by Equations (6.9) and (6.11), or in the linear time-invariant (LTI) case by Equations (6.24) and (6.25). This model is thus a discrete-time version of the linear Gaussian model analogously to Equation (10.51).

Assuming that $p(\mathbf{x}(t_0)) = \mathrm{N}(\mathbf{x}(t_0) \mid \mathbf{m}_0, \mathbf{P}_0)$, the filtering solution corresponding to Algorithm 10.15 is now given by the following Kalman filter (Kalman, 1960; Särkkä, 2013).

Algorithm 10.18 (Kalman filter). *The Kalman filter recursion is started from the prior mean* $\mathbf{m}_0$ *and covariance* $\mathbf{P}_0$. *For* $k = 1, 2, \ldots$, *we then perform the following steps:*

- Prediction step:

$$\begin{aligned} \mathbf{m}_k^- &= \mathbf{A}_{k-1}\,\mathbf{m}_{k-1}, \\ \mathbf{P}_k^- &= \mathbf{A}_{k-1}\,\mathbf{P}_{k-1}\,\mathbf{A}_{k-1}^\mathsf{T} + \boldsymbol{\Sigma}_{k-1}. \end{aligned} \tag{10.68}$$

- Update step:

$$\begin{aligned} \mathbf{v}_k &= \mathbf{y}_k - \mathbf{H}_k\,\mathbf{m}_k^-, \\ \mathbf{S}_k &= \mathbf{H}_k\,\mathbf{P}_k^-\,\mathbf{H}_k^\mathsf{T} + \mathbf{R}_k, \\ \mathbf{K}_k &= \mathbf{P}_k^-\,\mathbf{H}_k^\mathsf{T}\,\mathbf{S}_k^{-1}, \\ \mathbf{m}_k &= \mathbf{m}_k^- + \mathbf{K}_k\,\mathbf{v}_k, \\ \mathbf{P}_k &= \mathbf{P}_k^- - \mathbf{K}_k\,\mathbf{S}_k\,\mathbf{K}_k^\mathsf{T}. \end{aligned} \tag{10.69}$$

The Kalman filter gives the following distributions:

$$p(\mathbf{x}(t_k) \mid \mathbf{y}_1, \ldots, \mathbf{y}_{k-1}) = \mathrm{N}(\mathbf{x}(t_k) \mid \mathbf{m}_k^-, \mathbf{P}_k^-),$$
$$p(\mathbf{x}(t_k) \mid \mathbf{y}_1, \ldots, \mathbf{y}_k) = \mathrm{N}(\mathbf{x}(t_k) \mid \mathbf{m}_k, \mathbf{P}_k), \quad (10.70)$$
$$p(\mathbf{y}_k \mid \mathbf{y}_1, \ldots, \mathbf{y}_{k-1}) = \mathrm{N}(\mathbf{y}_k \mid \mathbf{H}_k \mathbf{m}_k^-, \mathbf{S}_k).$$

If we wish to predict the state from time t_k to some time $t = t_k + \Delta t$, then we just need to form the discretization matrices for the time step $t_k \to t_k + \Delta t$ and then perform the previous prediction step. Missing measurements can be handled just by skipping the update step. The Kalman filter equations are illustrated in a one-dimensional example presented as Example 10.19.

Example 10.19 (Kalman filter for the Ornstein–Uhlenbeck model). *Recall from Example 6.2 that the equivalent discretization of the Ornstein–Uhlenbeck model is*

$$x(t_{k+1}) = a_k \, x(t_k) + q_k,$$
$$y_k = x(t_k) + r_k, \quad (10.71)$$

with $q_k \sim \mathrm{N}(0, \Sigma_k)$, $r_k \sim \mathrm{N}(0, R)$, $a_k = \exp(-\lambda \, \Delta t_k)$, and $\Sigma_k = \frac{q}{2\lambda}[1 - \exp(-2\lambda \, \Delta t_k)]$. The Kalman filter for this model is then the following:

1. *Start from the initial mean m_0 and variance P_0.*
2. *For each $k = 1, 2, \ldots$, compute the following using the measurement sequence $y_1, y_2, \ldots$ as the input:*

$$m_k^- = \exp(-\lambda \, \Delta t_{k-1}) \, m_{k-1},$$
$$P_k^- = \exp(-2\lambda \, \Delta t_{k-1}) \, P_{k-1} + \frac{q}{2\lambda} [1 - \exp(-2\lambda \, \Delta t_{k-1})],$$
$$m_k = m_k^- + \left(\frac{P_k^-}{P_k^- + R} \right) (y_k - m_k^-), \quad (10.72)$$
$$P_k = P_k^- - \frac{[P_k^-]^2}{P_k^- + R}.$$

To obtain predicted distributions between the measurements, we can replace the measurement-time grid with a denser grid of times and perform only prediction steps (the first two equations) at the times where we have no measurement. Figure 10.3 shows the result of this kind of Kalman filter when run on simulated data.

It is also possible to form the Kalman filter corresponding to the continuous-discrete Bayesian filter formulation in Algorithm 10.16. The result is the following.

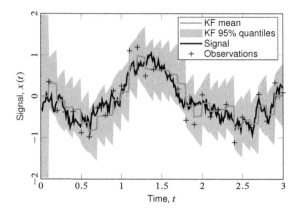

Figure 10.3 Results of the Kalman filters in Examples 10.19 and 10.21.

Algorithm 10.20 (Continuous-discrete Kalman filter). *The continuous-discrete Kalman filter recursion is started from the prior mean* $\mathbf{m}_0$ *and covariance* $\mathbf{P}_0$*. For* $k = 1, 2, \ldots$ *we then perform the following steps:*

- Prediction step: *The differential equations*

$$
\begin{aligned}
\frac{d\mathbf{m}}{dt} &= \mathbf{F}(t)\,\mathbf{m}, \\
\frac{d\mathbf{P}}{dt} &= \mathbf{F}(t)\,\mathbf{P} + \mathbf{P}\,\mathbf{F}^{\mathsf{T}}(t) + \mathbf{L}(t)\,\mathbf{Q}(t)\,\mathbf{L}^{\mathsf{T}}(t),
\end{aligned}
\tag{10.73}
$$

are integrated from the initial conditions $\mathbf{m}(t_{k-1}) = \mathbf{m}_{k-1}$, $\mathbf{P}(t_{k-1}) = \mathbf{P}_{k-1}$ *to time instant* t_k*. The predicted mean and covariance are given as* $\mathbf{m}_k^- = \mathbf{m}(t_k)$ *and* $\mathbf{P}_k^- = \mathbf{P}(t_k)$, *respectively.*

- Update step: *The update step is the same as the discrete-time Kalman filter update step given in Equations* (10.69).

Example 10.21 (Continuous-discrete Kalman filter for the Ornstein–Uhlenbeck model). *The continuous-discrete Kalman filter for the Ornstein–Uhlenbeck model in Example 10.19 is now the same as before except that* m_k^- *and* P_k^- *are computed by solving the differential equations*

$$
\begin{aligned}
\frac{dm}{dt} &= -\lambda\, m, \\
\frac{dP}{dt} &= -2\lambda\, P + q,
\end{aligned}
\tag{10.74}
$$

from initial conditions $m(t_{k-1}) = m_{k-1}$, $P(t_{k-1}) = P_{k-1}$ *up to time* t_k*.*

This filter produces the same result as the filter in Example 10.19 (see Figure 10.3).

10.7 Approximate Continuous-Discrete Filtering

For nonlinear filtering problems, there exists various approximations (see, e.g., Jazwinski, 1970; Maybeck, 1982a; Särkkä, 2006, 2013). One approach to approximate nonlinear continuous-discrete filtering is to approximate the transition density using Itô–Taylor or stochastic Runge–Kutta approximations or some of the other approximations discussed in the previous chapters (see Section 9.2 especially). For example, from the Euler–Maruyama discretization in the model (10.12), we get the approximate model

$$\mathbf{x}(t_{k+1}) = \mathbf{x}(t_k) + \mathbf{f}(\mathbf{x}(t_k), t_k)\, \Delta t + \mathbf{L}(\mathbf{x}(t_k), t_k)\, \Delta \boldsymbol{\beta}_k,$$
$$\mathbf{y}_k \sim p(\mathbf{y}_k \mid \mathbf{x}(t_k)), \tag{10.75}$$

which corresponds to a discrete-time model (10.51) with

$$p(\mathbf{x}(t_{k+1}) \mid \mathbf{x}(t_k)) = \mathrm{N}(\mathbf{x}(t_{k+1}) \mid \mathbf{x}(t_k) + \mathbf{f}(\mathbf{x}(t_k), t_k)\, \Delta t,$$
$$\mathbf{L}(\mathbf{x}(t_k), t_k)\, \mathbf{Q}\, \mathbf{L}^{\mathsf{T}}(\mathbf{x}(t_k), t_k)\, \Delta t). \tag{10.76}$$

Weak Itô–Taylor series and Runge–Kutta approximations similarly correspond to transition density approximations with either Gaussian or binomial/trinomial noises. Once we have the transition density approximation, then we can use standard discrete-time Bayesian filter approximations such as extended Kalman filters, unscented Kalman filters, or particle filters reviewed in Särkkä (2013).

To illustrate what the algorithms look like, let us take a look at the extended Kalman filter (EKF) solution to the problem. It requires the measurement model to have the specific form $\mathbf{y}_k = \mathbf{h}(\mathbf{x}_k) + \mathbf{r}_k$ (although extensions are possible; see Särkkä, 2013). Assuming that the approximate discrete-time model has the form

$$\mathbf{x}_k = \mathbf{f}_{k-1}^{\Delta t}(\mathbf{x}_{k-1}) + \mathbf{q}_{k-1},$$
$$\mathbf{y}_k = \mathbf{h}(\mathbf{x}_k) + \mathbf{r}_k, \tag{10.77}$$

where $\mathbf{x}_k \triangleq \mathbf{x}(t_k)$, $\mathbf{q}_{k-1} \sim \mathrm{N}(\mathbf{0}, \boldsymbol{\Sigma}_{k-1}(\mathbf{x}_{k-1}))$, and $\mathbf{r}_k \sim \mathrm{N}(\mathbf{0}, \mathbf{R}_k)$, then the EKF has the form indicated in Algorithm 10.22. The matrices $\mathbf{F}_{\mathbf{x}}^{\Delta t}(\bullet)$ and $\mathbf{H}_{\mathbf{x}}(\bullet)$ denote the Jacobians of $\mathbf{f}_{k-1}^{\Delta t}(\bullet)$ and $\mathbf{h}(\bullet)$, respectively.

Algorithm 10.22 (EKF). *The prediction and update steps of the EKF are as follows:*

- Prediction step:

$$\mathbf{m}_k^- = \mathbf{f}_{k-1}^{\Delta t}(\mathbf{m}_{k-1}),$$
$$\mathbf{P}_k^- = \mathbf{F}_\mathbf{x}^{\Delta t}(\mathbf{m}_{k-1}) \, \mathbf{P}_{k-1} \left[\mathbf{F}_\mathbf{x}^{\Delta t}\right]^\mathsf{T} (\mathbf{m}_{k-1}) + \mathbf{\Sigma}_{k-1}(\mathbf{m}_{k-1}). \tag{10.78}$$

- Update step:

$$\mathbf{v}_k = \mathbf{y}_k - \mathbf{h}(\mathbf{m}_k^-),$$
$$\mathbf{S}_k = \mathbf{H}_\mathbf{x}(\mathbf{m}_k^-) \, \mathbf{P}_k^- \, \mathbf{H}_\mathbf{x}^\mathsf{T}(\mathbf{m}_k^-) + \mathbf{R}_k,$$
$$\mathbf{K}_k = \mathbf{P}_k^- \, \mathbf{H}_\mathbf{x}^\mathsf{T}(\mathbf{m}_k^-) \, \mathbf{S}_k^{-1}, \tag{10.79}$$
$$\mathbf{m}_k = \mathbf{m}_k^- + \mathbf{K}_k \, \mathbf{v}_k,$$
$$\mathbf{P}_k = \mathbf{P}_k^- - \mathbf{K}_k \, \mathbf{S}_k \, \mathbf{K}_k^\mathsf{T}.$$

A more general Gaussian assumed density filter or Gaussian filter has the following form.

Algorithm 10.23 (Gaussian assumed density filter). *The prediction and update steps of the assumed density Gaussian (Kalman) filter are as follows:*

- Prediction step:

$$\mathbf{m}_k^- = \int \mathbf{f}_{k-1}^{\Delta t}(\mathbf{x}_{k-1}) \, \mathrm{N}(\mathbf{x}_{k-1} \mid \mathbf{m}_{k-1}, \mathbf{P}_{k-1}) \, \mathrm{d}\mathbf{x}_{k-1},$$

$$\mathbf{P}_k^- = \int \left[(\mathbf{f}_{k-1}^{\Delta t}(\mathbf{x}_{k-1}) - \mathbf{m}_k^-) \, (\mathbf{f}_{k-1}^{\Delta t}(\mathbf{x}_{k-1}) - \mathbf{m}_k^-)^\mathsf{T} \right. \tag{10.80}$$
$$\left. \times \, \mathrm{N}(\mathbf{x}_{k-1} \mid \mathbf{m}_{k-1}, \mathbf{P}_{k-1}) + \mathbf{\Sigma}_{k-1}(\mathbf{x}_{k-1}) \right] \mathrm{d}\mathbf{x}_{k-1}.$$

- Update step:

$$\boldsymbol{\mu}_k = \int \mathbf{h}(\mathbf{x}_k) \, \mathrm{N}(\mathbf{x}_k \mid \mathbf{m}_k^-, \mathbf{P}_k^-) \, \mathrm{d}\mathbf{x}_k,$$

$$\mathbf{S}_k = \int (\mathbf{h}(\mathbf{x}_k) - \boldsymbol{\mu}_k) \, (\mathbf{h}(\mathbf{x}_k) - \boldsymbol{\mu}_k)^\mathsf{T} \, \mathrm{N}(\mathbf{x}_k \mid \mathbf{m}_k^-, \mathbf{P}_k^-) \, \mathrm{d}\mathbf{x}_k + \mathbf{R}_k,$$

$$\mathbf{C}_k = \int (\mathbf{x}_k - \mathbf{m}_k^-) \, (\mathbf{h}(\mathbf{x}_k) - \boldsymbol{\mu}_k)^\mathsf{T} \, \mathrm{N}(\mathbf{x}_k \mid \mathbf{m}_k^-, \mathbf{P}_k^-) \, \mathrm{d}\mathbf{x}_k,$$

$$\mathbf{K}_k = \mathbf{C}_k \, \mathbf{S}_k^{-1},$$

$$\mathbf{m}_k = \mathbf{m}_k^- + \mathbf{K}_k \, (\mathbf{y}_k - \boldsymbol{\mu}_k),$$

$$\mathbf{P}_k = \mathbf{P}_k^- - \mathbf{K}_k \, \mathbf{S}_k \, \mathbf{K}_k^\mathsf{T}.$$

$$\tag{10.81}$$

In implementation of the preceding filter, we can numerically compute the Gaussian integrals by using sigma-point methods of the form given in Equation (9.16). This leads to a large class of filters called sigma-point filters (e.g., Särkkä, 2013).

The preceding algorithms correspond to the Bayesian filter formulation in Algorithm 10.15. As the algorithm is just a discrete-time filter, we do not give more examples of algorithms here, but the reader is referred to Särkkä (2013) for various algorithms for the problem. Algorithm 10.23 also corresponds to something called prior linearization, but it is also possible to use linearization with respect to other distributions as well. For further details, see García-Fernández et al. (2015), García-Fernández et al. (2017), and Tronarp et al. (2018). The choice of linearization is also closely related to Gauss–Newton interpretations of the iterated extended Kalman filters and smoothers (Bell and Cathey, 1993; Bell, 1994).

It is further possible to formulate a continuous-discrete version of the filter as in Algorithm 10.16, which is no longer equivalent to a discrete-time filtering problem.

Algorithm 10.24 (Continuous-discrete extended Kalman filter). *The prediction and update steps of the continuous-discrete extended Kalman filter (CDEKF) are as follows:*

- Prediction step: *Integrate the following differential equations from initial conditions* $\mathbf{m}(t_{k-1}) = \mathbf{m}_{k-1}$, $\mathbf{P}(t_{k-1}) = \mathbf{P}_{k-1}$ *to time instant* t_k:

$$\frac{\mathrm{d}\mathbf{m}}{\mathrm{d}t} = \mathbf{f}(\mathbf{m}, t),$$
$$\frac{\mathrm{d}\mathbf{P}}{\mathrm{d}t} = \mathbf{P}\,\mathbf{F}_x^{\mathsf{T}}(\mathbf{m}, t) + \mathbf{F}_x(\mathbf{m}, t)\,\mathbf{P} + \mathbf{L}(\mathbf{m}, t)\,\mathbf{Q}\,\mathbf{L}^{\mathsf{T}}(\mathbf{m}, t). \tag{10.82}$$

The predicted mean and covariance are then given as $\mathbf{m}_k^- = \mathbf{m}(t_k)$, $\mathbf{P}_k^- = \mathbf{P}(t_k)$, *respectively.*
- Update step: *The update step is the same as in Equation (10.79).*

We can also replace the Taylor series expansion behind Algorithm 10.24 with a more general Gaussian assumed density approximation (cf. Särkkä, 2013; Särkkä and Sarmavuori, 2013), which results in the following algorithm.

Algorithm 10.25 (Continuous-discrete Gaussian assumed density filter). *The prediction and update steps of the continuous-discrete assumed density Gaussian (Kalman) filter are as follows:*

- Prediction step: *Integrate the following differential equations from initial conditions* $\mathbf{m}(t_{k-1}) = \mathbf{m}_{k-1}$, $\mathbf{P}(t_{k-1}) = \mathbf{P}_{k-1}$ *to time instant* t_k:

$$\frac{d\mathbf{m}}{dt} = \int \mathbf{f}(\mathbf{x}, t) \, N(\mathbf{x} \mid \mathbf{m}, \mathbf{P}) \, d\mathbf{x}, \tag{10.83}$$

$$\frac{d\mathbf{P}}{dt} = \int \mathbf{f}(\mathbf{x}, t) \, (\mathbf{x} - \mathbf{m})^\mathsf{T} \, N(\mathbf{x} \mid \mathbf{m}, \mathbf{P}) \, d\mathbf{x}$$
$$+ \int (\mathbf{x} - \mathbf{m}) \, \mathbf{f}^\mathsf{T}(\mathbf{x}, t) \, N(\mathbf{x} \mid \mathbf{m}, \mathbf{P}) \, d\mathbf{x}$$
$$+ \int \mathbf{L}(\mathbf{x}, t) \, \mathbf{Q} \, \mathbf{L}^\mathsf{T}(\mathbf{x}, t) \, N(\mathbf{x} \mid \mathbf{m}, \mathbf{P}) \, d\mathbf{x}. \tag{10.84}$$

  The predicted mean and covariance are then given as $\mathbf{m}_k^- = \mathbf{m}(t_k)$, $\mathbf{P}_k^- = \mathbf{P}(t_k)$, *respectively.*
- Update step: *The update step is the same as in Equation* (10.81).

If we use a sigma-point approximation to the mean and covariance differential equations, we end up to the sigma-point SDE approximation used in Algorithm 9.5.

For additional approximate Bayesian filters, the reader is referred to Jazwinski (1970), Gelb (1974), Maybeck (1982a), Särkkä (2006), Särkkä (2007), Särkkä and Sottinen (2008), Särkkä and Sarmavuori (2013), Särkkä (2013), and the references therein.

Example 10.26 (Continuous-discrete EKF solution to the Beneš–Daum problem). *The continuous-discrete extended Kalman filter to the Beneš–Daum filtering problem presented in Example 10.17 is the following.*

- Prediction step:

$$\frac{dm}{dt} = \tanh(m),$$
$$\frac{dP}{dt} = 2\left(1 - \tanh^2(m)\right) P + 1. \tag{10.85}$$

- Update step:

$$m_k = m_k^- + \frac{P_k^-}{(\sigma^2 + P_k^-)} (y_k - m_k^-),$$
$$P_k = P_k^- - \frac{(P_k^-)^2}{(\sigma^2 + P_k^-)}. \tag{10.86}$$

An example of a filtering result using this EKF is shown in Figure 10.4.

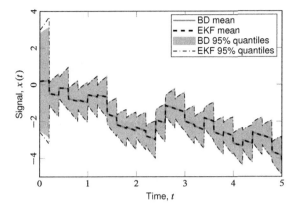

Figure 10.4 Illustration of the EKF approximation to the Beneš–Daum filtering problem given in Example 10.26. In this case, the EKF produces very accurate approximation to the exact Beneš–Daum filter (marked as "BD").

10.8 Smoothing in Continuous-Discrete and Continuous Time

The smoothing problem is concerned with computation of

$$p(\mathbf{x}(t) \mid \mathbf{y}_1, \dots, \mathbf{y}_T), \qquad t \in [t_0, t_T], \qquad (10.87)$$

or in the continuous case

$$p(\mathbf{x}(t) \mid \mathcal{Z}_T), \qquad t \in [t_0, T]. \qquad (10.88)$$

That is, smoothing is concerned with computation of the posterior distributions of the states within the range of all measurements. For example, in the car tracking problem, this corresponds to determination of the position and velocity of the car during some past points of time, but still by using information from all the measurements. That is, the smoothing problem is roughly equivalent to batch estimation of the state, although the algorithm is formulated a bit differently to gain better computational scaling. However, from smoothing algorithms we can only get the posterior distribution of single time instant $\mathbf{x}(t)$ at a time, whereas a full batch solution would correspond to a joint distribution of all times.

For mathematical simplicity, let us start with the continuous-discrete problem. Analogously to Section 10.5 we can, in principle, solve the transition density of the SDE to form a discrete-time model:

$$\begin{aligned} \mathbf{x}(t_{k+1}) &\sim p(\mathbf{x}(t_{k+1}) \mid \mathbf{x}(t_k)), \\ \mathbf{y}_k &\sim p(\mathbf{y}_k \mid \mathbf{x}(t_k)). \end{aligned} \tag{10.89}$$

The smoothing solution for the time instants t_k can now be computed with the following discrete-time smoothing recursion (e.g., Särkkä, 2013).

Algorithm 10.27 (Bayesian smoother). *The backward recursive equations (the Bayesian smoother) for computing the smoothed distributions $p(\mathbf{x}(t_k) \mid \mathbf{y}_{1:T})$ for any $k < T$ are given by the following Bayesian (fixed-interval) smoothing equations:*

$$p(\mathbf{x}(t_{k+1}) \mid \mathbf{y}_{1:k})$$
$$= \int p(\mathbf{x}(t_{k+1}) \mid \mathbf{x}(t_k))\, p(\mathbf{x}(t_k) \mid \mathbf{y}_{1:k})\, \mathrm{d}\mathbf{x}(t_k),$$
$$p(\mathbf{x}(t_k) \mid \mathbf{y}_{1:T})$$
$$= p(\mathbf{x}(t_k) \mid \mathbf{y}_{1:k}) \int \left[\frac{p(\mathbf{x}(t_{k+1}) \mid \mathbf{x}(t_k))\, p(\mathbf{x}(t_{k+1}) \mid \mathbf{y}_{1:T})}{p(\mathbf{x}(t_{k+1}) \mid \mathbf{y}_{1:k})} \right] \mathrm{d}\mathbf{x}(t_{k+1}), \tag{10.90}$$

where $p(\mathbf{x}(t_k) \mid \mathbf{y}_{1:k})$ is the filtering density of the distribution of the time step k. Note that the term $p(\mathbf{x}(t_{k+1}) \mid \mathbf{y}_{1:k})$ is simply the predicted distribution of time step $k + 1$ as in Algorithm 10.15. The recursion is started from the final step, where the filtering and smoothing densities are the same: $p(\mathbf{x}(t_T) \mid \mathbf{y}_{1:T})$.

Algorithm 10.27 can be used to compute the smoothing solution at any given time t by including an extra measurement-free time instant to the filter (e.g., Särkkä, 2006).

The Rauch–Tung–Striebel (RTS) smoother is the closed-form solution to the linear Gaussian smoothing problem. Assuming that we have formed an equivalent discrete-time model of the form (10.67), then we can use the following RTS smoother to compute the means and covariances of the smoothing distributions.

Algorithm 10.28 (RTS smoother). *The backward recursion equations for* the RTS smoother are given as

$$\begin{aligned}
\mathbf{m}_{k+1}^- &= \mathbf{A}_k \, \mathbf{m}_k, \\
\mathbf{P}_{k+1}^- &= \mathbf{A}_k \, \mathbf{P}_k \, \mathbf{A}_k^\mathsf{T} + \mathbf{\Sigma}_k, \\
\mathbf{G}_k &= \mathbf{P}_k \, \mathbf{A}_k^\mathsf{T} \, [\mathbf{P}_{k+1}^-]^{-1}, \\
\mathbf{m}_k^{\mathrm{s}} &= \mathbf{m}_k + \mathbf{G}_k \, [\mathbf{m}_{k+1}^{\mathrm{s}} - \mathbf{m}_{k+1}^-], \\
\mathbf{P}_k^{\mathrm{s}} &= \mathbf{P}_k + \mathbf{G}_k \, [\mathbf{P}_{k+1}^{\mathrm{s}} - \mathbf{P}_{k+1}^-] \, \mathbf{G}_k^\mathsf{T},
\end{aligned} \tag{10.91}$$

where $\mathbf{m}_k$ and $\mathbf{P}_k$ are the mean and covariance computed by the Kalman filter. The recursion is started from the last time step T, with $\mathbf{m}_T^{\mathrm{s}} = \mathbf{m}_T$ and $\mathbf{P}_T^{\mathrm{s}} = \mathbf{P}_T$. Note that the first two of the equations are simply the Kalman filter prediction equations.

The algorithm computes

$$p(\mathbf{x}(t_k) \mid \mathbf{y}_{1:T}) = \mathrm{N}(\mathbf{x}(t_k) \mid \mathbf{m}_k^{\mathrm{s}}, \mathbf{P}_k^{\mathrm{s}}) \tag{10.92}$$

for all $k = 1, 2, \ldots, T$.

Example 10.29 (RTS smoother for the Ornstein–Uhlenbeck model). *The RTS smoother for the Ornstein–Uhlenbeck model in Example 10.19 is the following:*

1. *Start from the filter mean $m_T^{\mathrm{s}} = m_T$ and variance $P_T^{\mathrm{s}} = P_T$.*
2. *For each $k = T - 1, T - 2, \ldots$, compute the following using the means m_k and variances P_k computed by the Kalman filter in Example 10.19:*

$$\begin{aligned}
m_{k+1}^- &= \exp(-\lambda \, \Delta t_k) \, m_k, \\
P_{k+1}^- &= \exp(-2\lambda \, \Delta t_k) \, P_k + \frac{q}{2\lambda} \, [1 - \exp(-2\lambda \, \Delta t_k)], \\
m_k^{\mathrm{s}} &= m_k + \left(\frac{P_k \, \exp(-\lambda \, \Delta t_k)}{P_{k+1}^-} \right) [m_{k+1}^{\mathrm{s}} - m_{k+1}^-], \\
P_k^{\mathrm{s}} &= P_k + \left(\frac{P_k \, \exp(-\lambda \, \Delta t_k)}{P_{k+1}^-} \right)^2 [P_{k+1}^{\mathrm{s}} - P_{k+1}^-].
\end{aligned} \tag{10.93}$$

The result of this smoother when run on the same data as in Example 10.19 is shown in Figure 10.5. For visualization, we have run the smoother in a denser grid than the measurements to obtain intermediate smoothing results between the measurements.

It is also possible to formulate the smoothing solution directly in continuous time. The following smoothing solution by Leondes et al. (1970) is a backward partial differential equation that can be seen as a continuous time

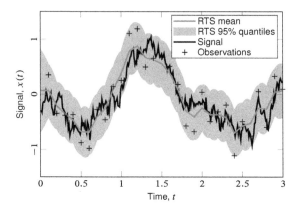

Figure 10.5 Result of the Rauch–Tung–Striebel smoother in Example 10.29.

version of Equation (10.90). It turns out the same equation works for both continuous-discrete and continuous-time problems. All we need to do is to use the correct filtering/predicted density in the equations – the smoother itself does not depend on the measurement model at all.

Algorithm 10.30 (Continuous-discrete/continuous Bayesian smoother). *Let us denote the filtering density as*

$$q(\mathbf{x}, t) \triangleq p(\mathbf{x}(t) \mid \mathbf{y}_{1:k}), \quad \text{(in the continuous-discrete case),}$$
$$q(\mathbf{x}, t) \triangleq p(\mathbf{x}(t) \mid \mathcal{Z}_t), \quad \text{(in the continuous-time case),} \tag{10.94}$$

where k is the largest integer such that $t_k \leq t$. Also denote the smoothing density as

$$s(\mathbf{x}, t) \triangleq p(\mathbf{x}(t) \mid \mathbf{y}_{1:T}), \quad \text{(in the continuous-discrete case),}$$
$$s(\mathbf{x}, t) \triangleq p(\mathbf{x}(t) \mid \mathcal{Z}_T), \quad \text{(in the continuous-time case).} \tag{10.95}$$

The smoothing solution is then given by solution to the following backward partial differential equation

$$\frac{\partial s(\mathbf{x}, t)}{\partial t} = s(\mathbf{x}, t) \sum_j \frac{\partial f_j(\mathbf{x}, t)}{\partial x_j} - \sum_j f_j(\mathbf{x}, t) \frac{\partial s(\mathbf{x}, t)}{\partial x_j}$$

$$+ \frac{1}{2} s(\mathbf{x}, t) \sum_{j,k} \frac{\partial^2 \left[\mathbf{L}(\mathbf{x}, t) \mathbf{Q} \mathbf{L}^\mathsf{T}(\mathbf{x}, t) \right]_{jk}}{\partial x_j \partial x_k}$$

$$- \frac{1}{2} \sum_{j,k} \left[\mathbf{L}(\mathbf{x}, t) \mathbf{Q} \mathbf{L}^\mathsf{T}(\mathbf{x}, t) \right]_{jk} \frac{\partial^2 s(\mathbf{x}, t)}{\partial x_j \partial x_k}$$

$$+ \frac{1}{q(\mathbf{x}, t)} \sum_{j,k} \left[\mathbf{L}(\mathbf{x}, t) \mathbf{Q} \mathbf{L}^\mathsf{T}(\mathbf{x}, t) \right]_{jk} \frac{\partial s(\mathbf{x}, t)}{\partial x_j} \frac{\partial q(\mathbf{x}, t)}{\partial x_k}$$

$$- \frac{s(\mathbf{x}, t)}{q^2(\mathbf{x}, t)} \sum_{j,k} \left[\mathbf{L}(\mathbf{x}, t) \mathbf{Q} \mathbf{L}^\mathsf{T}(\mathbf{x}, t) \right]_{jk} \frac{\partial q(\mathbf{x}, t)}{\partial x_j} \frac{\partial q(\mathbf{x}, t)}{\partial x_k}$$

$$+ \frac{s(\mathbf{x}, t)}{q(\mathbf{x}, t)} \sum_{j,k} \frac{\partial \left[\mathbf{L}(\mathbf{x}, t) \mathbf{Q} \mathbf{L}^\mathsf{T}(\mathbf{x}, t) \right]_{jk}}{\partial x_j} \frac{\partial q(\mathbf{x}, t)}{\partial x_k}$$

$$+ \frac{s(\mathbf{x}, t)}{q(\mathbf{x}, t)} \sum_{j,k} \left[\mathbf{L}(\mathbf{x}, t) \mathbf{Q} \mathbf{L}^\mathsf{T}(\mathbf{x}, t) \right]_{jk} \frac{\partial^2 q(\mathbf{x}, t)}{\partial x_j \partial x_k} \tag{10.96}$$

with the terminal condition $q(\mathbf{x}, t_T) = s(\mathbf{x}, t_T)$ at the final step.

Remark 10.31. *As pointed out by Anderson (1972), the smoothing equations in Algorithm 10.30 can also be written in simpler form as*

$$\frac{\partial s(\mathbf{x}, t)}{\partial t} = \mathcal{A}^*[q(\mathbf{x}, t)] \frac{s(\mathbf{x}, t)}{q(\mathbf{x}, t)} - q(\mathbf{x}, t) \mathcal{A} \left[\frac{s(\mathbf{x}, t)}{q(\mathbf{x}, t)} \right]. \tag{10.97}$$

The RTS smoother for the linear Gaussian models corresponding to the preceding formulation is as follows.

Algorithm 10.32 (Continuous-discrete/continuous RTS smoother). *The differential backward equations of the smoothing solution (characterized by the mean $\mathbf{m}^s$ and covariance $\mathbf{P}^s$) for linear Gaussian models are*

$$\frac{d\mathbf{m}^s}{dt} = \mathbf{F}(t) \mathbf{m}^s + \mathbf{L}(t) \mathbf{Q}(t) \mathbf{L}^\mathsf{T}(t) \mathbf{P}^{-1} [\mathbf{m}^s - \mathbf{m}],$$

$$\frac{d\mathbf{P}^s}{dt} = \left[\mathbf{F}(t) + \mathbf{L}(t) \mathbf{Q}(t) \mathbf{L}(t) \mathbf{P}^{-1} \right] \mathbf{P}^s \tag{10.98}$$

$$+ \mathbf{P}^s \left[\mathbf{F}(t) + \mathbf{L}(t) \mathbf{Q}(t) \mathbf{L}^\mathsf{T}(t) \mathbf{P}^{-1} \right]^\mathsf{T} - \mathbf{L}(t) \mathbf{Q}(t) \mathbf{L}^\mathsf{T}(t).$$

The means $\mathbf{m}(t)$ *and covariances* $\mathbf{P}(t)$ *can be from either the Kalman–Bucy filter in Algorithm 10.9 or one of the continuous-discrete Kalman filters in Algorithms 10.18 and 10.20. In the latter case, the means and covariances between the measurements should be their predicted values.*

Example 10.33 (Continuous-discrete/continuous RTS smoother for the Ornstein–Uhlenbeck model). *The continuous-time version of the smoother in Example 10.29 is defined via the differential equations*

$$\begin{aligned}
\frac{dm^s}{dt} &= -\lambda\, m^s + \left(\frac{q}{P}\right)(m^s - m), \\
\frac{dP^s}{dt} &= 2\left(-\lambda + \frac{q}{P}\right)P^s - q.
\end{aligned} \tag{10.99}$$

The result produced by this smoother is the same as of the RTS smoother in Example 10.29 and Figure 10.5.

10.9 Approximate Smoothing Algorithms

We can form approximate smoothing algorithms corresponding to both the discretized and continuous smoothers in Algorithms 10.27 and 10.30. Provided that we have discretized the model into form (10.77), then we can directly apply off-the-shelf nonlinear smoothers (e.g., Särkkä, 2013). For example, the extended RTS smoother looks like the following.

Algorithm 10.34 (Discrete-time extended RTS smoother). *The equations for the discrete-time extended RTS (ERTS) smoother are*

$$\begin{aligned}
\mathbf{m}_{k+1}^- &= \mathbf{f}_k^{\Delta t}(\mathbf{m}_k), \\
\mathbf{P}_{k+1}^- &= \mathbf{F}_\mathbf{x}^{\Delta t}(\mathbf{m}_k)\,\mathbf{P}_k\,\left[\mathbf{F}_\mathbf{x}^{\Delta t}\right]^\mathsf{T}(\mathbf{m}_k) + \boldsymbol{\Sigma}_k(\mathbf{m}_k), \\
\mathbf{G}_k &= \mathbf{P}_k\,\mathbf{F}_\mathbf{x}^\mathsf{T}(\mathbf{m}_k)\,[\mathbf{P}_{k+1}^-]^{-1}, \\
\mathbf{m}_k^s &= \mathbf{m}_k + \mathbf{G}_k\,[\mathbf{m}_{k+1}^s - \mathbf{m}_{k+1}^-], \\
\mathbf{P}_k^s &= \mathbf{P}_k + \mathbf{G}_k\,[\mathbf{P}_{k+1}^s - \mathbf{P}_{k+1}^-]\,\mathbf{G}_k^\mathsf{T},
\end{aligned} \tag{10.100}$$

where the matrix $\mathbf{F}_\mathbf{x}(\mathbf{m}_k)$ *is the Jacobian matrix of* $\mathbf{f}(\mathbf{x})$ *evaluated at* $\mathbf{m}_k$.

The preceding procedure is a recursion that can be used for computing the smoothing distribution of time step k from the smoothing distribution of time step $k + 1$. Because the smoothing distribution and filtering distribution of the last time step T are the same, we have $\mathbf{m}_T^s = \mathbf{m}_T$ *and* $\mathbf{P}_T^s = \mathbf{P}_T$, *and thus the recursion can be used for computing the smoothing distributions of all time steps by starting from the last step $k = T$ and proceeding backward to the initial step $k = 0$.*

Various other discrete-time smoother approximations based on, for example, sigma-point approximations and Monte Carlo can be found in the book of Särkkä (2013). Some iterated sigma-point smoothers have also been introduced by García-Fernández et al. (2017) and Tronarp et al. (2018), which in turn are based on the Gauss–Newton method interpretation of extended Kalman smoothers discussed in Bell (1994).

A review of Gaussian approximation–based nonlinear smoothers corresponding to the smoother formulation in Algorithm 10.30 can be found in the article of Särkkä and Sarmavuori (2013). In the following, we briefly express the ideas of the Type I, II, and III smoothers introduced in Särkkä and Sarmavuori (2013) and give Taylor series–based implementation examples of them. In the following, for notational convenience, we assume that $\mathbf{L}(t)$ does not depend on the state.

A "Type I" smoother refers to a smoother that is derived as a Gaussian approximation to the smoothing equations of Leondes et al. (1970) given in Algorithm 10.30. The first-order Taylor series–based smoother of this type is the following.

Algorithm 10.35 (Type I extended RTS smoother). *The smoother consists of the following equations which are integrated backward starting from the filtering solution* $\mathbf{m}^s(t_T) = \mathbf{m}(t_T)$, $\mathbf{P}^s(t_T) = \mathbf{P}(t_T)$ *at the end point:*

$$
\begin{aligned}
\frac{d\mathbf{m}^s}{dt} &= \mathbf{f}(\mathbf{m}^s, t) + \mathbf{L}(t)\,\mathbf{Q}\,\mathbf{L}^\mathsf{T}(t)\,\mathbf{P}^{-1}\,[\mathbf{m}^s - \mathbf{m}], \\
\frac{d\mathbf{P}^s}{dt} &= \mathbf{F_x}(\mathbf{m}^s, t)\,\mathbf{P}^s + \mathbf{P}^s\,\mathbf{F_x}^\mathsf{T}(\mathbf{m}^s, t) - \mathbf{L}(t)\,\mathbf{Q}\,\mathbf{L}^\mathsf{T}(t) \qquad (10.101) \\
&\quad + \mathbf{L}(t)\,\mathbf{Q}\,\mathbf{L}^\mathsf{T}(t)\,\mathbf{P}^{-1}\,\mathbf{P}^s + \mathbf{P}^s\,\mathbf{P}^{-1}\,\mathbf{L}(t)\,\mathbf{Q}\,\mathbf{L}^\mathsf{T}(t).
\end{aligned}
$$

A "Type II" smoother refers to a smoother that can be derived by taking the formal limit of the corresponding discrete-time smoother (cf. Särkkä, 2013). A characteristic of these kinds of smoothers is that they are linear in the smoothing solution. The first-order Taylor series–based approximation of it is the following.

Algorithm 10.36 (Type II extended RTS smoother). *In the smoother, we*

integrate the following equations backward starting from the filtering solu-
tion $\mathbf{m}^s(t_T) = \mathbf{m}(t_T)$, $\mathbf{P}^s(t_T) = \mathbf{P}(t_T)$ *at the end point:*

$$
\begin{aligned}
\frac{d\mathbf{m}^s}{dt} &= \mathbf{f}(\mathbf{m}, t) \\
&\quad + \left[\mathbf{F_x}(\mathbf{m}, t) + \mathbf{L}(t)\,\mathbf{Q}\,\mathbf{L}^\mathsf{T}(t)\,\mathbf{P}^{-1}\right][\mathbf{m}^s - \mathbf{m}], \\
\frac{d\mathbf{P}^s}{dt} &= \mathbf{F_x}(\mathbf{m}, t)\,\mathbf{P}^s + \mathbf{P}^s\,\mathbf{F_x^\mathsf{T}}(\mathbf{m}, t) - \mathbf{L}(t)\,\mathbf{Q}\,\mathbf{L}^\mathsf{T}(t) \\
&\quad + \mathbf{L}(t)\,\mathbf{Q}\,\mathbf{L}^\mathsf{T}(t)\,\mathbf{P}^{-1}\,\mathbf{P}^s + \mathbf{P}^s\,\mathbf{P}^{-1}\,\mathbf{L}(t)\,\mathbf{Q}\,\mathbf{L}^\mathsf{T}(t).
\end{aligned}
\tag{10.102}
$$

"Type III" refers to a smoother that is mathematically equivalent to the
Type II smoother, but the backward differential equation is rewritten as
two forward differential equations and a discrete-time backward recursion.
This formulation is advantageous from a numerical point of view. The algorithm resulting from a first-order Taylor series expansion is given in Algorithm 10.37.

Algorithm 10.37 (Type III extended RTS smoother). *The smoother con-*
sists of three forward differential equations along with a discrete-time
backward recursion:

$$
\begin{aligned}
\frac{d\mathbf{m}}{dt} &= \mathbf{f}(\mathbf{m}, t), \\
\frac{d\mathbf{P}}{dt} &= \mathbf{F_x}(\mathbf{m}, t)\,\mathbf{P} + \mathbf{P}\,\mathbf{F_x^\mathsf{T}}(\mathbf{m}, t) + \mathbf{L}(t)\,\mathbf{Q}\,\mathbf{L}^\mathsf{T}(t), \\
\frac{d\mathbf{C}_k(t)}{dt} &= \mathbf{C}_k(t)\,\mathbf{F_x^\mathsf{T}}(\mathbf{m}, t), \\
\mathbf{G}_{k+1} &= \mathbf{C}_k(t_{k+1}^-)\,\mathbf{P}^{-1}(t_{k+1}^-), \\
\mathbf{m}^s(t_k) &= \mathbf{m}(t_k) + \mathbf{G}_{k+1}\left[\mathbf{m}^s(t_{k+1}) - \mathbf{m}(t_{k+1}^-)\right], \\
\mathbf{P}^s(t_k) &= \mathbf{P}(t_k) + \mathbf{G}_{k+1}\left[\mathbf{P}^s(t_{k+1}) - \mathbf{P}(t_{k+1}^-)\right]\mathbf{G}_{k+1}^\mathsf{T}.
\end{aligned}
\tag{10.103}
$$

The smoother is started from the filtering solution $\mathbf{m}^s(t_T) = \mathbf{m}(t_T)$,
$\mathbf{P}^s(t_T) = \mathbf{P}(t_T)$. *We also need to put* $\mathbf{C}_k(t_k) = \mathbf{P}(t_k)$ *always before solv-*
ing the differential equation of $\mathbf{C}_k(t)$ *for* $t \in [t_k, t_{k+1}]$.

The following example shows a Type II smoother.

Example 10.38 (Extended RTS solution to Beneš and Beneš–Daum filtering problems). *The equations of the Type II extended RTS smoother that*
can be used both for the Beneš filtering problem in Example 10.8 as well

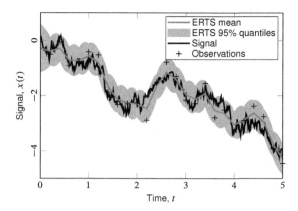

Figure 10.6 Illustration of the result of the ERTS solution to the
Beneš–Daum (smoothing) problem in Example 10.38. The
estimate and the uncertainty estimates are now conditioned on all
the data, which makes the discontinuities seen in Figures 10.2 and
10.4 disappear.

as for the Beneš–Daum filtering problem in the Example 10.17 are

$$
\frac{dm^{\mathrm{s}}}{dt} = \tanh(m) + \left(\frac{(1 - \tanh^2(m)) \, P + 1}{P} \right) (m^{\mathrm{s}} - m),
$$
$$
\frac{dP^{\mathrm{s}}}{dt} = 2 \left(\frac{(1 - \tanh^2(m)) \, P + 1}{P} \right) P^{\mathrm{s}} - 1,
$$

(10.104)

*where $m(t)$ and $P(t)$ are the mean and covariance from the extended
Kalman filter. Figure 10.6 shows the result of this smoother.*

10.10 Exercises

10.1 Consider the Beneš filtering problem in Example 10.8:

(a) Compute the normalized posterior distribution by evaluating the normalization constant in Equation (10.36).

(b) Show that the normalized posterior distribution solves the Kushner–
Stratonovich equation.

(c) Form the corresponding unnormalized distribution $q(x, t)$ and show that
it solves the Zakai equation.

10.2 Consider the Kushner–Stratonovich equation for the Ornstein–Uhlenbeck
model:

(a) Write down the Kushner–Stratonovich equation for the model

$$dx = -\lambda x \, dt + d\beta,$$
$$dz = x \, dt + d\eta,$$

where β and η are independent standard Brownian motions with diffusion coefficients q and r.

(b) Show that the Kalman–Bucy filter to the Ornstein–Uhlenbeck model in Example 10.10 solves the Kushner–Stratonovich equation.

10.3 Write down the equations of the Kalman–Bucy filter for the car tracking model in Example 10.3.

10.4 Continuous-time approximate nonlinear filtering: Consider the model

$$dx = \tanh(x) \, dt + d\beta,$$
$$dz = \sin(x) \, dt + d\eta,$$

where β and η are independent Brownian motions with diffusion coefficients $q = 1$ and $r = 0.01$, respectively.

(a) Write down the extended Kalman–Bucy filter for this model.

(b) Simulate data from the model over a time span $[0, 5]$ with $\Delta t = 1/100$, and implement the filtering method numerically. How does it work, and how could you improve the filter?

10.5 Show that the Kalman filters in Examples 10.19 and 10.21 are equivalent.

10.6 Find the Kalman filter for the car tracking model in the resources of the book Särkkä (2013). Implement the continuous-discrete Kalman filter for the same problem and check that the results match.

10.7 Recall the smartphone tracking Example 3.5, where the three-dimensional orientation of the device was modeled with the help of observations of acceleration and angular velocity in the local coordinate frame of the device. Consider that you have (noisy) observations provided by a gyroscope and accelerometer at 100 Hz (you can assume Δt to be constant).

(a) Write the corresponding state-space model for the smartphone tracking problem. How would you formulate the model in terms of the two data sources. Also recall that the SDE model was solved in Exercise 6.2.

(b) Formulate the filtering problem that provides a solution to the tracking problem.

(c) Implement the filter numerically and check your solution by simulating data (you can assume the device to rotate with a constant speed around one axis in your simulation).

10.8 Consider the Black–Scholes model from Example 3.8:

(a) Formulate the corresponding state-space model, where the state is observed through a log-normal measurement model.

(b) Formulate the filtering solution to the state-space model.

(c) Derive the corresponding smoother.

10.9 Show that the Rauch–Tung–Striebel smoothers in Examples 10.29 and 10.33 are equivalent.

10.10 Show that the smoother for the Ornstein–Uhlenbeck model given in Example 10.33 solves the Leondes equation (Algorithm 10.30).

10.11 Write down the Type I and Type III smoothers for Example 10.38.

11

Parameter Estimation in SDE Models

An issue that often arises in the context of practical modeling with SDEs is the problem of parameter estimation. In that context, we might know the parametric form of the SDE, but the parameters of the SDE have unknown values. However, we might have a set of experimental data that we wish to use for determining the values of the parameters. The aim of this chapter is to give an overview of solutions to these kinds of problems. We specifically consider statistical likelihood-based inference methods, but we also give pointers to other types of methods.

The problem of parameter estimation in SDE models has a long history, and overviews of parameter estimation methods in different types of models can be found, for example, in the articles of Nielsen et al. (2000), Sørensen (2004), and Särkkä et al. (2015b) and in the theses of Jeisman (2005) and Mbalawata (2014) as well as in the books of Rao (1999) and Iacus (2008). In the case of partially observed systems, the problem is closely related to parameter estimation in (discrete-time) state-space models (or hidden Markov models), and parameter estimation methods for these kinds of models have been summarized in the books of Särkkä (2013) and Cappé et al. (2005).

11.1 Overview of Parameter Estimation Methods

In the typical setting that we consider here, we have an SDE with a vector of (unknown) parameters $\theta = (\theta_1, \theta_2, \ldots, \theta_d)$:

$$d\mathbf{x} = \mathbf{f}(\mathbf{x}, t; \theta) \, dt + \mathbf{L}(\mathbf{x}, t; \theta) \, d\beta, \quad \mathbf{x}(t_0) = \mathbf{x}_0. \qquad (11.1)$$

The diffusion matrix of the Brownian motion $\mathbf{Q}(\theta)$ might also depend on the parameters. Additionally, we have a set of observations of the SDE. For example, we might have a set of known values of the state $\mathbf{x}(t)$ at a certain finite number of time points. Alternatively, we might only have partial observations of the state, and these observations might also be corrupted by noise.

Example 11.1 (Parameters in the Ornstein–Uhlenbeck model). *Consider, for example, the Ornstein–Uhlenbeck process*

$$\mathrm{d}x = -\lambda\, x\, \mathrm{d}t + \mathrm{d}\beta, \quad x(0) = x_0. \tag{11.2}$$

For the sake of parameter estimation, we can assume that the parameters λ *and* q *are unknown, that is, we have the unknown parameter vector* $\boldsymbol{\theta} = (\lambda, q)$. *Furthermore, we could assume that we know the values of the SDE at times* $x(\Delta t), x(2\,\Delta t), \dots, x(T\,\Delta t)$ *which forms the data that are used for estimating the parameters.*

In the case that we observe a finite number of values of the SDE, say, $\mathbf{x}(t_1), \mathbf{x}(t_2), \dots, \mathbf{x}(t_T)$, a classical method for SDE parameter estimation is the maximum likelihood (ML) method. Due to the Markov properties of SDEs (cf. Section 5.4), we can write down the likelihood of the observed values given the parameters as follows:

$$p(\mathbf{x}(t_1), \dots, \mathbf{x}(t_T) \mid \boldsymbol{\theta}) = \prod_{k=0}^{T-1} p(\mathbf{x}(t_{k+1}) \mid \mathbf{x}(t_k), \boldsymbol{\theta}), \tag{11.3}$$

where $p(\mathbf{x}(t_{k+1}) \mid \mathbf{x}(t_k), \boldsymbol{\theta})$ are the transition densities of the SDE that we discussed in Section 5.4. In the ML method, we wish to maximize the preceding likelihood expression or, equivalently, minimize the negative log-likelihood:

$$\ell(\boldsymbol{\theta}) = -\log p(\mathbf{x}(t_1), \dots, \mathbf{x}(t_T) \mid \boldsymbol{\theta})$$

$$= -\sum_{k=0}^{T} \log p(\mathbf{x}(t_{k+1}) \mid \mathbf{x}(t_k), \boldsymbol{\theta}). \tag{11.4}$$

Alternately, given the likelihood we can use Bayesian methods that directly perform inference on the posterior distribution,

$$p(\boldsymbol{\theta} \mid \mathbf{x}(t_1), \dots, \mathbf{x}(t_T)) = \frac{p(\mathbf{x}(t_1), \dots, \mathbf{x}(t_T) \mid \boldsymbol{\theta})\, p(\boldsymbol{\theta})}{p(\mathbf{x}(t_1), \dots, \mathbf{x}(t_T))}$$

$$\propto p(\boldsymbol{\theta}) \prod_{k=0}^{T-1} p(\mathbf{x}(t_{k+1}) \mid \mathbf{x}(t_k), \boldsymbol{\theta}), \tag{11.5}$$

where $p(\boldsymbol{\theta})$ is the prior distribution and $\propto$ denotes a constant proportionality. Typical Bayesian methods include maximum a posteriori (MAP) estimation, Laplace approximations, Markov chain Monte Carlo, and other Monte Carlo methods.

In order to use ML methods or Bayesian methods, we need to be able to evaluate the likelihood, which in general is hard. This is because the likelihood depends on the transition densities, which are solutions to the Fokker–Planck–Kolmogorov (FPK) equation (see Section 5.4) and are thus hard to compute. If we know the transition densities, then we can explicitly evaluate the likelihood, which is indeed the case for linear SDEs. Parameter estimation in these kinds of models is considered in Section 11.3.

In the case of general multivariate nonlinear SDEs, we cannot solve the FPK and thus the transition density is unknown. In that case, the typical approach is to replace the SDE or its transition density in the likelihood with a tractable approximation. We can, for example, use the various SDE discretization methods that we have already covered (Itô–Taylor, stochastic Runge–Kutta, linearization) for forming a discrete-time SDE approximation whose transition density we can evaluate. Another way is to directly approximate the transition density of the SDE using Gaussian approximations, Hermite expansions, or other approximations that we saw in the previous chapters.

In this chapter, we also discuss partially and noisily observed SDE models. In those models, we do not observe the values $\mathbf{x}(t_1), \mathbf{x}(t_2), \ldots, \mathbf{x}(t_T)$ directly, but instead we only observe noisy versions of them, such as $\mathbf{y}_k = \mathbf{H}\,\mathbf{x}(t_k) + \mathbf{r}_k$, where $\mathbf{H}$ is some (possibly singular) matrix and $\mathbf{r}_k \sim \mathrm{N}(0, \mathbf{R})$ is a Gaussian noise. More generally, as in the previous chapter, the observations might come from a conditional distribution $p(\mathbf{y}_k \mid \mathbf{x}(t_k))$. In these models, we cannot use the previous likelihood expressions, but instead we need to compute the marginal likelihood of the measurements given the parameters $p(\mathbf{y}_1, \mathbf{y}_2, \ldots, \mathbf{y}_T \mid \boldsymbol{\theta})$ or the corresponding posterior distributions $p(\boldsymbol{\theta} \mid \mathbf{y}_1, \mathbf{y}_2, \ldots, \mathbf{y}_T)$ at least up to an unknown constant factor. In order to do parameter estimation in these models, we need to use the filtering and smoothing methods from the previous chapter as parts of the parameter estimation methods for evaluating the marginal likelihood or posterior distribution.

11.2 Computational Methods for Parameter Estimation

As discussed in the previous section, provided that we can evaluate the likelihood $p(\mathbf{x}(t_1), \mathbf{x}(t_2), \ldots, \mathbf{x}(t_T) \mid \boldsymbol{\theta})$ or its negative logarithm $\ell(\boldsymbol{\theta}) =$

$-\log p(\mathbf{x}(t_1), \mathbf{x}(t_2), \ldots, \mathbf{x}(t_T) \mid \boldsymbol{\theta})$, we have a wide range of off-the-shelf computational methods that we can use for estimating the parameters $\boldsymbol{\theta}$. The aim of this section is to discuss a couple of them, optimization-based methods and Markov chain Monte Carlo (MCMC) methods. For more details and more advanced computational methods, the reader is referred to the books of Luenberger and Ye (2008), Liu (2001), Brooks et al. (2011), and Gelman et al. (2013), along with Särkkä (2013).

The simplest approach to estimate parameters is to maximize the likelihood of the measured values with respect to the parameter values. This can be seen as a method that finds the parameters that best fit the data. The method can be written in algorithmic form as follows.

Algorithm 11.2 (ML estimate). *The maximum likelihood estimate of SDE parameters can be obtained by finding the vector of parameters $\boldsymbol{\theta}_{\mathrm{ML}}$ that minimizes the negative log-likelihood $\ell(\boldsymbol{\theta})$ defined in Equation* (11.4):

$$\boldsymbol{\theta}_{\mathrm{ML}} = \arg\min_{\boldsymbol{\theta}} \ell(\boldsymbol{\theta}). \tag{11.6}$$

In practice, the minimum can be computed either analytically by setting derivatives to zero or by using numerical optimization methods (see, e.g., Luenberger and Ye, 2008).

In the Bayesian setting, we might also have a prior distribution $p(\boldsymbol{\theta})$, which restricts or weights the possible parameter values. Then we can define the unnormalized negative log-posterior by using the following:

$$\ell_p(\boldsymbol{\theta}) = \ell(\boldsymbol{\theta}) - \log p(\boldsymbol{\theta}). \tag{11.7}$$

With this prior information, the ML estimate is generalized to the maximum a posteriori estimate shown in Algorithm 11.3.

Algorithm 11.3 (MAP estimate). *The MAP estimate $\boldsymbol{\theta}_{\mathrm{MAP}}$ of the SDE parameters can be found by minimizing the unnormalized negative log-posterior defined in Equation* (11.7):

$$\boldsymbol{\theta}_{\mathrm{MAP}} = \arg\min_{\boldsymbol{\theta}} \ell_p(\boldsymbol{\theta}). \tag{11.8}$$

A completely different class of algorithms for likelihood-based inference are MCMC methods. These kinds of methods are particularly commonly used in Bayesian analysis (Gelman et al., 2013; Särkkä, 2013), but MCMC methods have a multitude of other applications as well (Liu, 2001; Brooks et al., 2011).

The aim of MCMC in the SDE context is to generate samples from the posterior distribution (11.5). That is, instead of summarizing the posterior distribution via its single maximum (the MAP estimate), we generate a set of samples from the distribution. These samples can then be used for computing the best parameter estimates (e.g., the posterior mean) as well as uncertainty in the parameter estimates (e.g., the posterior covariance).

The most common MCMC method is the Metropolis–Hastings algorithm in Algorithm 11.4. In order to implement it, we only need to be able to evaluate the unnormalized negative log-posterior (11.7).

Algorithm 11.4 (Metropolis–Hastings). *The Metropolis–Hastings (MH) algorithm for generating samples from a distribution $p(\boldsymbol{\theta}) \propto \exp(-\ell_p(\boldsymbol{\theta}))$ is the following.*

1. *Draw the starting point, $\boldsymbol{\theta}^{(0)}$ from an arbitrary initial distribution.*
2. *For each iteration $i = 1, 2, \ldots, N$, do the following steps:*

   a. *Sample a candidate point $\boldsymbol{\theta}^*$ from the proposal distribution:*

   $$\boldsymbol{\theta}^* \sim q(\boldsymbol{\theta}^* \mid \boldsymbol{\theta}^{(i-1)}). \tag{11.9}$$

   b. *Evaluate the acceptance probability:*

   $$\alpha_i = \min\left\{1, \exp(\ell_p(\boldsymbol{\theta}^{(i-1)}) - \ell_p(\boldsymbol{\theta}^*)) \frac{q(\boldsymbol{\theta}^{(i-1)} \mid \boldsymbol{\theta}^*)}{q(\boldsymbol{\theta}^* \mid \boldsymbol{\theta}^{(i-1)})}\right\}. \tag{11.10}$$

   c. *Generate a uniform random variable $u \sim \mathrm{U}(0, 1)$ and set the following:*

   $$\boldsymbol{\theta}^{(i)} = \begin{cases} \boldsymbol{\theta}^*, & \text{if } u \leq \alpha_i, \\ \boldsymbol{\theta}^{(i-1)}, & \text{otherwise.} \end{cases} \tag{11.11}$$

The Metropolis algorithm is a commonly used special case of MH, where the proposal distribution is symmetric, $q(\boldsymbol{\theta}^{(i-1)} \mid \boldsymbol{\theta}^{(i)}) = q(\boldsymbol{\theta}^{(i)} \mid \boldsymbol{\theta}^{(i-1)})$. In this case, the acceptance probability reduces to the following:

$$\alpha_i = \min\left\{1, \exp(\ell_p(\boldsymbol{\theta}^{(i-1)}) - \ell_p(\boldsymbol{\theta}^*))\right\}. \tag{11.12}$$

The MH algorithm basically has a single design parameter, the proposal distribution $q(\boldsymbol{\theta}^* \mid \boldsymbol{\theta})$. However, this distribution almost completely defines the algorithm operation, and by selecting it in specific ways we get different brands of MCMC methods. For details, the reader is referred to Brooks et al. (2011).

11.3 Parameter Estimation in Linear SDE Models

In this section, the aim is to consider ML and Bayesian inference in linear SDEs. What makes linear SDEs special is that their transition densities are Gaussian and hence can be efficiently evaluated. In certain simple cases, we can also compute the ML estimates (or MAP estimates) of the parameters in closed form. However, more generally we need to resort to the computational methods outlined in the previous section.

Let us start by a considering the Ornstein–Uhlenbeck process which we already saw in Example 11.1:

$$\mathrm{d}x = -\lambda\,x\,\mathrm{d}t + \mathrm{d}\beta, \quad x(0) = x_0, \tag{11.13}$$

where λ is unknown and β has an unknown diffusion constant q. The vector of unknown parameters is thus $\boldsymbol{\theta} = (\lambda, q)$, and we assume that we have observed the SDE trajectory at $x(\Delta t), x(2\,\Delta t), \ldots, x(T\,\Delta t)$.

The transition density of the SDE is now given as (recall Example 6.2)

$$p(x(t + \Delta t) \mid x(t))$$
$$= \mathrm{N}\left(x(t + \Delta t) \mid \exp(-\lambda\,\Delta t)\,x(t), \frac{q}{2\lambda}\,[1 - \exp(-2\lambda\,\Delta t)]\right). \tag{11.14}$$

Thus the negative log-likelihood can be written as

$$\ell(\lambda, q) = \sum_{k=0}^{T-1}\left[\frac{1}{2}\log\left(2\pi\,\frac{q}{2\lambda}\,[1 - \exp(-2\lambda\,\Delta t)]\right)\right.$$
$$\left. + \frac{\lambda}{q\,[1 - \exp(-2\lambda\,\Delta t)]}\,(x(t_{k+1}) - \exp(-\lambda\,\Delta t)\,x(t_k))^2\right]. \tag{11.15}$$

However, for practical computation of the ML estimate it is more convenient to reparametrize the negative log-likelihood in terms of

$$\begin{aligned} a &= \exp(-\lambda\,\Delta t), \\ \Sigma &= \frac{q}{2\lambda}\,[1 - \exp(-2\lambda\,\Delta t)], \end{aligned} \tag{11.16}$$

which thus gives

$$\ell(a, \Sigma) = \sum_{k=0}^{T-1}\left[\frac{1}{2}\log(2\pi\,\Sigma) + \frac{1}{2\Sigma}\,(x(t_{k+1}) - a\,x(t_k))^2\right]. \tag{11.17}$$

Setting derivatives with respect to a and Σ to zero then gives

$$a_{\text{ML}} = \frac{\sum_{k=0}^{T-1} x(t_k) \, x(t_{k+1})}{\sum_{k=0}^{T-1} x(t_k) \, x(t_k)},$$

$$\Sigma_{\text{ML}} = \frac{1}{T} \sum_{k=0}^{T-1} (x(t_{k+1}) - a_{\text{ML}} \, x(t_k))^2, \tag{11.18}$$

which in terms of original λ and q gives the final ML estimates of the parameters:

$$\lambda_{\text{ML}} = -\frac{1}{\Delta t} \log \left[\frac{\sum_{k=0}^{T-1} x(t_k) \, x(t_{k+1})}{\sum_{k=0}^{T-1} x(t_k) \, x(t_k)} \right],$$

$$q_{\text{ML}} = \frac{1}{T} \left(\frac{2 \lambda_{\text{ML}}}{1 - \exp(-2 \lambda_{\text{ML}} \, \Delta t)} \right) \sum_{k=0}^{T-1} (x(t_{k+1}) - \exp(-\lambda_{\text{ML}} \, \Delta t) \, x(t_k))^2. \tag{11.19}$$

We can also, in principle, do the similar inference for a more general LTI SDE

$$d\mathbf{x} = \mathbf{F}(\boldsymbol{\theta}) \, \mathbf{x} \, dt + \mathbf{L}(\boldsymbol{\theta}) \, d\boldsymbol{\beta}, \quad \mathbf{x}(0) = \mathbf{x}_0, \tag{11.20}$$

where the vector of Brownian motions $\boldsymbol{\beta}$ has the diffusion matrix $\mathbf{Q}(\boldsymbol{\theta})$. With sampling at $\mathbf{x}(\Delta t), \mathbf{x}(2 \, \Delta t), \dots, \mathbf{x}(T \, \Delta t)$, we get the negative log-likelihood

$$\ell(\mathbf{A}, \boldsymbol{\Sigma}) = \sum_{k=0}^{T-1} \left[\frac{1}{2} \log |2\pi \, \boldsymbol{\Sigma}| \right.$$
$$\left. + \frac{1}{2} (\mathbf{x}(t_{k+1}) - \mathbf{A} \, \mathbf{x}(t_k))^\mathsf{T} \, \boldsymbol{\Sigma}^{-1} \, (\mathbf{x}(t_{k+1}) - \mathbf{A} \, \mathbf{x}(t_k)) \right], \tag{11.21}$$

which we have already written in terms of

$$\mathbf{A} = \exp(\mathbf{F}(\boldsymbol{\theta}) \, \Delta t),$$

$$\boldsymbol{\Sigma} = \int_0^{\Delta t} \exp(\mathbf{F}(\boldsymbol{\theta}) \, (\Delta t - \tau)) \, \mathbf{L}(\boldsymbol{\theta}) \, \mathbf{Q}(\boldsymbol{\theta}) \, \mathbf{L}^\mathsf{T}(\boldsymbol{\theta}) \, \exp(\mathbf{F}(\boldsymbol{\theta}) \, (\Delta t - \tau))^\mathsf{T} \, d\tau. \tag{11.22}$$

When we set the derivatives with respect to $\mathbf{A}$ and $\mathbf{\Sigma}$ to zero, we get

$$\mathbf{\Sigma}_{\text{ML}} = \frac{1}{T} \sum_{k=0}^{T-1} (\mathbf{x}(t_{k+1}) - \mathbf{A}\,\mathbf{x}(t_k))\,(\mathbf{x}(t_{k+1}) - \mathbf{A}\,\mathbf{x}(t_k))^{\mathsf{T}},$$

$$\mathbf{A}_{\text{ML}} = \left(\sum_{k=0}^{T-1} \mathbf{x}(t_{k+1})\,\mathbf{x}^{\mathsf{T}}(t_k) \right) \left(\sum_{k=0}^{T-1} \mathbf{x}(t_k)\,\mathbf{x}^{\mathsf{T}}(t_k) \right)^{-1}, \tag{11.23}$$

after which we still need to solve $\boldsymbol{\theta}$ from Equation (11.23). Unfortunately, this solution is rarely possible and seldom even exists when the parameters theta appear nontrivially in $\mathbf{A}$ and $\mathbf{Q}$.

In the preceding derivation for general $\mathbf{A}$ and $\mathbf{Q}$, we have failed to take into account that $\boldsymbol{\theta}$ might have significantly lower dimensionality than $\mathbf{A}$ and $\mathbf{Q}$ – thus it does not lead to the correct ML estimate unless we can uniquely solve the parameters given $\mathbf{A}$ and $\mathbf{Q}$. Furthermore, we often do not have a constant sampling period Δt, and the SDE might be time-varying – and we might also have an unknown offset function in the SDE.

For general linear SDEs of the general form

$$\mathrm{d}\mathbf{x} = \mathbf{F}(t;\boldsymbol{\theta})\,\mathbf{x}\,\mathrm{d}t + \mathbf{u}(t;\boldsymbol{\theta})\,\mathrm{d}t + \mathbf{L}(t;\boldsymbol{\theta})\,\mathrm{d}\boldsymbol{\beta}, \tag{11.24}$$

it is advisable to directly consider the negative log-likelihood, which is more generally given as

$$\begin{aligned} \ell(\boldsymbol{\theta}) = \sum_{k=0}^{T-1} \Bigg[&\frac{1}{2} \log |2\pi\,\mathbf{\Sigma}_k(\boldsymbol{\theta})| \\ &+ \frac{1}{2} (\mathbf{x}(t_{k+1}) - \mathbf{A}_k(\boldsymbol{\theta})\,\mathbf{x}(t_k) - \mathbf{u}_k(\boldsymbol{\theta}))^{\mathsf{T}}\,\mathbf{\Sigma}_k^{-1}(\boldsymbol{\theta}) \\ &\times (\mathbf{x}(t_{k+1}) - \mathbf{A}_k(\boldsymbol{\theta})\,\mathbf{x}(t_k) - \mathbf{u}_k(\boldsymbol{\theta})) \Bigg], \end{aligned} \tag{11.25}$$

where $\mathbf{A}_k(\boldsymbol{\theta})$, $\mathbf{u}_k(\boldsymbol{\theta})$, and $\mathbf{\Sigma}_k(\boldsymbol{\theta})$ are given by Equations (6.9), (6.10), and (6.11), respectively. We can now numerically find the minimum of the negative log-likelihood by using numerical optimization methods (e.g., Luenberger and Ye, 2008). In order to do that, we also need to compute the derivatives with respect to the parameters, where the difficulty arises from deriving the partial derivatives $\partial \mathbf{A}_k(\boldsymbol{\theta})/\partial \theta_i$, $\partial \mathbf{u}_k(\boldsymbol{\theta})/\partial \theta_i$, and $\partial \mathbf{\Sigma}_k(\boldsymbol{\theta})/\partial \theta_i$. However, this can be done by using the matrix fraction decomposition (see, e.g., Mbalawata et al., 2013).

The negative log-likelihood expression in Equation (11.25) also allows for the use of Bayesian methods for parameter estimation. This is because

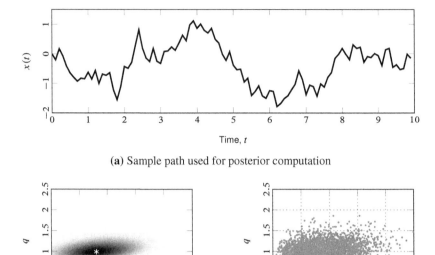

(**a**) Sample path used for posterior computation

(**b**) Posterior distribution   (**c**) MCMC samples from posterior

Figure 11.1 Illustration of the posterior computation for the Ornstein–Uhlenbeck model in Example 11.5. Subfigure (a) shows the simulated data (with $\lambda = 1/2$, $q = 1$), (b) shows the posterior distribution along with the ML/MAP estimates $\lambda_{\mathrm{MAP}} = 0.80$ and $q_{\mathrm{MAP}} = 0.99$, and (c) shows MCMC samples from the posterior generated with Metropolis–Hastings.

by Equation (11.7), the posterior distribution of the parameters can be written as

$$p(\boldsymbol{\theta} \mid \mathbf{x}(t_1), \mathbf{x}(t_2), \dots, \mathbf{x}(t_T)) \propto p(\boldsymbol{\theta}) \, \exp(-\ell(\boldsymbol{\theta}))$$
$$= \exp(-\ell_p(\boldsymbol{\theta})), \qquad (11.26)$$

where $p(\boldsymbol{\theta})$ is the prior distribution. This unnormalized posterior distribution can now be plugged into various MCMC methods (see, e.g., Brooks et al., 2011), or we can compute MAP estimates (or Laplace approximations; see Gelman et al., 2013) by minimizing the negative logarithm of it by numerical optimization (e.g., Luenberger and Ye, 2008).

Example 11.5 (Exact parameter estimation in the Ornstein–Uhlenbeck model). *The posterior distribution of the parameters λ and q of the*

Ornstein–Uhlenbeck model

$$dx = -\lambda\, x\, dt + d\beta, \quad x(0) = 0, \quad (11.27)$$

using the exact negative log-likelihood (11.15) *and a uniform prior, is shown in Figure 11.1. The data used for the posterior distribution are shown as well. The data were generated using the parameter values* $\lambda = 1/2$, $q = 1$, *and the sampling period in the data was* $\Delta t = 1/10$ *with a total of* 100 *points.*

11.4 Approximated-Likelihood Methods

With nonlinear SDEs of the generic form given in Equation (11.1), we have an additional challenge that the evaluation of the transition density is intractable. This also makes parameter estimation harder, because we cannot evaluate the likelihood term in Equation (11.3) nor its negative logarithm in Equation (11.4).

In approximated-likelihood methods, we replace the likelihood, or more specifically the transition densities used for computing the likelihood, with approximations. One approach is to approximate the SDE with a continuous- or discrete-time system whose transition density we can evaluate. For that purpose, we can use the various SDE simulation and discretization methods that we discussed in Chapter 8.

For example, recall that one step of the Euler–Maruyama method in Algorithm 8.1 is

$$\hat{\mathbf{x}}(t_{k+1}) = \hat{\mathbf{x}}(t_k) + \mathbf{f}(\hat{\mathbf{x}}(t_k), t_k; \boldsymbol{\theta})\, \Delta t + \mathbf{L}(\hat{\mathbf{x}}(t_k), t_k; \boldsymbol{\theta})\, \Delta\boldsymbol{\beta}_k, \quad (11.28)$$

for which $\Delta\boldsymbol{\beta}_k \sim \mathrm{N}(\mathbf{0}, \mathbf{Q}(\boldsymbol{\theta})\, \Delta t)$. It has a Gaussian transition density and hence leads to the approximation

$$
\begin{aligned}
p(\mathbf{x}(t_{k+1}) \mid \mathbf{x}(t_k), \boldsymbol{\theta}) \\
\approx \mathrm{N}(\mathbf{x}(t_{k+1}) \mid \mathbf{x}(t_k) + \mathbf{f}(\mathbf{x}(t_k), t_k; \boldsymbol{\theta})\, \Delta t, \\
\mathbf{L}(\mathbf{x}(t_k), t_k; \boldsymbol{\theta})\, \mathbf{Q}(\boldsymbol{\theta})\, \mathbf{L}^{\mathsf{T}}(\mathbf{x}(t_k), t_k; \boldsymbol{\theta})\, \Delta t). \quad (11.29)
\end{aligned}
$$

The approximation to the likelihood in Equation (11.3) is then given as

$$
\begin{aligned}
p(\mathbf{x}(t_1), \ldots, \mathbf{x}(t_T) \mid \boldsymbol{\theta}) \\
= \prod_{k=0}^{T-1} \mathrm{N}(\mathbf{x}(t_{k+1}) \mid \mathbf{x}(t_k) + \mathbf{f}(\mathbf{x}(t_k), t_k; \boldsymbol{\theta})\, \Delta t, \\
\mathbf{L}(\mathbf{x}(t_k), t_k; \boldsymbol{\theta})\, \mathbf{Q}(\boldsymbol{\theta})\, \mathbf{L}^{\mathsf{T}}(\mathbf{x}(t_k), t_k; \boldsymbol{\theta})\, \Delta t). \quad (11.30)
\end{aligned}
$$

and the approximation to the negative log-likelihood (11.4) will be

$$
\ell(\boldsymbol{\theta}) = \sum_{k=0}^{T-1} \left[\frac{1}{2} \log |2\pi \, \mathbf{L}(\mathbf{x}(t_k), t_k; \boldsymbol{\theta}) \, \mathbf{Q}(\boldsymbol{\theta}) \, \mathbf{L}^{\mathsf{T}}(\mathbf{x}(t_k), t_k; \boldsymbol{\theta}) \, \Delta t | \right.
$$
$$
+ \frac{1}{2} (\mathbf{x}(t_{k+1}) - \mathbf{x}(t_k) - \mathbf{f}(\mathbf{x}(t_k), t_k; \boldsymbol{\theta}) \, \Delta t)^{\mathsf{T}}
$$
$$
\times (\mathbf{L}(\mathbf{x}(t_k), t_k; \boldsymbol{\theta}) \, \mathbf{Q}(\boldsymbol{\theta}) \, \mathbf{L}^{\mathsf{T}}(\mathbf{x}(t_k), t_k; \boldsymbol{\theta}) \, \Delta t)^{-1}
$$
$$
\left. \times (\mathbf{x}(t_{k+1}) - \mathbf{x}(t_k) - \mathbf{f}(\mathbf{x}(t_k), t_k; \boldsymbol{\theta}) \, \Delta t) \right], \quad (11.31)
$$

which is going to be as (in)accurate as the Euler–Maruyama method is – thus we expect this approximation to work only with small Δt. Similarly, we can approximate the SDE with a step of the Milstein method or higher-order Itô–Taylor expansion-based methods. We can also use strong or weak Runge–Kutta and related methods. The only limitation is that we need to be able to evaluate the transition density corresponding to the discrete-time approximation. In algorithm form, we have the following.

Algorithm 11.6 (Discretization-based approximated-likelihood estimation). *SDE discretization-based approximated-likelihood parameter estimation can be done as follows.*

1. *Use an SDE discretization such as Euler–Maruyama, Itô–Taylor expansions, or stochastic Runge–Kutta to form a discrete-time approximation to the SDE.*
2. *Let $\hat{p}(\mathbf{x}(t_{k+1}) \mid \mathbf{x}(t_k), \boldsymbol{\theta})$ be the transition density of the discrete-time approximation. Approximate the negative log-likelihood as*

$$
\hat{\ell}(\boldsymbol{\theta}) = - \sum_{k=0}^{T-1} \log \hat{p}(\mathbf{x}(t_{k+1}) \mid \mathbf{x}(t_k), \boldsymbol{\theta}). \quad (11.32)
$$

3. *Perform maximum likelihood estimation or any form of Bayesian estimation by replacing the exact likelihood ℓ with the preceding approximation $\hat{\ell}$.*

Instead of approximating the SDE as such, we can also approximate its transition density. One general approach to transition density approximation is to approximate is as Gaussian. For this, we can use, for example, the Gaussian and linearization approximations considered in Sections 9.1, 9.2, and 9.3 or alternatively using the Taylor series expansions for the first two moments as described in Section 9.4. Even the Euler–Maruyama happens to have this form, although higher-order Ito–Taylor based methods typically do not. Thus in the end we get a transition density approximation of the form

$$p(\mathbf{x}(t_{k+1}) \mid \mathbf{x}(t_k), \boldsymbol{\theta}) \approx \mathrm{N}(\mathbf{x}(t_{k+1}) \mid \boldsymbol{\mu}(\mathbf{x}(t_k), \Delta t; \boldsymbol{\theta}), \boldsymbol{\Sigma}(\mathbf{x}(t_k), \Delta t; \boldsymbol{\theta})),$$
(11.33)

which can be then further plugged into the likelihood expression. Thus we get the following algorithm.

Algorithm 11.7 (Gaussian approximated-likelihood parameter estimation). *Gaussian approximation-based approximated-likelihood parameter estimation can be done as follows:*

1. *Use linearization, moment approximation, or any other methods to form the approximations to the mean $\boldsymbol{\mu}(\mathbf{x}(t_k), \Delta t; \boldsymbol{\theta})$ and covariance $\boldsymbol{\Sigma}(\mathbf{x}(t_k), \Delta t; \boldsymbol{\theta})$ of the Gaussian approximation on $p(\mathbf{x}(t_{k+1}) \mid \mathbf{x}(t_k), \boldsymbol{\theta})$.*

2. *Approximate the negative log likelihood as*

$$\hat{\ell}(\boldsymbol{\theta}) = -\sum_{k=0}^{T-1} \log \mathrm{N}(\mathbf{x}(t_{k+1}) \mid \boldsymbol{\mu}(\mathbf{x}(t_k), \Delta t; \boldsymbol{\theta}), \boldsymbol{\Sigma}(\mathbf{x}(t_k), \Delta t; \boldsymbol{\theta}))$$

$$= \sum_{k=0}^{T-1} \Bigg[\frac{1}{2} \log |2\pi \, \boldsymbol{\Sigma}(\mathbf{x}(t_k), \Delta t; \boldsymbol{\theta})|$$

$$+ \frac{1}{2}(\mathbf{x}(t_{k+1}) - \boldsymbol{\mu}(\mathbf{x}(t_k), \Delta t; \boldsymbol{\theta}))^\mathsf{T} \, \boldsymbol{\Sigma}^{-1}(\mathbf{x}(t_k), \Delta t; \boldsymbol{\theta})$$

$$\times (\mathbf{x}(t_{k+1}) - \boldsymbol{\mu}(\mathbf{x}(t_k), \Delta t; \boldsymbol{\theta})) \Bigg].$$
(11.34)

3. *Perform maximum likelihood estimation or any form of Bayesian estimation by replacing the exact likelihood with the preceding approximation.*

Note that one way to approximate the mean and covariance of the preceding Gaussian approximation is as the conditional mean $\mathrm{E}[\mathbf{x}(t_{k+1}) \mid \mathbf{x}(t_k)]$ and covariance $\mathrm{Cov}[\mathbf{x}(t_{k+1}) \mid \mathbf{x}(t_k)]$ of the SDE, that is, by *moment matching*. However, this might not always lead to the best possible approximation on the likelihood as whole (cf. Archambeau and Opper, 2011; García-Fernández et al., 2017; Tronarp et al., 2018).

Not all transition densities can be approximated as Gaussian, for example, when they are multimodal. Then we can use non-Gaussian approximations such as the Hermite expansions considered in Section 9.5. We can also numerically solve the Fokker–Planck–Kolmogorov partial differential equation as described in Section 9.6 using methods such as finite-differences or Galerkin methods, which also leads to approximations to the transition density. The Taylor series expansion considered in Section 9.4 can also be used to approximate the moments of the transition density, and given the moments we can, for example, form maximum entropy approximation to the density (Cover and Thomas, 2006). The simulated likelihood method in Section 9.7 was also originally proposed exactly for this purpose.

We get the following algorithm.

Algorithm 11.8 (Non-Gaussian approximated-likelihood estimation). *Transition density approximation-based non-Gaussian approximated-likelihood parameter estimation can be done as follows:*

1. *Use a suitable method to form a parametric approximation $\hat{p}(\mathbf{x}(t_{k+1}) \mid \mathbf{x}(t_k), \boldsymbol{\theta})$ to the transition density using some of the previously discussed methods. Then approximate the negative log-likelihood as*

$$\hat{\ell}(\boldsymbol{\theta}) = - \sum_{k=0}^{T-1} \log \hat{p}(\mathbf{x}(t_{k+1}) \mid \mathbf{x}(t_k), \boldsymbol{\theta}). \qquad (11.35)$$

2. *Perform maximum likelihood estimation or any form of Bayesian estimation by replacing the exact likelihood ℓ with the preceding approximation $\hat{\ell}$.*

Example 11.9 (Approximate parameter estimation in the Ornstein–Uhlenbeck model). *The parameter estimation problem in Example 11.5 was repeated using an approximated likelihood method using the Euler–Maruyama transition density approximation. Figure 11.2 shows a comparison of the exact and approximate posteriors. As can be seen in the figure, the effect of the approximation is the loss of the correlation between the parameters.*

11.5 Likelihood Methods for Indirectly Observed SDEs

So far in this chapter, we have assumed that the states $\mathbf{x}(t_1), \mathbf{x}(t_2), \mathbf{x}(t_3), \ldots$ have been perfectly measured. However, as we saw in Chapter 10, in many real-world problems we do not directly observe the states, but we only get to see measurements $\mathbf{y}_1, \mathbf{y}_2, \ldots$, which are indirectly related to the state and contain noise. However, it turns out

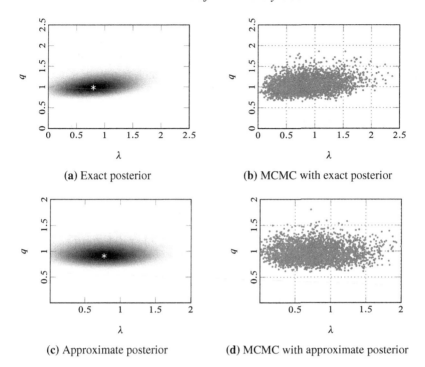

(a) Exact posterior

(b) MCMC with exact posterior

(c) Approximate posterior

(d) MCMC with approximate posterior

Figure 11.2 Illustration of the effect of the Euler–Maruyama
approximation to the posterior distribution of
Ornstein–Uhlenbeck model parameters from Example 11.9. The
exact posterior and its MCMC samples are shown in subfigures
(a) and (b), and the approximations in (c) and (d), respectively.
The point estimates of the parameters with Euler–Maryuama
approximation are $\hat{\lambda}_{\mathrm{MAP}} = 0.77$ and $\hat{q}_{\mathrm{MAP}} = 0.91$ when with the
exact posterior they were $\lambda_{\mathrm{MAP}} = 0.80$ and $q_{\mathrm{MAP}} = 0.99$.

that the filtering and smoothing methods can be combined with parameter
estimation methods to allow for parameter estimation from such indirect,
noisy measurements. These kinds of methods for SDE models have been
presented, for example, in Singer (2002), Mbalawata et al. (2013), and
Särkkä et al. (2015b) and their discrete-time analogs in Cappé et al.
(2005) and Särkkä (2013). However, the idea itself dates back at least to
Schweppe (1965) and Jazwinski (1970).

Let us assume that our model has the form

$$
\begin{aligned}
\mathrm{d}\mathbf{x} &= \mathbf{f}(\mathbf{x}, t; \boldsymbol{\theta})\,\mathrm{d}t + \mathbf{L}(\mathbf{x}, t; \boldsymbol{\theta})\,\mathrm{d}\boldsymbol{\beta}(t), \\
\mathbf{y}_k &\sim p(\mathbf{y}_k \mid \mathbf{x}(t_k); \boldsymbol{\theta}),
\end{aligned}
\tag{11.36}
$$

where $\boldsymbol{\theta}$ is a vector of unknown parameters of the system. This model is a continuous-discrete model for which we developed two types of Bayesian filters in Section 10.5. The key to the solution is to notice that the normalization constant Z_k in Equation (10.54), which is a byproduct of the Bayesian filtering equations, is actually

$$Z_k(\boldsymbol{\theta}) = \int p(\mathbf{y}_k \mid \mathbf{x}(t_k); \boldsymbol{\theta}) \, p(\mathbf{x}(t_k) \mid \mathbf{y}_1, \dots, \mathbf{y}_{k-1}; \boldsymbol{\theta}) \, d\mathbf{x}(t_k)$$

$$= p(\mathbf{y}_k \mid \mathbf{y}_1, \dots, \mathbf{y}_{k-1}; \boldsymbol{\theta}). \tag{11.37}$$

We can now express the marginal likelihood by using the prediction error decomposition

$$p(\mathbf{y}_1, \dots, \mathbf{y}_T \mid \boldsymbol{\theta}) = \prod_{k=1}^{T} p(\mathbf{y}_k \mid \mathbf{y}_1, \dots, \mathbf{y}_{k-1}; \boldsymbol{\theta}) = \prod_{k=1}^{T} Z_k(\boldsymbol{\theta}). \tag{11.38}$$

Thus, provided that we can compute the terms $Z_k(\boldsymbol{\theta})$ or approximate them accurately, we can obtain the expression for the (negative logarithm of marginal) likelihood as

$$\ell(\boldsymbol{\theta}) = -\log p(\mathbf{y}_1, \dots, \mathbf{y}_T \mid \boldsymbol{\theta}) = -\sum_{k=1}^{T} \log Z_k(\boldsymbol{\theta}). \tag{11.39}$$

For linear Gaussian models, we can compute these likelihoods exactly with Kalman filters, whereas for nonlinear and non-Gaussian models we need to approximate them. Fortunately, the same approximations that can be used for approximate continuous-discrete filtering also provide approximations to the terms $Z_k(\boldsymbol{\theta})$.

The resulting likelihood expressions or approximations can be further optimized or sampled using methods such as MCMC. For more information on these kinds of methods, the reader is referred to Särkkä (2013), Mbalawata et al. (2013), and Särkkä et al. (2015b).

11.6 Expectation–Maximization, Variational Bayes, and Other Methods

There is also a wide range of other parameter estimation methods that we have not discussed here. We have not considered, for example, estimating function methods, generalized method of moments, nor methods based on approximating the continuous-time estimators. For more details about these methods, the reader is referred to the books of Rao (1999) and Iacus

(2008). We also have only considered parametric models, although non-parametric estimation of, for example, drift functions is possible as well (see Section 12.6 and Rao, 1999; Ruttor et al., 2013).

The indirectly observed case in the previous section has also spanned many other families of methods for parameter estimation in SDEs. For example, expectation–maximization methods can be used to approximate the maximum likelihood methods of discrete-time state-space models (e.g., Shumway and Stoffer, 1982; Särkkä, 2013; Kokkala et al., 2016). However, provided that we discretize the model first, these methods are directly applicable to continuous-discrete SDE models as well. So-called variational Bayes methods (e.g., Šmídl and Quinn, 2006) can also be used in the context of SDEs after the SDE has been discretized. Archambeau and Opper (2011) have also developed variational Bayes methods that can directly be used to estimate parameters in SDEs.

11.7 Exercises

11.1 Consider the following problem with unknown parameter θ:

$$x_k = \theta + r_k, \quad k = 1, 2, \ldots, T, \tag{11.40}$$

where $r_k \sim N(0, \sigma^2)$ are independent. Then accomplish the following:

(a) Derive the ML estimate of θ given the measurements $x_1, x_2, \ldots, x_T$.
(b) Fix $\theta = 1$, $\sigma = 1$ and simulate a set of data from the preceding model. Then compute the ML estimate. How close is it to the truth?
(c) Plot the negative log-likelihood as function of the parameter θ. Is the maximum close to the true value?

11.2 Assume that in Equation (11.40) we have a Gaussian prior density $p(\theta) = N(\theta \mid 0, \lambda^2)$. Then accomplish the following:

(a) Derive the MAP estimate of θ.
(b) How does the estimate behave as function of λ?
(c) Fix $\theta = 1$, $\sigma = 1$, $\lambda = 2$, and simulate a set of data from the preceding model. Then compute the MAP estimate. How close is it to the truth?
(d) Plot the posterior distribution of the parameter. Is the true parameter value well within the support of the distribution?

11.3 Estimate the parameter θ in Equation (11.40) using MCMC:

(a) Fix $\theta = 1$, $\sigma = 1$, $\lambda = 2$, and simulate a set of data from Equation 11.40 (with the prior).
(b) Implement an MH algorithm for sampling the parameter θ. Use a Gaussian proposal distribution $q(\theta^* \mid \theta) = N(\theta^* \mid \theta, \gamma^2)$, and select the parameter γ such that about $1/4$ of the proposals are accepted.

    (c) Plot the true posterior distribution and the histogram of samples. Do the results match?

11.4   Fill in the details in the derivation of the Ornstein–Uhlenbeck ML estimate:

    (a) Derive Equations (11.18) from (11.17).

    (b) Derive Equation (11.19).

11.5   Consider the parameter estimates (11.23) in the context of the Wiener velocity model. Assume that the only unknown is q:

    (a) Recall the expressions for $\mathbf{A}(\Delta t)$ and $\mathbf{\Sigma}(\Delta t)$ from Example 6.3.

    (b) Choose some Δt and q, and simulate some data from the exact discrete-time model in the example.

    (c) Compute estimates $\mathbf{A}_{\mathrm{ML}}$ and $\mathbf{\Sigma}_{\mathrm{ML}}$ using Equations 11.23.

    (d) Does the estimate of $\mathbf{A}$ match the correct one? Could you determine the parameter q from the estimate of $\mathbf{\Sigma}$?

11.6   Using the data from the preceding exercise, estimate q by numerically finding the ML estimate using the negative log-likelihood expression (11.25).

11.7   Still using the same data, estimate the posterior distribution of the parameter q by generating samples from the posterior distribution with uniform prior $p(q) \propto 1$ for $q > 0$ using the MH algorithm. Use a log transform $\theta = \log q$, which transforms the uniform prior to $p(\theta) \propto \exp(\theta)$.

11.8   Derive the derivatives of Equation (11.25) using the matrix fraction decomposition (see, e.g., Mbalawata et al., 2013). Check numerically using the Wiener velocity model that they are correct.

11.9   Consider the following model, where the parameters θ_1 and θ_2 are unknown:

$$\mathrm{d}x = \theta_1 \sin(x - \theta_2)\, \mathrm{d}t + \mathrm{d}\beta, \quad x(0) = 0,$$

where β is a standard Brownian motion.

    (a) Simulate data from the model using the Euler–Maruyama method.

    (b) Compute the maximum likelihood estimates of the parameters by using Euler–Maruyama approximation to the transition density.

    (c) Compute the MAP estimates of the parameters with independent $N(0, 1)$ priors on them both.

    (d) Simulate samples from the posterior distribution of parameters using the MH algorithm.

11.10 Repeat Exercise 11.9 with the Itô–Taylor method in Algorithm 8.4.

11.11 Repeat Exercise 11.9 by replacing the transition density with a linearization approximation given in Algorithm 9.4.

11.12 Extend the Kalman filter equations in Example 10.19 so that they can be used to compute the marginal likelihood of parameters λ and q. Using simulated data, estimate the parameters from noisy observations by maximizing the marginal likelihood.

12

Stochastic Differential Equations in Machine Learning

In this chapter, we aim to present an overview of how the topics in the previous chapters are related to concepts in probabilistic machine learning. This link is presented with the focus on Gaussian processes, which are stochastic processes already familiar from the previous chapters. In particular, it is useful to notice that solutions of linear SDEs are always Gaussian processes. An interesting question to ask is whether all Gaussian processes can be constructed as solutions to SDEs – the answer is negative, but we can get quite close to that.

In probabilistic machine learning, Gaussian process (GP) regression (O'Hagan, 1978; MacKay, 1998; Rasmussen and Williams, 2006) is a probabilistic paradigm where instead of using a fixed-form parametric model for regression such as a multi layer perceptron (MLP) neural network (Bishop, 2006), one postulates a Gaussian process prior over the model functions. Learning in Gaussian process regression means computing the posterior Gaussian process, which is conditioned to observed measurements. The prediction of unobserved values amounts to computing predictive distributions and their statistics. Although here we restrict our discussion to regression problems, we can similarly do Gaussian process classification (Rasmussen and Williams, 2006), provided that we use a logit or probit transformation in the measurement model (for further connections to general non-Gaussian likelihoods, see Nickisch et al., 2018).

It turns out that filtering and smoothing on linear stochastic differential equations are more or less equivalent to Gaussian process regression. The presentation in this chapter summarizes the presentation in Solin (2016) on the connections between (infinite-dimensional) Kalman filtering and GP regression (see also Hartikainen and Särkkä, 2010, 2011; Särkkä and Hartikainen, 2012; Särkkä et al., 2013, for further results and analysis). Additionally, we discuss learning of drift functions by using GP regression as well as latent force models and GP regression on SDE trajectories.

12.1 Gaussian Processes

A Gaussian process $\mathbf{x}(\boldsymbol{\xi})$ is a random function on a d-dimensional input space. Note that in the previous chapters we have always assumed that the input $\boldsymbol{\xi}$ is one-dimensional time t, but here we allow for multiple input dimensions.

Definition 12.1 (Gaussian process (GP)). *A Gaussian process $\mathbf{x}(\boldsymbol{\xi})$ is a random function $\mathbf{x}(\boldsymbol{\xi})$ with d-dimensional input $\boldsymbol{\xi}$ such that any finite collection of random variables $\mathbf{x}(\boldsymbol{\xi}_1), \mathbf{x}(\boldsymbol{\xi}_2), \ldots, \mathbf{x}(\boldsymbol{\xi}_n)$ has a multidimensional Gaussian distribution.*

As we already saw in Chapter 6, solutions of linear SDEs are Markov processes with Gaussian transition densities. Thus their finite-dimensional distributions are Gaussian and they are Gaussian processes by the preceding definition. Gaussian processes with $d > 1$ are also sometimes called *Gaussian fields*, but here we just call them *processes* as is commonly done in machine learning (Rasmussen and Williams, 2006).

A GP is commonly defined through the following kernel (covariance function) formalism (Rasmussen and Williams, 2006) denoted by

$$\mathbf{x}(\boldsymbol{\xi}) \sim \text{GP}(\mathbf{m}(\boldsymbol{\xi}), \mathbf{C}(\boldsymbol{\xi}, \boldsymbol{\xi}')), \tag{12.1}$$

which means that $\mathbf{x}(\boldsymbol{\xi})$ is a Gaussian process with mean and covariance functions $\mathbf{m}(\boldsymbol{\xi})$ and $\mathbf{C}(\boldsymbol{\xi}, \boldsymbol{\xi}')$, respectively. The mean and covariance functions also completely characterize a Gaussian process.

Definition 12.2 (Mean and covariance functions of GPs). *A Gaussian process can be defined in terms of a* mean $\mathbf{m}(\boldsymbol{\xi})$ *and* covariance function *(kernel)* $\mathbf{C}(\boldsymbol{\xi}, \boldsymbol{\xi}')$:

$$\begin{aligned}
\mathbf{m}(\boldsymbol{\xi}) &= \mathrm{E}[\mathbf{x}(\boldsymbol{\xi})], \\
\mathbf{C}(\boldsymbol{\xi}, \boldsymbol{\xi}') &= \mathrm{E}[(\mathbf{x}(\boldsymbol{\xi}) - \mathbf{m}(\boldsymbol{\xi}))(\mathbf{x}(\boldsymbol{\xi}') - \mathbf{m}(\boldsymbol{\xi}'))^{\mathsf{T}}].
\end{aligned} \tag{12.2}$$

The joint distribution of an arbitrary finite collection of random variables $\mathbf{x}(\boldsymbol{\xi}_1), \mathbf{x}(\boldsymbol{\xi}_2), \ldots, \mathbf{x}(\boldsymbol{\xi}_n)$ is then multidimensional Gaussian:

$$\begin{pmatrix} \mathbf{x}(\boldsymbol{\xi}_1) \\ \vdots \\ \mathbf{x}(\boldsymbol{\xi}_n) \end{pmatrix} \sim \mathrm{N} \left(\begin{pmatrix} \mathbf{m}(\boldsymbol{\xi}_1) \\ \vdots \\ \mathbf{m}(\boldsymbol{\xi}_n) \end{pmatrix}, \begin{pmatrix} \mathbf{C}(\boldsymbol{\xi}_1, \boldsymbol{\xi}_1) & \cdots & \mathbf{C}(\boldsymbol{\xi}_1, \boldsymbol{\xi}_n) \\ \vdots & \ddots & \vdots \\ \mathbf{C}(\boldsymbol{\xi}_n, \boldsymbol{\xi}_1) & \cdots & \mathbf{C}(\boldsymbol{\xi}_n, \boldsymbol{\xi}_n) \end{pmatrix} \right). \tag{12.3}$$

Even though the model is built upon functions depending on continuously defined inputs, data are always finite and allow the computations to be performed on finite index sets. In machine learning, GP models are often referred to as *nonparametric* because the number of parameters in the model is not fixed but rather spanned by the number of data points.

Special structure in data or the covariance function leads to useful special cases and formulations that bridge concepts across stochastic differential equations and GP regression.

Definition 12.3 (Stationary Gaussian process). *A process is said to be sta-tionary or* homogeneous *if its mean is constant and the covariance function is of the form*

$$\mathbf{C}(\boldsymbol{\xi}, \boldsymbol{\xi}') = \mathbf{C}(\boldsymbol{\xi} - \boldsymbol{\xi}'). \tag{12.4}$$

The covariance functions of stationary processes are often written as func-tions of variable $\boldsymbol{\xi}$ only, that is, $\mathbf{C}(\boldsymbol{\xi})$. Here we have formally made the substitution $\boldsymbol{\xi} \rightarrow \boldsymbol{\xi} - \boldsymbol{\xi}'$.

GP models with a stationary covariance function can be represented in terms of their spectral density. This stems from *Bochner's theorem* (see, e.g., Akhiezer and Glazman, 1993; Da Prato and Zabczyk, 2014), which states that an arbitrary positive definite function can be represented as a Fourier transform of a positive measure. If the measure has a density, it is called the *spectral density* of the process. This relation is the Fourier duality of covariance and spectral density, which is known as the *Wiener–Khinchin theorem* (we saw it already in Section 6.6). In its matrix form, it can be written as

$$\begin{aligned} \mathbf{S}(\boldsymbol{\omega}) &= \mathfrak{F}[\mathbf{C}(\boldsymbol{\xi})] \\ &= \int \mathbf{C}(\boldsymbol{\xi}) \, \exp(-i \, \boldsymbol{\omega}^{\mathsf{T}} \boldsymbol{\xi}) \, d\boldsymbol{\xi}. \end{aligned} \tag{12.5}$$

Definition 12.4 (Isotropic Gaussian process). *An* isotropic *covariance function has the form*

$$\mathbf{C}(\boldsymbol{\xi}, \boldsymbol{\xi}') = \mathbf{C}(\|\boldsymbol{\xi} - \boldsymbol{\xi}'\|). \tag{12.6}$$

That is, it is only a function of the distance of two input vectors, which means that the covariance is both translation and rotation invariant.

Definition 12.5 (Separable Gaussian process). *A* separable *covariance function can be rewritten as a product of two covariance functions with no shared inputs. For example, if $\boldsymbol{\xi} = (\boldsymbol{\xi}_1, \boldsymbol{\xi}_2)$, the covariance function $\mathbf{C}(\boldsymbol{\xi}, \boldsymbol{\xi}') = \mathbf{C}_1(\boldsymbol{\xi}_1, \boldsymbol{\xi}_1') \, \mathbf{C}_2(\boldsymbol{\xi}_2, \boldsymbol{\xi}_2')$ is separable.*

12.2 Gaussian Process Regression

In its simplest form, Gaussian process regression (Rasmussen and Williams, 2006) considers predicting values of an unknown scalar function

$$y = x(\boldsymbol{\xi}) \tag{12.7}$$

at certain test points $\{(\boldsymbol{\xi}_{\star,i}, y_{\star,i}) \mid i = 1, 2, \ldots, m\}$, based on a finite number of samples $\{(\boldsymbol{\xi}_k, y_k) \mid k = 1, 2, \ldots, n\}$ observed from it. Instead of postulating a parametric form of the function $x(\boldsymbol{\xi}; \boldsymbol{\theta})$ as in parametric regression and estimating the parameters $\boldsymbol{\theta}$, it is assumed that the function $x(\boldsymbol{\xi})$ is a sample from a zero mean Gaussian process with a given covariance function $C(\boldsymbol{\xi}, \boldsymbol{\xi}')$. A typical choice for the covariance function is the *squared exponential* (Rasmussen and Williams, 2006) or *exponentiated quadratic* (Álvarez et al., 2012) covariance function

$$C(\boldsymbol{\xi}, \boldsymbol{\xi}') = \sigma^2 \, \exp\left(-\frac{1}{2\ell^2}\|\boldsymbol{\xi} - \boldsymbol{\xi}'\|^2\right). \tag{12.8}$$

Note that although the common name for the covariance function is the squared exponential covariance function, the exponentiated quadratic covariance function is a far more descriptive name for it – anyway, $\exp(-x)^2 = \exp(-2x)$, which not equal to $\exp(-x^2)$.

If the vector of observed points is denoted as $\mathbf{y}_o = (y_1, \ldots, y_n)^\mathsf{T}$ and the vector of test points as $\mathbf{y}_\star = (y_{\star,1}, \ldots, y_{\star,m})^\mathsf{T}$, then based on the Gaussian assumption their joint distribution is

$$\begin{pmatrix} \mathbf{y}_o \\ \mathbf{y}_\star \end{pmatrix} = \mathrm{N}\left(\begin{pmatrix} \mathbf{0} \\ \mathbf{0} \end{pmatrix}, \begin{pmatrix} \mathbf{C}_{oo} & \mathbf{C}_{\star o}^\mathsf{T} \\ \mathbf{C}_{\star o} & \mathbf{C}_{\star\star} \end{pmatrix}\right), \tag{12.9}$$

where $[\mathbf{C}_{oo}]_{ij} = C(\boldsymbol{\xi}_i, \boldsymbol{\xi}_j)$ is the joint covariance matrix of observed points, $[\mathbf{C}_{\star o}]_{ij} = C(\boldsymbol{\xi}_{\star,i}, \boldsymbol{\xi}_j)$ is the cross-covariance of the test and observed points, and $[\mathbf{C}_{\star\star}]_{ij} = C(\boldsymbol{\xi}_{\star,i}, \boldsymbol{\xi}_{\star,j})$ is the joint covariance of test points.

The conditional distribution of test points given the observed values is now Gaussian, and by using the computation rules of Gaussian distribution, we can compute the minimum mean squared estimate of the test points, which is given by the conditional mean

$$\mathrm{E}[\mathbf{y}_\star \mid \mathbf{y}_o] = \mathbf{C}_{\star o} \, \mathbf{C}_{oo}^{-1} \, \mathbf{y}_o. \tag{12.10}$$

The covariance of the corresponding Gaussian posterior distribution is

$$\mathrm{Cov}[\mathbf{y}_\star \mid \mathbf{y}_o] = \mathbf{C}_{\star\star} - \mathbf{C}_{\star o} \, \mathbf{C}_{oo}^{-1} \, \mathbf{C}_{\star o}^\mathsf{T}. \tag{12.11}$$

Equations (12.10) and (12.11) are useful in applications where we can assume that the observed values are error-free and we want to interpolate the test point values using the observed values.

In statistical estimation problems, it is often assumed that the measurements are not perfect, but corrupted by additive Gaussian noise. That is, the measurements are modeled as

$$y_k = x(\boldsymbol{\xi}_k) + \varepsilon_k, \qquad \varepsilon_k \sim N(0, \sigma_n^2), \tag{12.12}$$

where ε_k are independent random variables, a priori independent of the Gaussian process $x(\boldsymbol{\xi})$. Now we are interested in computing the mean squared estimate of the values of the "clean" function $x(\boldsymbol{\xi})$ at a set of test points $\boldsymbol{\xi}_\star = \{\boldsymbol{\xi}_{\star,i} \mid i = 1, 2, \dots, m\}$ given the measurements at the observed points $\boldsymbol{\xi}_o = \{\boldsymbol{\xi}_i \mid i = 1, 2, \dots, n\}$. If we denote the vector of values of the function at the test points as $\mathbf{x}_\star = (x(\boldsymbol{\xi}_{\star,1}), \dots, x(\boldsymbol{\xi}_{\star,m}))^\mathsf{T}$, then the joint distribution of observed measurement and function values at test points is

$$\begin{pmatrix} \mathbf{y}_o \\ \mathbf{x}_\star \end{pmatrix} = N\left(\begin{pmatrix} \mathbf{0} \\ \mathbf{0} \end{pmatrix}, \begin{pmatrix} \mathbf{C}_{oo} + \sigma_n^2 \mathbf{I} & \mathbf{C}_{\star o}^\mathsf{T} \\ \mathbf{C}_{\star o} & \mathbf{C}_{\star\star} \end{pmatrix} \right), \tag{12.13}$$

and the conditional mean and covariance are given as

$$\begin{aligned} E[\mathbf{x}_\star \mid \mathbf{y}_o] &= \mathbf{C}_{\star o} (\mathbf{C}_{oo} + \sigma_n^2 \mathbf{I})^{-1} \mathbf{y}_o, \\ \mathrm{Cov}[\mathbf{x}_\star \mid \mathbf{y}_o] &= \mathbf{C}_{\star\star} - \mathbf{C}_{\star o} (\mathbf{C}_{oo} + \sigma_n^2 \mathbf{I})^{-1} \mathbf{C}_{\star o}^\mathsf{T}. \end{aligned} \tag{12.14}$$

The model can be easily generalized to vector processes and processes, including a prior mean function in the GP such as

$$\begin{aligned} \mathbf{y}_k &= \mathbf{x}(\boldsymbol{\xi}_k) + \boldsymbol{\varepsilon}_k, \qquad \boldsymbol{\varepsilon}_k \sim N(\mathbf{0}, \boldsymbol{\Sigma}), \\ \mathbf{m}(\boldsymbol{\xi}) &= E[\mathbf{x}(\boldsymbol{\xi})], \\ \mathbf{C}(\boldsymbol{\xi}, \boldsymbol{\xi}') &= E[(\mathbf{x}(\boldsymbol{\xi}) - \mathbf{m}(\boldsymbol{\xi})) (\mathbf{x}(\boldsymbol{\xi}) - \mathbf{m}(\boldsymbol{\xi}))^\mathsf{T}]. \end{aligned} \tag{12.15}$$

The joint distribution still has a quite familiar form

$$\begin{pmatrix} \mathbf{y}_o \\ \mathbf{x}_\star \end{pmatrix} = N\left(\begin{pmatrix} \mathbf{m}_o \\ \mathbf{m}_\star \end{pmatrix}, \begin{pmatrix} \mathbf{C}_{oo} + \hat{\boldsymbol{\Sigma}} & \mathbf{C}_{\star o}^\mathsf{T} \\ \mathbf{C}_{\star o} & \mathbf{C}_{\star\star} \end{pmatrix} \right), \tag{12.16}$$

where matrices have been formed from the blocks $[\mathbf{C}_{oo}]_{ij} = \mathbf{C}(\boldsymbol{\xi}_i, \boldsymbol{\xi}_j)$, $[\mathbf{C}_{\star o}]_{ij} = \mathbf{C}(\boldsymbol{\xi}_{\star,i}, \boldsymbol{\xi}_j)$, and $[\mathbf{C}_{\star\star}]_{ij} = \mathbf{C}(\boldsymbol{\xi}_{\star,i}, \boldsymbol{\xi}_{\star,j})$ analogously to the one-dimensional case. The mean vectors have been formed by similar stacking, $\mathbf{m}_{o,i} = \mathbf{m}(\boldsymbol{\xi}_i)$ and $\mathbf{m}_{\star,i} = \mathbf{m}(\boldsymbol{\xi}_{\star,i})$. The matrix $\hat{\boldsymbol{\Sigma}} = \mathbf{I} \otimes \boldsymbol{\Sigma}$, is a block-diagonal matrix containing matrices $\boldsymbol{\Sigma}$ as the diagonal blocks.

The conditional mean and covariance then have almost the same functional form as before:

$$E[\mathbf{x}_\star \mid \mathbf{y}_o] = \mathbf{m}_\star + \mathbf{C}_{\star o}\,(\mathbf{C}_{oo} + \hat{\mathbf{\Sigma}})^{-1}\,(\mathbf{y}_o - \mathbf{m}_o),$$
$$\text{Cov}[\mathbf{x}_\star \mid \mathbf{y}_o] = \mathbf{C}_{\star\star} - \mathbf{C}_{\star o}\,(\mathbf{C}_{oo} + \hat{\mathbf{\Sigma}})^{-1}\,\mathbf{C}_{\star o}^\mathsf{T}. \tag{12.17}$$

A useful way of interpreting the preceding equations is that they actually define a posterior Gaussian process that has the mean and covariance functions

$$\hat{\mathbf{m}}(\boldsymbol{\xi} \mid \mathbf{y}_o) = \mathbf{m}(\boldsymbol{\xi}) + \mathbf{C}(\boldsymbol{\xi}, \boldsymbol{\xi}_o')\,(\mathbf{C}(\boldsymbol{\xi}_o, \boldsymbol{\xi}_o') + \hat{\mathbf{\Sigma}})^{-1}\,(\mathbf{y}_o - \mathbf{m}(\boldsymbol{\xi}_o)),$$
$$\hat{\mathbf{C}}(\boldsymbol{\xi}, \boldsymbol{\xi}' \mid \mathbf{y}_o) = \mathbf{C}(\boldsymbol{\xi}, \boldsymbol{\xi}') - \mathbf{C}(\boldsymbol{\xi}, \boldsymbol{\xi}_o')\,(\mathbf{C}(\boldsymbol{\xi}_o, \boldsymbol{\xi}_o') + \hat{\mathbf{\Sigma}})^{-1}\,\mathbf{C}^\mathsf{T}(\boldsymbol{\xi}', \boldsymbol{\xi}_o'). \tag{12.18}$$

Here the apostrophe is used for emphasizing that the index sets such as $\boldsymbol{\xi}_o$ and $\boldsymbol{\xi}_o'$ are independent such that the matrix is formed from all the pairs of indexes. In particular, the posterior distribution of an arbitrary function value $\mathbf{x}(\boldsymbol{\xi})$ is

$$p(\mathbf{x}(\boldsymbol{\xi}) \mid \mathbf{y}_o) = \mathrm{N}(\mathbf{x}(\boldsymbol{\xi}) \mid \hat{\mathbf{m}}(\boldsymbol{\xi} \mid \mathbf{y}_o), \hat{\mathbf{C}}(\boldsymbol{\xi}, \boldsymbol{\xi} \mid \mathbf{y}_o)). \tag{12.19}$$

We can also define the observation model through a linear operator $\mathcal{H}$ that evaluates the latent function values $\mathbf{x}(\boldsymbol{\xi})$ at the observation inputs such that $(\mathbf{x}(\boldsymbol{\xi}_1), \ldots, \mathbf{x}(\boldsymbol{\xi}_n))^\mathsf{T} = \mathcal{H}\,\mathbf{x}(\boldsymbol{\xi})$. Similarly, we can consider more general measurement operators and allow $\mathcal{H}$ to be a more general linear operator, such as a differential or pseudodifferential operator (e.g., Särkkä, 2011; Särkkä et al., 2013).

The generalized GP regression problem thus takes the following form:

$$\mathbf{x}(\boldsymbol{\xi}) \sim \mathrm{GP}(\mathbf{m}(\boldsymbol{\xi}), \mathbf{C}(\boldsymbol{\xi}, \boldsymbol{\xi}')),$$
$$\mathbf{y}_o = \mathcal{H}\,\mathbf{x}(\boldsymbol{\xi}) + \boldsymbol{\varepsilon}, \tag{12.20}$$

for which the solution is given by

$$\hat{\mathbf{m}}(\boldsymbol{\xi} \mid \mathbf{y}_o) =$$
$$\mathbf{m}(\boldsymbol{\xi}) + [\mathbf{C}(\boldsymbol{\xi}, \boldsymbol{\xi}')\,\mathcal{H}^*]\,[\mathcal{H}\,\mathbf{C}(\boldsymbol{\xi}, \boldsymbol{\xi}')\,\mathcal{H}^* + \hat{\mathbf{\Sigma}}]^{-1}\,[\mathbf{y}_o - \mathcal{H}\,\mathbf{m}(\boldsymbol{\xi})],$$
$$\hat{\mathbf{C}}(\boldsymbol{\xi}, \boldsymbol{\xi}' \mid \mathbf{y}_o) =$$
$$\mathbf{C}(\boldsymbol{\xi}, \boldsymbol{\xi}') - [\mathbf{C}(\boldsymbol{\xi}, \boldsymbol{\xi}')\,\mathcal{H}^*]\,[\mathcal{H}\,\mathbf{C}(\boldsymbol{\xi}, \boldsymbol{\xi}')\,\mathcal{H}^* + \hat{\mathbf{\Sigma}}]^{-1}\,\mathcal{H}\,\mathbf{C}(\boldsymbol{\xi}, \boldsymbol{\xi}'), \tag{12.21}$$

where $\mathcal{H}^*$ denotes the adjoint of $\mathcal{H}$, effectively reversing the direction of the operation (cf. Särkkä, 2011; Särkkä et al., 2013).

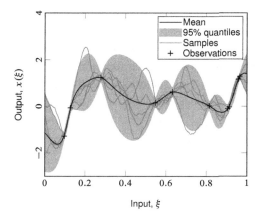

Figure 12.1 Batch GP regression solution from Example 12.6 with the data shown by black crosses, and the posterior mean and marginal variance visualized together with random realizations of the posterior process.

Example 12.6 (Batch GP regression). *Consider the following one-dimensional GP regression model:*

$$x(\xi) \sim \mathrm{GP}(0, C(\xi, \xi')),$$
$$y_k = x(\xi_k) + \varepsilon_k, \qquad \varepsilon_k \sim \mathrm{N}(0, \sigma_{\mathrm{n}}^2), \tag{12.22}$$

where the first equation defines the GP prior and second the observation model. Let the a priori covariance function be a Matérn covariance function:

$$C_{\mathrm{Mat.}}(\xi, \xi') = \sigma^2 \left(1 + \frac{\sqrt{3}\,|\xi - \xi'|}{\ell} \right) \exp\left(-\frac{\sqrt{3}\,|\xi - \xi'|}{\ell} \right), \tag{12.23}$$

where ℓ now stands for a characteristic "length scale" and σ^2 for a magnitude scale (hyper) parameter of the model.

Figure 12.1 shows an example where the data $\{(\xi_k, y_k) \mid k = 1, 2, \ldots\}$ are shown with black crosses, and the predicted marginal posterior process on the entire interval is visualized by the shaded patch.

12.3 Converting between Covariance Functions and SDEs

Consider now a special case of Gaussian process regression, where the input is one-dimensional and denoted as $\xi = t$. Thus we have a relationship of the form

$$y = x(t), \tag{12.24}$$

and the a priori mean and covariance functions are given functions $m(t)$ and $C(t, t')$. In practice, we usually observe the process only at discrete points of time, and the measurements are corrupted by some noise, for example,

$$y_k = x(t_k) + \varepsilon_k, \qquad \varepsilon_k \sim N(0, \sigma^2). \tag{12.25}$$

The aim of Gaussian process regression is (in a simple case) to infer the posterior distribution

$$p(x(t) \mid \mathbf{y}_o) = N(x(t) \mid \hat{m}(t \mid \mathbf{y}_o), \hat{C}(t, t \mid \mathbf{y}_o)). \tag{12.26}$$

for an arbitrary t. It is now easy to see that if the process $x(t)$ was represented in terms of an SDE (i.e., state-space model), this would exactly correspond to the smoothing distribution discussed in Section 10.8. It turns out that this is indeed the case provided that we can find an SDE that "generates" the mean and covariance functions $m(t)$ and $C(t, t')$, respectively. In the following, we take $m(t) = 0$ without loss of generality.

To illustrate the main idea, assume that the a priori GP has zero mean and the covariance function

$$C(t, t') = \frac{q}{2\lambda} \exp(-\lambda \,|t - t'|), \tag{12.27}$$

which is the covariance function of the Ornstein–Uhlenbeck process (see Example 6.8). Thus the SDE having this covariance function is

$$dx = -\lambda \, x \, dt + d\beta. \tag{12.28}$$

In machine learning, this covariance function is also known as the exponential covariance function. This implies that the smoothers in Examples 10.29 and 10.33 actually compute the Gaussian process regression solution (12.26) for the covariance function (12.27).

Many other widely used covariance functions can also be converted to SDE models. When the spectral density of the covariance function is a rational function, this conversion can be done exactly. For many other covariance functions, we can also form SDE representations that accurately approximate the covariance function, for example, by approximating the spectral density as a rational function.

In the following sections, the interest will be in classes of Gaussian process regression problems, which can be represented in terms of a linear time-invariant SDE model and a linear measurement model of the following form:

$$
\begin{aligned}
d\mathbf{x} &= \mathbf{F}\,\mathbf{x}\,dt + \mathbf{L}\,d\boldsymbol{\beta}, \\
y_k &= \mathbf{H}\,\mathbf{x}(t_k) + \varepsilon_k,
\end{aligned}
\tag{12.29}
$$

where $\mathbf{x}(t) = (x_1(t),\, x_2(t),\, \ldots,\, x_D(t))$ contains the D stochastic processes, and the multidimensional Brownian motion $\boldsymbol{\beta}$ has a diffusion matrix $\mathbf{Q} \in \mathbb{R}^{S \times S}$. The model is defined by the matrices $\mathbf{F} \in \mathbb{R}^{D \times D}$ and $\mathbf{L} \in \mathbb{R}^{D \times S}$, and the initial state covariance $\mathbf{P}_0$. The observation model is defined by the measurement model matrix $\mathbf{H} \in \mathbb{R}^{1 \times D}$. The measurements are assumed to be corrupted by independent Gaussian noises, $\varepsilon_k \sim \mathrm{N}(0, \sigma_n^2)$.

The basic idea is to construct $\mathbf{F}$, $\mathbf{L}$, and $\mathbf{P}_0$ such that the process

$$
x(t) = \mathbf{H}\,\mathbf{x}(t)
\tag{12.30}
$$

has a given covariance function $C(t, t')$ that we prescribed for the GP regression problem. Then a smoothing algorithm will produce the distributions $p(\mathbf{x}(t) \mid \mathbf{y}_o)$, where we can recover $p(x(t) \mid \mathbf{y}_o)$ in Equation (12.26) by a marginalization.

Some examples of widely used covariance functions of the LTI SDE kind include the ones listed in the following. Rasmussen and Williams (2006) provide a detailed presentation of most of the covariance functions in the GP context. For the details of the corresponding state-space representations, see, for example, Solin (2016). At the end of this section, we also show how state-space models can be combined to produce sums and products of the corresponding covariance functions.

Exponential: The exponential (or Ornstein–Uhlenbeck) covariance function defines a stationary process, where the model functions are continuous but not differentiable:

$$
C_{\exp}(t, t') = \sigma^2 \exp\left(-\frac{|t - t'|}{\ell} \right),
\tag{12.31}
$$

where σ^2 is a magnitude scale parameter and ℓ the characteristic length-scale parameter. The exponential covariance function is the covariance function of the *Ornstein–Uhlenbeck process*, which was already presented earlier with the parameterization $\lambda = 1/\ell$ and $q = 2\sigma^2/\ell$.

Thus the exponential covariance function has the following LTI SDE representation:

$$F = -\frac{1}{\ell}, \quad L = 1, \quad q = \frac{2\sigma^2}{\ell}, \quad \text{and} \quad H = 1. \qquad (12.32)$$

The initial variance is given by the stationary state covariance $P_0 = P_\infty = \sigma^2$.

Matérn: A wide class of covariance functions can be defined by the Matérn class, which is given in its general form as

$$C_{\text{Mat.}}(t, t') = \sigma^2 \frac{2^{1-\nu}}{\Gamma(\nu)} \left(\frac{\sqrt{2\nu}\,|t - t'|}{\ell} \right)^\nu K_\nu \left(\frac{\sqrt{2\nu}\,|t - t'|}{\ell} \right), \qquad (12.33)$$

where σ^2 is a magnitude scale hyperparameter, ℓ the characteristic length scale, and ν a smoothness parameter. $K_\nu(\bullet)$ denotes the modified Bessel function of the second kind. For this class, the corresponding process is k-times differentiable if $\nu > k$. For half-integer values of ν, the expression simplifies and it has an exact representation in terms of a LTI SDE. For example, if $\nu = 1/2$, the exponential covariance function is recovered. For higher-order half-integer Matérn covariance functions, the LTI SDE representation is given as follows (Hartikainen and Särkkä, 2010): let $\lambda = \sqrt{2\nu}/\ell$ and the state dimensionality $D = \nu + 1/2$, then the model is given in the so-called companion form as

$$F = \begin{pmatrix} 0 & 1 & & \\ \vdots & & 0 & 1 & \\ & & & \ddots & \ddots \\ -a_1 \lambda^P & -a_2 \lambda^{D-1} & \cdots & & -a_D \lambda \end{pmatrix}, \quad L = \begin{pmatrix} 0 \\ \vdots \\ 0 \\ 1 \end{pmatrix}, \qquad (12.34)$$

where $a_i = \binom{D}{i-1}$ are the binomial coefficients. The measurement model matrix is $H = \begin{pmatrix} 1 & 0 & \cdots & 0 \end{pmatrix}$, the diffusion coefficient is

$$q = \sigma^2 \frac{[(D-1)!]^2}{(2D-2)!} (2\lambda)^{2D-1}, \qquad (12.35)$$

and the initial state covariance is the steady-state covariance P_∞ of the model. For arbitrary values of the smoothness parameter ν, the spectral density can be approximated by series expansions giving a rational representation.

Squared exponential: The squared exponential (also known as the Gaussian, radial basis function, RBF, or exponentiated quadratic) covariance function is given by

$$C_{\mathrm{se}}(t,t') = \sigma^2 \exp\left(-\frac{|t-t'|^2}{2\ell^2}\right), \qquad (12.36)$$

where σ^2 is a magnitude scale hyperparameter and ℓ is the characteristic length scale. It can be recovered in the limit of the Matérn smoothness parameter ν going to infinity. The model functions are thus infinitely smooth (infinitely differentiable), which would require an infinite number of derivatives stacked in the state for an exact SDE representation. However, the spectral density of the squared exponential can be efficiently approximated by a Taylor expansion (see Hartikainen and Särkkä, 2010; Särkkä et al., 2013) or Padé approximants (Särkkä and Piché, 2014) which leads to a finite-dimensional SDE representation.

In particular, as discussed in Särkkä and Piché (2014), a general class of approximations can be constructed as

$$S_{\mathrm{se}}(\omega) = \sigma^2 \sqrt{2\pi}\, \ell \exp\left(-\frac{\ell^2 \omega^2}{2}\right)$$

$$\approx \sigma^2 \sqrt{2\pi}\, \ell \left(\frac{1 + b_1\left(\frac{\ell^2 \omega^2}{2}\right) + \cdots + b_L\left(\frac{\ell^2 \omega^2}{2}\right)^L}{1 + a_1\left(\frac{\ell^2 \omega^2}{2}\right) + \cdots + a_M\left(\frac{\ell^2 \omega^2}{2}\right)^M}\right), \quad (12.37)$$

where a_i for $i = 1, \ldots, M$ and b_j for $j = 1, \ldots, L$ are the coefficients of a Padé approximant for the exponential function $\exp(-x)$ for $x \geq 0$ such that $L < M$. The corresponding SDE can constructed by using spectral factorization.

Rational quadratic: The rational quadratic (RQ) covariance function

$$C_{\mathrm{rq}}(t,t') = \sigma^2 \left(1 + \frac{|t-t'|^2}{2\alpha\ell^2}\right)^{-\alpha}, \qquad (12.38)$$

where $\alpha > 0$ is a shape parameter that defines the decay of the tail. The RQ covariance function corresponds to a scale-mixture model of squared exponential covariance functions with a gamma prior distribution on the inverse squared length scale. The Cauchy covariance is a special case for $\alpha = 1$. As $\alpha \to \infty$, the covariance function converges to the squared exponential.

The scale-mixture representation of the rational quadratic covariance function can be approximated in terms of (Solin and Särkkä, 2014b)

$$C_{\mathrm{rq}}(t - t') \approx \sum_{i=1}^{n} C_{\mathrm{se}}(t, t' \mid \sigma_i^2, \ell_i), \tag{12.39}$$

where the squared exponentials are evaluated with magnitudes $\sigma_i^2 = \sigma_{\mathrm{rq}}^2 \, w_i / \Gamma(\alpha)$ and length scales $\ell_i^2 = \ell_{\mathrm{rq}}^2 \, \alpha / x_i$. The points x_i and weights w_i are given such that $x_i, i = 1, 2, \ldots, n$, are the roots of the generalized Laguerre polynomial $\mathrm{L}_n^{\alpha-1}(x)$, and the weights w_i are given as follows:

$$w_i = \frac{\Gamma(n + \alpha) \, x_i}{n! \, (n + 1)^2 \left[\mathrm{L}_{n+1}^{\alpha-1}(x_i) \right]^2}. \tag{12.40}$$

The state-space representation of the RQ covariance function can then be constructed as the sum of squared-exponential SDE models with hyperparameters σ_i^2 and ℓ_i^2.

Periodic: The canonical periodic covariance function often encountered in machine learning applications can be given as

$$C_{\mathrm{periodic}}(t, t') = \sigma^2 \exp\left(-\frac{2 \sin^2\left(\frac{\omega_0 \, |t - t'|}{2} \right)}{\ell^2} \right), \tag{12.41}$$

where σ^2 is the magnitude scale and ℓ the characteristic length scale. The period length is determined by $\omega_0 = 2\pi / t_{\mathrm{period}}$. The periodic covariance function has a spectrum with a set of delta peaks appearing at the harmonic frequencies.

An approximative SDE model can be given in terms of a sum of n two-dimensional oscillatory SDE models of the following form. The feedback matrices are (Solin and Särkkä, 2014a)

$$\mathbf{F}_j = \begin{pmatrix} 0 & -\omega_0 \, j \\ \omega_0 \, j & 0 \end{pmatrix}, \tag{12.42}$$

dispersion matrix $\mathbf{L}_j = \mathbf{I}$, and the stationary covariances $\mathbf{P}_{\infty, j} = q_j^2 \, \mathbf{I}_2$. The process does not have a diffusion term, so $\mathbf{Q} = \mathbf{0}$. The measurement models are $\mathbf{H}_j = \begin{pmatrix} 1 & 0 \end{pmatrix}$. The coefficients are given by $q_j^2 = 2\,\mathrm{I}_j(\ell^{-2}) / \exp(\ell^{-2})$, for $j = 1, 2, \ldots, n$, and $q_0^2 = \mathrm{I}_0(\ell^{-2}) / \exp(\ell^{-2})$, where $\mathrm{I}_\alpha(z)$ is the modified Bessel function of the first kind of order α.

In practical modeling, it is common to allow for seasonable variation (quasiperiodicity) by considering a product of a long length scale stationary covariance function and a periodic covariance function. This allows the model to decay away from exact periodicity. Forming the product of covariance functions in SDE formulation is explained in Algorithm 12.10. For the product formulation in Algorithm 12.10 to hold, the process noise spectral density of the periodic SDE needs to be redefined as $\mathbf{Q}_j = q_j^2 \mathbf{I}_2$ (see Solin and Särkkä, 2014a).

Wiener process: As already discussed in Section 7.2, the Wiener process (Brownian motion) with a diffusion constant q has the covariance function

$$C_{\text{WP}}(t, t') = q \, \min(t, t'), \tag{12.43}$$

which is valid on the domain $t, t' \geq 0$.

Wiener velocity: The Wiener velocity model is a commonly used model in tracking applications (see Example 6.3). It corresponds to a once integrated Wiener process, where the derivative (velocity) is modeled as a Wiener process. The covariance function is (for $t, t' \geq 0$)

$$C_{\text{WV}}(t, t') = q \left(\frac{\min^3(t, t')}{3} + |t - t'| \frac{\min^2(t, t')}{2} \right). \tag{12.44}$$

Polynomial: The polynomial covariance function defines a nonstationary kernel that produces polynomials

$$C_{\text{poly}}(t, t') = \sigma^2 \, (t \, t')^p, \tag{12.45}$$

where σ^2 is a magnitude scale hyperparameter and p is a positive integer. Example 12.8 and Exercise 12.6 show examples of state-space models for this class of covariance functions.

Constant: Constant bias in the outputs can be modeled by including a constant covariance function in the covariance structure of the model. The constant covariance function is given as

$$C_{\text{const.}}(t, t') = \sigma^2, \tag{12.46}$$

where σ^2 is a magnitude scale hyperparameter. The covariance function is degenerate, it has a rank one. The corresponding degenerate state-space model has the feedback $F = 0$, dispersion $L = 1$, diffusion coefficient $q = 0$, observation model $H = 1$, and initial state variance $P_0 = \sigma^2$.

Noise: The (white) noise covariance function is

$$C_{\text{noise}}(t, t') = \sigma^2 \, \delta(t - t'), \tag{12.47}$$

where σ^2 is a magnitude scale hyperparameter and $\delta(t - t')$ represents the Dirac delta function. For an indexed set of inputs, the covariance is given by $C(t_i, t_j) = \sigma^2 \delta_{ij}$, where δ_{ij} is the Kronecker delta function. This covariance function has a flat frequency spectrum. It can be seen as the limit of other stationary kernels, when the characteristic length scale tends to zero, $\ell \to 0$ (see Exercise 3.5).

Example 12.7 (Matérn SDE model). *For covariance functions of the Matérn family for which* $v = 3/2$, *the processes are continuous and once differentiable. The covariance function becomes*

$$C_{\mathrm{Mat.}}(t, t') = \sigma^2 \left(1 + \frac{\sqrt{3}\,|t - t'|}{\ell} \right) \exp\left(-\frac{\sqrt{3}\,|t - t'|}{\ell} \right). \quad (12.48)$$

This model has the SDE representation

$$\mathbf{F} = \begin{pmatrix} 0 & 1 \\ -\lambda^2 & -2\lambda \end{pmatrix}, \quad \mathbf{L} = \begin{pmatrix} 0 \\ 1 \end{pmatrix}, \quad and \quad \mathbf{P}_\infty = \begin{pmatrix} \sigma^2 & 0 \\ 0 & \lambda^2 \sigma^2 \end{pmatrix}, \quad (12.49)$$

where $\lambda = \sqrt{3}/\ell$. *The diffusion constant of the Brownian motion is* $q = 4\lambda^3 \sigma^2$. *The measurement model matrix is* $\mathbf{H} = (1\ 0)$.

Example 12.8 (Linear covariance function). *In the polynomial covariance function, when* $p = 1$, *it becomes the linear covariance function*

$$C_{\mathrm{linear}}(t, t') = \sigma^2\, t\, t', \quad (12.50)$$

where σ^2 *is a magnitude scale hyperparameter. This is a degenerate covariance function, whose corresponding LTI SDE model is derived through the corresponding ODE model defining the evolution of the linear basis functions and the covariance of the Gaussian initial state. Thus the SDE is*

$$\mathbf{F} = \begin{pmatrix} 0 & 1 \\ 0 & 0 \end{pmatrix}, \quad \mathbf{L} = \begin{pmatrix} 0 \\ 1 \end{pmatrix}, \quad and \quad \mathbf{P}_0 = \sigma^2 \begin{pmatrix} t_0^2 & t_0 \\ t_0 & 1 \end{pmatrix}, \quad (12.51)$$

where t_0 *is the time at the initial point. The spectral density of the Brownian motion is* $Q = 0$ *(zero diffusion) and observation model* $\mathbf{H} = (1\ 0)$.

In GP modeling, common practice of encoding prior beliefs is to build more complicated covariance functions from sums and products of simpler ones. Given the SDE representations for the covariance functions $C_1(t, t')$ and $C_2(t, t')$, we can form the SDE representations of the sums and products, as we see in the following.

The sum $C(t, t') = C_1(t, t') + C_2(t, t')$ of two covariance functions is a valid covariance function (see, e.g., Rasmussen and Williams, 2006).

Algorithm 12.9 (Sums of stationary covariance functions as SDEs). *In terms of the state-space SDE model, this corresponds to a stacked model* $\mathbf{x}(t) = (\mathbf{x}_1(t), \mathbf{x}_2(t))$ *of the two SDE models:*

$$
\begin{aligned}
\mathbf{F} &= \text{blkdiag}(\mathbf{F}_1, \mathbf{F}_2), & \mathbf{L} &= \text{blkdiag}(\mathbf{L}_1, \mathbf{L}_2), \\
\mathbf{Q} &= \text{blkdiag}(\mathbf{Q}_1, \mathbf{Q}_2), & \mathbf{P}_0 &= \text{blkdiag}(\mathbf{P}_{0,1}, \mathbf{P}_{0,2}),
\end{aligned}
\tag{12.52}
$$

where the indices 1 and 2 refer to $C_1(t, t')$ *and* $C_2(t, t')$, *respectively. The dynamics of the two state-space models are independent, but coupled by the measurement model* $\mathbf{H} = (\mathbf{H}_1 \ \mathbf{H}_2)$. *The state dimension becomes* $D = D_1 + D_2$.

The product $C(t, t') = C_1(t, t') C_2(t, t')$ of two covariance functions is also a valid covariance function (see, e.g., Rasmussen and Williams, 2006). The corresponding product of two state-space SDEs must preserve the dynamics of both the models. The SDE corresponding to the product of two covariance functions is given by the following algorithm.

Algorithm 12.10 (Products of stationary covariance functions as SDEs). *The feedback matrix is given by*

$$
\mathbf{F} = \mathbf{F}_1 \otimes \mathbf{I} + \mathbf{I} \otimes \mathbf{F}_2,
\tag{12.53}
$$

where $\otimes$ *denotes the Kronecker product. The dispersion, spectral density, initial state covariance, and measurement model matrices are then given as follows:*

$$
\begin{aligned}
\mathbf{L} &= \mathbf{L}_1 \otimes \mathbf{L}_2, & \mathbf{Q} &= \mathbf{Q}_1 \otimes \mathbf{Q}_2, \\
\mathbf{P}_0 &= \mathbf{P}_{0,1} \otimes \mathbf{P}_{0,2}, & \mathbf{H} &= \mathbf{H}_1 \otimes \mathbf{H}_2.
\end{aligned}
\tag{12.54}
$$

The state dimension becomes $D = D_1 D_2$.

Equation (12.53) is known as the Kronecker sum of matrices $\mathbf{F}_1$ and $\mathbf{F}_2$, which makes the matrix exponential factor into the Kronecker product of the corresponding two matrix exponentials (see Higham, 2008; Solin and Särkkä, 2014a).

12.4 GP Regression via Kalman Filtering and Smoothing

In Chapter 6, we saw that linear stochastic differential equations can be solved for discrete time instants, which corresponds to conversion to an equivalent discrete-time state-space model. If the observations are linear and corrupted by additive Gaussian noise, the state-space inference problem can be solved by the Kalman filter (Algorithms 10.18 and 10.20) and

Rauch–Tung–Striebel smoother (Algorithms 10.28 and 10.32). This implies that if the GP covariance function has an equivalent representation in terms of a linear stochastic differential equation, then the batch GP regression problem can be solved using a Kalman filter and Rauch–Tung–Striebel smoother (Hartikainen and Särkkä, 2010, 2011; Särkkä and Hartikainen, 2012; Särkkä et al., 2013; Solin, 2016).

While a naïve implementation of solving the batch GP regression problem in Equation (12.18) scales cubically with respect to the number of observations, solving the inference problem sequentially with a Kalman filter and RTS smoother scales linearly. This makes the sequential approach for GP regression appealing for long (or even unbounded) temporal datasets.

The Kalman filter and RTS smoother only return the marginal means and variances. For recovering the full posterior GP covariance $\mathrm{Cov}[\mathbf{x}_\star \mid \mathbf{y}_\mathrm{o}]$, the smoother outcome can be used as follows. The diagonal elements (marginal variances) are given by $\boldsymbol{\Sigma}_{k,k} = \mathbf{H}\,\mathbf{P}_k^\mathrm{s}\,\mathbf{H}^\mathsf{T}$, for $k = 1, 2, \ldots, n$, and the lower triangle of the symmetric covariance matrix is given by

$$\boldsymbol{\Sigma}_{i,j} = \mathbf{H}\left(\prod_{k=j}^{i-1} \mathbf{G}_k \right) \mathbf{P}_i^\mathrm{s}\,\mathbf{H}^\mathsf{T}, \quad \text{for} \quad i > j, \qquad (12.55)$$

where $\mathbf{P}_k^\mathrm{s}$ is the smoother covariance of the state and $\mathbf{G}_k$ the smoother gain in Equation (10.91). However, for drawing posterior samples from the latent process, the full covariance does not have to be formed. This can be done directly by using the smoother solution as explained in Doucet (2010) (cf. also Särkkä et al., 2015b).

For training the hyperparameters, the parameter estimation methods discussed in Sections 11.5 and 11.6 apply as such.

Example 12.11 (Sequential solution to GP regression)**.** *Figure 12.2 demonstrates the Kalman filtering and RTS smoothing approach to solving the GP regression problem from Example 12.6. The filtering outcome shows the sequential nature of the filter, where the marginal variance drops as new data points are encountered. The smoother outcomes on the right side correspond exactly (up to floating point inaccuracy) to the batch solution given in Example 12.6.*

12.5 Spatiotemporal Gaussian Process Models

The connection between batch GP regression models and SDEs can be extended from purely temporal models to spatiotemporal models. Modeling of spatiotemporal data is straightforward under the GP regression

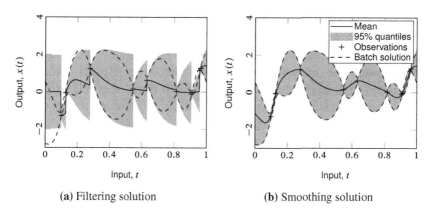

(a) Filtering solution   **(b)** Smoothing solution

Figure 12.2 Sequential solution to GP regression from Example 12.11: (a) the result after the (forward) filtering step, and (b) the (backward) smoother outcome that corresponds to the batch GP regression solution.

paradigm. The space $\mathbf{r} \in \mathbb{R}^d$ and time $t \in \mathbb{R}$ variables jointly take the roles of input variables of the regression function: $\boldsymbol{\xi} = (\mathbf{r}, t)$. While the batch GP regression formalism in Section 12.2 can be employed for solving models of this type, in practical applications the cubic computational scaling easily becomes an issue.

Kalman filtering and smoothing solutions to spatiotemporal GP regression problems were introduced in the article by Särkkä et al. (2013). The idea is to let the state be a function in an infinite-dimensional Hilbert space. The rationale is that similarly as for the finite-dimensional state variables in Equation (12.29), the evolution of the infinite-dimensional quantities can be encoded into a state-space model. For spatiotemporal models, this is interpreted as a Gaussian process (Gaussian random field) that evolves over time. The GP regression problem can be reformulated into a stochastic partial differential equation (SPDE; see Da Prato and Zabczyk, 2014) describing the temporal dynamics of the spatial process:

$$\frac{\partial \mathbf{x}(\mathbf{r}, t)}{\partial t} = \mathscr{F}\, \mathbf{x}(\mathbf{r}, t) + \mathbf{L}\, \mathbf{w}(\mathbf{r}, t),$$
$$\mathbf{y}_k = \mathscr{H}_k\, \mathbf{x}(\mathbf{r}, t_k) + \boldsymbol{\varepsilon}_k, \tag{12.56}$$

where the process is driven by a spatiotemporal white noise process $\mathbf{w}(\mathbf{r}, t)$ with a spectral density function $\mathbf{Q}(\mathbf{r}, \mathbf{r}')$.

The dynamical model in Equation (12.56) is an infinite-dimensional, time-invariant linear stochastic partial differential equation. If the operator $\mathcal{F}$ is a differential operator, the dynamic model is an evolution type SPDE (see, e.g., Chow, 2007). The model formulation generalizes to a setup, where the operators are integrodifferential operators or pseudodifferential operators (Shubin, 1987). A pseudodifferential operator refers to an operator that is not a finite linear combination of differential operators, but instead can be defined as an infinite series of them.

Analogously to the one-dimensional SDE case, the infinite-dimensional SPDE model can be solved by a Kalman filter and RTS smoother, though now by using their *infinite-dimensional* counterparts. We point the reader to Särkkä et al. (2013) and Solin (2016) for a more detailed overview and examples of this approach.

12.6 Gaussian Process Approximation of Drift Functions

Gaussian processes can also be used as nonparametric priors in inferring the unknown drift function in an SDE by observing a realization path of the process. Ruttor et al. (2013) present an approach that we simplify here to one-dimensional problems:

$$dx = f(x)\,dt + d\beta, \tag{12.57}$$

where $\beta(t)$ is a Brownian motion with diffusion constant q. We assume that $f(x) \sim \mathrm{GP}(0, C(x, x'))$ is given a Gaussian process prior with a known covariance function, and we have observed a realization of the SDE.

The general construction for the GP approximation is as follows. Consider approximating an observed path by the Euler–Maruyama method:

$$\hat{x}(t_{k+1}) - \hat{x}(t_k) = f(\hat{x}(t_k))\Delta t + \Delta\beta_k, \tag{12.58}$$

where $\Delta\beta_k \sim \mathrm{N}(0, q\,\Delta t)$ and $\hat{x}(t)$ stands for the observed path seen at discrete time instants. As in Section 11.4, we now discretize the solution trajectory using Euler–Maruyama and use it to approximate the distribution of the trajectory. Since the transition probabilities in the model are Gaussian, the preceding model can be read as

$$p(\hat{x}(t_1), \ldots, \hat{x}(t_T) \mid f)$$
$$\propto \exp\left(-\frac{1}{2q\,\Delta t}\sum_{k=0}^{T-1}|\hat{x}(t_{k+1}) - \hat{x}(t_k) - f(\hat{x}(t_k))\Delta t|^2\right). \tag{12.59}$$

We can now consider a Gaussian process model as the conjugate prior for the drift function f.

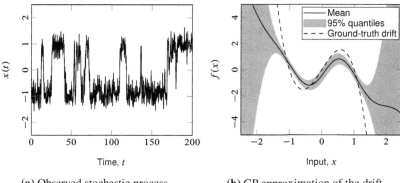

(a) Observed stochastic process   **(b)** GP approximation of the drift

Figure 12.3 An example (see Example 12.12) of learning the double-well drift model from an observed stochastic process shown in (a) using a Gaussian process prior. The resulting drift estimate is shown in (b).

By directly applying the GP regression formulation from Section 12.2, we get that the posterior process over drift functions f has a posterior mean and covariance:

$$\mathrm{E}[f(x_\star) \mid \hat{x}(t_1), \dots, \hat{x}(t_T)] = \mathbf{C}_{\star 0} \left(\mathbf{C}_{00} + \frac{q}{\Delta t} \mathbf{I} \right)^{-1} \mathbf{d},$$

$$\mathrm{Cov}[f(x_\star) \mid \hat{x}(t_1), \dots, \hat{x}(t_T)] = \mathbf{C}_{\star\star} - \mathbf{C}_{\star 0} \left(\mathbf{C}_{00} + \frac{q}{\Delta t} \mathbf{I} \right)^{-1} \mathbf{C}_{0\star},$$

$$(12.60)$$

where the elements of $\mathbf{d}$ are $d_k = (\hat{x}(t_{k+1}) - \hat{x}(t_k))/\Delta t$, for $k = 1, 2, \dots, T - 1$. However, in practice this approach becomes computationally heavy for larger datasets and multidimensional models, and more complicated if the data are not uniformly sampled. For more details, the reader is directed to Ruttor et al. (2013).

Methods for inferring the nonlinear dynamics in discrete-time models – so-called Gaussian process state-space (GPSS) models – have been considered, for example, by Turner et al. (2010), Frigola et al. (2013), and Svensson et al. (2016).

Example 12.12 (GP approximation of the double-well model). *Consider the following double-well model*

$$\mathrm{d}x = 4\,(x - x^3)\,\mathrm{d}t + \mathrm{d}\beta, \tag{12.61}$$

where $\beta(t)$ is a standard Brownian motion ($q = 1$). We simulate 20,000 *samples using Euler–Maruyama with $\Delta t = 0.01$ and keep every 20th sample point as observation, which leads to $n = 1,000$ observations. We aim to infer the drift function by using a GP prior with the following covariance function*

$$C(x, x') = 10^2 \, x \, x' + 20^2 \, \exp\left(-|x - x'|^2\right),\qquad(12.62)$$

which corresponds to a sum of a linear and squared-exponential covariance function with fixed hyperparameters. Figure 12.3 shows the original data, the GP solution, and the ground-truth drift function.

12.7 SDEs with Gaussian Process Inputs

In this book so far we have assumed that the SDEs are driven with Brownian motions (or equivalently white noises). However, from a modeling point of view, it is sometimes convenient to assume a correlation structure in the driving noise. In machine learning, these kinds of models are called latent force models (LFMs; Álvarez et al., 2009, 2013; Hartikainen and Särkkä, 2011; Hartikainen et al., 2012). To illustrate the main principles, consider the following spring model (cf. Equation 2.1) with a Gaussian process $g(t)$ as its input:

$$\frac{d^2 x}{dt^2} + \gamma \frac{dx}{dt} + v^2 x = g(t),\qquad(12.63)$$

which is equivalent to the state-space model (cf. Equation 2.4)

$$\frac{d\mathbf{x}}{dt} = \mathbf{F}\,\mathbf{x} + \mathbf{L}\,g(t),\qquad(12.64)$$

where

$$\mathbf{x} = \begin{pmatrix} x \\ \frac{dx}{dt} \end{pmatrix}, \quad \mathbf{F} = \begin{pmatrix} 0 & 1 \\ -v^2 & -\gamma \end{pmatrix}, \quad \text{and} \quad \mathbf{L} = \begin{pmatrix} 0 \\ 1 \end{pmatrix}.\qquad(12.65)$$

Recall from Chapter 2 that the solution is (assuming zero initial conditions) as follows:

$$\mathbf{x}(t) = \int_0^t \exp(\mathbf{F}\,(t - \tau))\,\mathbf{L}\,g(\tau)\,d\tau.\qquad(12.66)$$

Now assume that $g(t)$ is a zero-mean Gaussian process with a given covariance function

$$C_g(t, t') = \mathrm{E}[g(t)\,g(t')].\qquad(12.67)$$

Then $\mathbf{x}(t)$ is a zero-mean Gaussian process as well, and we can compute its covariance function as

$$\mathbf{C}_{\mathbf{x}}(t, t') = \int_0^t \int_0^{t'} \exp(\mathbf{F}\,(t - \tau))\,\mathbf{L}\,C_g(\tau, \tau')\,\mathbf{L}^{\mathsf{T}}\,\exp(\mathbf{F}\,(t' - \tau'))^{\mathsf{T}}\,\mathrm{d}\tau\,\mathrm{d}\tau'.$$

(12.68)

That is, the Gaussian process–driven spring model is equivalent to a Gaussian process model with the preceding modified covariance function. In the LFM framework of Álvarez et al. (2009) and Álvarez et al. (2013) (see also Hartikainen and Särkkä, 2011; Hartikainen et al., 2012), methods from Gaussian process regression are then used to infer the parameters of both the ODE and driving Gaussian process as well as to do predictions of $x(t)$ given (noisy) observations of it.

As pointed out by Hartikainen and Särkkä (2011), it is also possible to use state-space representations discussed in the previous sections to reduce a Gaussian process driven ODE such as (12.64) into a larger dimensional ODE driven by white noise, that is, to a linear Itô SDE. This is possible when the driving Gaussian process has a state-space representation as well.

For example, assume that $g(t)$ has a covariance function

$$C_g(t, t') = \frac{q}{2\lambda}\,\exp(-\lambda\,|t - t'|),$$

(12.69)

which can be recognized as the steady-state covariance function of the Ornstein–Uhlenbeck process (see Example 6.8). In SDE representation, it thus corresponds to

$$\mathrm{d}g = -\lambda\,g\,\mathrm{d}t + \mathrm{d}\beta, \quad g(0) \sim \mathrm{N}(0, P_\infty),$$

(12.70)

where $P_\infty = C_g(0, 0) = q/(2\lambda)$ is the steady-state variance of $x(t)$. Thus the following system has the same covariance function in its (x_1, x_2) components as the GP-driven spring model (12.64), with $g(t)$ having the covariance $C_g(t, t')$ as in Equation (12.69):

$$\begin{aligned}
\mathrm{d}x_1 &= x_2\,\mathrm{d}t, \\
\mathrm{d}x_2 &= -\nu^2\,x_1\,\mathrm{d}t - \gamma\,x_2\,\mathrm{d}t + g\,\mathrm{d}t, \\
\mathrm{d}g &= -\lambda\,g\,\mathrm{d}t + \mathrm{d}\beta,
\end{aligned}$$

(12.71)

where $x_1(0) = 0$, $x_2(0) = 0$, and $g(0) \sim \mathrm{N}(0, P_\infty)$.

The preceding system can be written as a state-space SDE as follows:

$$\mathrm{d}\tilde{\mathbf{x}} = \tilde{\mathbf{F}}\,\tilde{\mathbf{x}}\,\mathrm{d}t + \tilde{\mathbf{L}}\,\mathrm{d}\beta,$$

(12.72)

where

$$\tilde{\mathbf{x}} = \begin{pmatrix} x_1 \\ x_2 \\ g \end{pmatrix}, \quad \tilde{\mathbf{F}} = \begin{pmatrix} 0 & 1 & 0 \\ -\nu^2 & -\gamma & 1 \\ -\lambda & 0 & 0 \end{pmatrix}, \quad \text{and} \quad \tilde{\mathbf{L}} = \begin{pmatrix} 0 \\ 0 \\ 1 \end{pmatrix}. \quad (12.73)$$

Clearly, the preceding procedure can be applied to any Gaussian process–driven ODE model, where we can represent the GP in state-space form. After this conversion, all the SDE methods presented in this book as well as the Kalman filtering and smoothing solutions are directly applicable. In particular, the computational methods for linear SDE given in Chapter 6 provide easy computational means to compute the exact discretizations and covariance functions.

The preceding state-space conversion also works for nonlinear ODE models, which then turns the GP-driven ODE model into a nonlinear SDE. Although we no longer get closed form solutions in the same sense as for linear SDE, we can still apply the various approximation methods discussed in the previous chapters for approximate inference and covariance function computation.

In mathematical sense, the LFMs are related to stochastic integrals with respect to fractional Brownian motions and more general Gaussian processes. For more information on the mathematical theory, the reader is referred to Nualart (2006) and the references therein. These kinds of driving processes might not have a state-space representation and often they cannot even sensibly be approximated with a state-space model.

12.8 Gaussian Process Approximation of SDE Solutions

Consider the following continuous-discrete system

$$\begin{aligned} d\mathbf{x} &= \mathbf{f}(\mathbf{x}, t)\, dt + \mathbf{L}(t)\, d\boldsymbol{\beta}, \\ \mathbf{y}_k &= \mathbf{h}(\mathbf{x}(t_k)) + \mathbf{r}_k, \end{aligned} \quad (12.74)$$

where $\boldsymbol{\beta}(t)$ is a Brownian motion with diffusion $\mathbf{Q}$ and $\mathbf{r}_k \sim \mathrm{N}(\mathbf{0}, \mathbf{R})$. The smoothing solution to problems of this kind was presented in Section 10.8. Archambeau et al. (2007) present an alternative approach for assessing the smoothing solution through a variational approximation to the problem. The method is based on approximating the model by a linear process

$$d\tilde{\mathbf{x}} = [-\mathbf{A}(t)\,\tilde{\mathbf{x}} + \mathbf{b}(t)]\, dt + \mathbf{L}(t)\, d\boldsymbol{\beta}, \quad (12.75)$$

that is, the actual drift function is replaced by an approximation $\tilde{\mathbf{f}}(\mathbf{x}, t) = -\mathbf{A}(t)\,\mathbf{x} + \mathbf{b}(t)$, where the time-dependent matrix $\mathbf{A}(t)$ and vector $\mathbf{b}(t)$ are parameters of the approximation.

As we saw in Chapter 6, the solutions to linear SDEs are Gaussian processes, and the marginal density given by the approximation is Gaussian $q(\mathbf{x}(t)) = \mathrm{N}(\mathbf{x}(t) \mid \mathbf{m}(t), \mathbf{P}(t))$. The mean and covariance obey the differential equations

$$\frac{\mathrm{d}\mathbf{m}(t)}{\mathrm{d}t} = -\mathbf{A}(t)\,\mathbf{m}(t) + \mathbf{b}(t),$$
$$\frac{\mathrm{d}\mathbf{P}(t)}{\mathrm{d}t} = -\mathbf{A}(t)\,\mathbf{P}(t) - \mathbf{P}(t)\,\mathbf{A}(t)^{\mathsf{T}} + \mathbf{L}(t)\,\mathbf{Q}\,\mathbf{L}^{\mathsf{T}}(t). \tag{12.76}$$

The parameters $\mathbf{A}(t)$ and $\mathbf{b}(t)$ can be estimated by minimizing the Kullback–Leibler (KL) divergence as follows (Archambeau et al., 2007)

$$\mathrm{KL}[q \parallel p] = \int_{t_0}^{T} \mathrm{E}_q\left[e(\mathbf{x}(t), t) + \sum_k u_k(\mathbf{x}(t))\,\delta(t - t_k)\right]\mathrm{d}t, \tag{12.77}$$

where the expectation is taken with respect to the marginal distribution q, $\delta(\bullet)$ denotes the Dirac delta function, and

$$e(\mathbf{x}(t), t) = \frac{1}{2}\,[\mathbf{f}(\mathbf{x}(t), t) + \mathbf{A}(t)\,\mathbf{x}(t) - \mathbf{b}(t)]^{\mathsf{T}}\,[\mathbf{L}(t)\,\mathbf{Q}\,\mathbf{L}^{\mathsf{T}}(t)]^{-1}$$
$$\times\,[\mathbf{f}(\mathbf{x}(t), t) + \mathbf{A}(t)\,\mathbf{x}(t) - \mathbf{b}(t)],$$
$$u_k(\mathbf{x}(t)) = \frac{1}{2}\,[\mathbf{y}_k - \mathbf{h}(\mathbf{x}(t))]^{\mathsf{T}}\,\mathbf{R}^{-1}\,[\mathbf{y}_k - \mathbf{h}(\mathbf{x}(t))]. \tag{12.78}$$

In order to find the parameters and the moments, the KL divergence (12.77) is minimized subject to the constraints given by Equation (12.76). This problem can be formulated through Lagrange multipliers $\boldsymbol{\lambda}(t)$ and $\boldsymbol{\Psi}(t)$, which gives (see Ala-Luhtala et al., 2015, for a detailed explanation)

$$\frac{\mathrm{d}\boldsymbol{\lambda}(t)}{\mathrm{d}t} = \mathbf{A}(t)\,\boldsymbol{\lambda}(t) - \nabla_{\mathbf{m}}\,\mathrm{E}_q[e(\mathbf{x}(t), t)],$$
$$\frac{\mathrm{d}\boldsymbol{\Psi}(t)}{\mathrm{d}t} = \boldsymbol{\Psi}(t)\,\mathbf{A}(t) + \mathbf{A}(t)^{\mathsf{T}}\,\boldsymbol{\Psi}(t) - \nabla_{\mathbf{P}}\,\mathrm{E}_q[e(\mathbf{x}(t), t)], \tag{12.79}$$

$$\mathbf{A}(t) = -\mathrm{E}_q[\mathbf{F}_x(\mathbf{x}(t), t)] + 2\,\mathbf{L}(t)\,\mathbf{Q}\,\mathbf{L}^{\mathsf{T}}(t)\,\boldsymbol{\Psi}(t),$$
$$\mathbf{b}(t) = \mathrm{E}_q[\mathbf{f}(\mathbf{x}(t), t)] + \mathbf{A}(t)\,\mathbf{m}(t) + \mathbf{L}(t)\,\mathbf{Q}\,\mathbf{L}^{\mathsf{T}}(t)\,\boldsymbol{\lambda}(t), \tag{12.80}$$

The variational Gaussian process approximation to the smoothing problem can thus be given in the form of the following algorithm (Archambeau et al., 2007).

Algorithm 12.13 (Variational approximation of continuous-discrete smoothing). *Initialize* $\mathbf{A}^{(0)}(t)$ *and* $\mathbf{b}^{(0)}(t)$ *suitably. For* $i = 1, 2, \ldots$ *iterate the following until convergence:*

1. *Given* $\mathbf{A}^{(i)}(t)$ *and* $\mathbf{b}^{(i)}(t)$, *solve the mean and covariance equations* *(12.76) forward in time to get* $\mathbf{m}^{(i+1)}(t)$ *and* $\mathbf{P}^{(i+1)}(t)$.
2. *Use* $\mathbf{m}^{(i+1)}(t)$ *and* $\mathbf{P}^{(i+1)}(t)$ *to solve the Lagrange differential equations* *(12.79) backward in time to get* $\boldsymbol{\lambda}^{(i+1)}(t)$ *and* $\boldsymbol{\Psi}^{(i+1)}(t)$.
3. *The new parameters estimates for the linearized model are given by*

$$
\begin{aligned}
\mathbf{A}^{(i+1)}(t) &= \mathbf{A}^{(i)}(t) + \gamma \left[\mathbf{A}(t) - \mathbf{A}^{(i)}(t)\right], \\
\mathbf{b}^{(i+1)}(t) &= \mathbf{b}^{(i)}(t) + \gamma \left[\mathbf{b}(t) - \mathbf{b}^{(i)}(t)\right],
\end{aligned}
\tag{12.81}
$$

where $\mathbf{A}(t)$ *and* $\mathbf{b}(t)$ *are solved from Equations (12.80) and* $\gamma \in [0, 1]$ *is a damping coefficient preventing instability from possibly too large update steps.*

It is worth noting that although the preceding method is based on variational Bayesian methodology, the resulting approximation is a Gaussian approximation to the SDE. The relationship with Gaussian assumed density approximations is discussed in Ala-Luhtala et al. (2015).

12.9 Exercises

12.1 Consider the operator formulation of batch GP regression (12.21), with the linear operator defined as

$$
\mathcal{H} x(t) = \left(\frac{\mathrm{d}x(t)}{\mathrm{d}t}\Big|_{t=t_1} \quad \cdots \quad \frac{\mathrm{d}x(t)}{\mathrm{d}t}\Big|_{t=t_n}\right)
$$

and the covariance functions being of the Matérn class as given in Equation (12.48) (for $\nu = 3/2$). Derive the covariance function corresponding to the "derivative observations" model defined by the operator.

12.2 The SDE corresponding to the Matérn covariance function with $\nu = 3/2$ was given in Example 12.7. Derive (starting from the spectral density function) the corresponding expression for the Matérn covariance function (12.33) with $\nu = 5/2$.

12.3 The squared exponential covariance function:

(a) Why is it not possible to write an exact finite-dimensional SDE model for the squared exponential covariance function?

(b) The squared exponential covariance function has the spectral density function

$$S(\omega) = \sigma^2 \sqrt{2\pi} \, \ell \exp\left(-\frac{\ell^2 \omega^2}{2}\right).$$

Approximate this function by a rational function by forming a Taylor series expansion in the denominator.

(c) Factor the rational function into an unstable and stable part corresponding to the transfer function (see, e.g., Särkkä et al., 2013).

(d) Form the so-called companion matrix corresponding to the linear SDE drift function by truncating the series and collecting the coefficients. Truncate your model so that you end up with a six-dimensional SDE model.

(e) Compare numerically the covariance function of the state-space model to the original squared exponential covariance function.

12.4   The periodic covariance function:

(a) Derive an expression for the spectral density of the canonical periodic covariance function given in Equation (12.41).

(b) Consider approximating the behavior of the periodic covariance function with a superposition of n stochastic oscillators with feedback matrices

$$\mathbf{F}_j = \begin{pmatrix} 0 & -\omega_0 \, j \\ \omega_0 \, j & 0 \end{pmatrix}, \tag{12.82}$$

zero diffusion, and a stationary covariances $\mathbf{P}_{\infty,j} = q_j^2 \, \mathbf{I}_2$. Derive the expression for q_j^2 for matching the spectral densities of the oscillators and the original covariance function.

(c) Simulate from the SDE you obtained by solving the LTI SDE model and compare to realizations directly simulated from a Gaussian with a covariance matrix formed by evaluating the original periodic covariance function.

12.5   Expand the model obtained in the previous exercise to produce quasiperiodic realizations such as what you would obtain by considering the covariance function given by

$$C_{\text{quasiperiodic}}(t, t') = C_{\text{periodic}}(t, t') \, C_{\text{exp}}(t, t'). \tag{12.83}$$

Derive a truncated LTI SDE approximation for the quasiperiodic covariance function. Simulate trajectories from the model that demonstrate the quasiperiodic behavior.

12.6   Consider the polynomial covariance function for $p = 2$. Derive the SDE model corresponding to this covariance function.

12.7   Gaussian process regression:

(a) Draw 10 random points from a Gaussian with an exponential (Ornstein–Uhlenbeck) covariance structure. Choose your parameters for the magnitude, length scale, and measurement noise variance as you see fit, and start by first drawing the 10 input locations uniformly from the interval [0, 1]. Add independent Gaussian measurement noise to your realization trajectory.

(b) Implement batch Gaussian process regression for an exponential covariance function model, where you use the data you just simulated. Visualize your result by plotting the mean and marginal variance over [01] with 100 discretization points.

(c) The corresponding SDE model was given in Example 12.7. Write the closed-form discrete-time state-space model for this SDE.

(d) Implement the sequential way of solving the GP regression problem by Kalman filtering and Rauch–Tung–Striebel smoothing (refer to Chapter 10 for implementation details). Check numerically that you get the same solution (up to minor numerical errors) as in the batch solution.

13

Epilogue

We are now reaching the end of this book and it is time to summarize and discuss what is there beyond this book. In this chapter, we first take a brief outlook to each of the topics, then discuss the selection of numerical methods, and finally discuss some more advanced aspects that we have not discussed here.

13.1 Overview of the Covered Topics

First of all, let us take another look at a generic form of SDEs that we have studied for the whole book:

$$d\mathbf{x} = \mathbf{f}(\mathbf{x}, t)\, dt + \mathbf{L}(\mathbf{x}, t)\, d\boldsymbol{\beta}. \tag{13.1}$$

In the early chapters of the book, we started by interpreting SDEs as white noise–driven differential equations, which corresponds to formally dividing the preceding equation with the differential dt:

$$\frac{d\mathbf{x}}{dt} = \mathbf{f}(\mathbf{x}, t) + \mathbf{L}(\mathbf{x}, t)\frac{d\boldsymbol{\beta}}{dt}, \tag{13.2}$$

and interpreting $\mathbf{w}(t) = d\boldsymbol{\beta}(t)/dt$ as white noise. It did work up to a point, but in the end we had to redefine SDEs as Itô integral equations, for which Equation (13.1) is a shorthand notation. However, until the end we were still allowed to build models of real-world phenomena by thinking of white noise–driven differential equations such as Equation (13.2), which is fine, provided that we realize that mathematically it means Equation (13.1).

After defining what we mean by an SDE, we then took a look at probability distributions, moments, Markov properties, and many other aspects of SDEs. One of the most important tools was the Itô formula that we could use to derive many of the results. We also took a more careful look at linear SDEs (with $\mathbf{f}(\mathbf{x}, t) = \mathbf{F}(t)\,\mathbf{x} + \mathbf{u}(t)$ and $\mathbf{L}(\mathbf{x}, t) = \mathbf{L}(t)$), which we can solve explicitly and whose statistics and probability distribution can be computed easily. Additionally, we discussed more advanced concepts related to transformations, construction, and advanced applications of SDEs.

For nonlinear SDEs, we had to conclude that they do not allow for any general solution methods and the only way to go is to numerically approximate them. We took a look at Itô–Taylor and stochastic Runge–Kutta methods for numerically simulating SDE solutions along with a few more specific methods (the Exact algorithm and Verlet methods). After that, we discussed various methods that can be used to approximate the statistics of SDE solutions. Especially Gaussian and linearization approximations turned out to be common in this context, but various series expansion methods are available as well.

In the final chapters, we discussed applications aspects of SDEs, which included filtering and smoothing theory, parameter estimation, and machine learning. These chapters very strongly leaned on the theory and methods that we discussed in the preceding chapters. Especially the numerical methods were in very intensive use, and we could even say that it is the performance of the SDE numerical methods that determines the success of these applications.

13.2 Choice of SDE Solution Method

We could now ask the important question, "Which numerical method should I choose to solve a given SDE?" but it is hard to answer. It not only depends on the SDE, but also on the application, that is, on what you want to do with the SDE. If the interest is in sample paths and their statistics as such, then one should use the numerical simulation methods. Probably the first attempt should always be the Euler–Maruyama method – it scales badly, but is easy to implement and provides a useful reference solution. With more sophisticated methods, it makes a huge difference whether the dispersion matrix $\mathbf{L}(\mathbf{x}, t)$ depends on the state $\mathbf{x}$ or not. Attempting to simulate strong solutions with state-dependent dispersion matrix is hard, because we encounter problematic iterated Itô integrals already in low-order numerical methods. However, if weak solutions are enough, then we have a selection of methods to apply even in the state-dependent case.

If we are only interested in approximating the statistics of SDEs, then the state dependence of the dispersion matrix is not so much of a problem. In many applications, (local) Gaussian approximations are enough, and we have a wide selection of them to choose from. A Gaussian approximation should indeed be the first attempt for this purpose – recall that Euler–Maruyama is a Gaussian approximation as well, and even that can be tried. However, for example, linearization methods are likely to perform better if we are only interested in the statistics, not sample paths of SDEs. If higher-order approximations are needed, then we can use Fourier–Hermite or Wong–Zakai series, or approximate the Fokker–Planck–Kolmogorov PDE using numerical methods for PDEs. But everything depends on the application.

13.3 Beyond the Topics

One important direction that we have not touched are jump process–driven SDEs. They result from replacing the Brownian motion in (13.1) with some other stochastic process $\boldsymbol{v}(t)$, which is not continuous, but has jumps:

$$d\mathbf{x} = \mathbf{f}(\mathbf{x}, t)\, dt + \mathbf{L}(\mathbf{x}, t)\, d\boldsymbol{v}. \tag{13.3}$$

We can, for example, model the process $\boldsymbol{v}(t)$ as a Lévy process, which then leads to Lévy process–driven SDEs and the related stochastic calculus. More information on this kind of processes and their numerical treatment can be found, for example, in the books of Applebaum (2009) and Platen and Bruti-Liberati (2010).

Instead of replacing the Brownian motion with a Lévy process, we can also replace it with, for example, fractional Brownian motion or some more general Gaussian process, as we already discussed in Section 12.7. For the related mathematical theory, the reader is referred to Nualart (2006). However, as discussed in Section 12.7, when the driving Gaussian process has a rational spectrum, then such an SDE is equivalent to an SDE with a few additional augmented states.

Another extension that we already considered in Section 12.5 is the extension to spatiotemporal models. One way to think about them is to assume that the state depends on a space location $\mathbf{r}$ as in the model (12.56). However, we can also directly consider abstract equations of the form

$$d\mathcal{X} = \mathcal{F}[\mathcal{X}, t]\, dt + \mathcal{L}[\mathcal{X}, t]\, d\mathcal{B}, \tag{13.4}$$

where the state $\mathcal{X}$ and the infinite-dimensional Brownian motion $\mathcal{B}$ live in a suitable Hilbert space. For the theory of this type of process, the reader is referred to Da Prato and Zabczyk (2014).

Finally, one possible extension is to consider stochastic control problems (e.g., Maybeck, 1982b; Stengel, 1994; Øksendal, 2003; Øksendal and Sulem, 2007) related to SDEs. Certainly, this can also be seen as application of SDEs, but stochastic control theory is an independent field as such.

The idea is to include control signal $\mathbf{u}(t)$ to the SDE

$$d\mathbf{x} = \mathbf{f}(\mathbf{x}, \mathbf{u}, t)\, dt + \mathbf{L}(\mathbf{x}, t)\, d\boldsymbol{\beta}, \qquad (13.5)$$

which we can then "connect" back to $\mathbf{x}$ directly or via noisy observations to yield a close-loop control for the process. The control can be chosen to, for example, make the SDE follow a given trajectory or hit a certain target stage while minimizing a cost function. These kinds of models have applications both in navigation and finance – in navigation, we might be attempting to steer a rocket to hit the moon with minimum fuel or to keep an autonomous car on the road, while in finance we might steer our portfolio to the maximum profitable stage.

References

Aihara, S. I. and Bagchi, A. 1999. On the Mortensen equation for maximum likelihood state estimate. *IEEE Transactions on Automatic Control*, **44**(10), 1955–1961. (Cited on page 205.)

Aït-Sahalia, Y. 2002. Maximum likelihood estimation of discretely sampled diffusions: A closed-form approximation approach. *Econometrica*, **70**(1), 223–262. (Cited on pages 179, 183, 184, and 185.)

Aït-Sahalia, Y. 2008. Closed-form likelihood expansions for multivariate diffusions. *Annals of Statistics*, **36**(2), 906–937. (Cited on pages 179, 183, and 184.)

Akhiezer, N. I. and Glazman, I. M. 1993. *Theory of Linear Operators in Hilbert Space*. New York, NY: Dover. (Cited on page 253.)

Ala-Luhtala, J., Särkkä, S., and Piché, R. 2015. Gaussian filtering and variational approximations for Bayesian smoothing in continuous-discrete stochastic dynamic systems. *Signal Processing*, **111**, 124–136. (Cited on pages 166, 273, and 274.)

Allen, M. P. and Tildesley, D. J. 1991. *Computer Simulation of Liquids*. New York, NY: Oxford University Press. (Cited on page 156.)

Álvarez, M., Luengo, D., and Lawrence, N. 2009. Latent force models. Pages 9–16 of: van Dyk, D. and Welling, M. (eds.), *Proceedings of the Twelfth International Conference on Artificial Intelligence and Statistics*. Proceedings of Machine Learning Research, vol. 5. Clearwater Beach, FL: PMLR. (Cited on pages 270 and 271.)

Álvarez, M. A., Rosasco, L., and Lawrence, N. D. 2012. Kernels for vector-valued functions: a review. *Foundations and Trends® in Machine Learning*, **4**(3), 195–266. (Cited on page 254.)

Álvarez, M. A., Luengo, D., and Lawrence, N. D. 2013. Linear latent force models using Gaussian processes. *IEEE Transactions on Pattern Analysis and Machine Intelligence*, **35**(11), 2693–2705. (Cited on pages 270 and 271.)

Anderson, B. D. O. 1972. Fixed interval smoothing for nonlinear continuous time systems. *Information and Control*, **20**(3), 294–300. (Cited on page 227.)

Applebaum, D. 2009. *Lévy Processes and Stochastic Calculus*. Second edn. Cambridge: Cambridge University Press. (Cited on pages 35 and 279.)

Arasaratnam, I. and Haykin, S. 2009. Cubature Kalman filters. *IEEE Transactions on Automatic Control*, **54**(6), 1254–1269. (Cited on page 169.)

Arasaratnam, I., Haykin, S., and Hurd, T. R. 2010. Cubature Kalman filtering for continuous-discrete systems: theory and simulations. *IEEE Transactions on Signal Processing*, **58**(10), 4977–4993. (Cited on pages 135 and 169.)

Archambeau, C. and Opper, M. 2011. Approximate inference for continuous-time Markov processes. Pages 125–140 of: *Bayesian Time Series Models*. New York, NY: Cambridge University Press. (Cited on pages 166, 168, 245, and 249.)

Archambeau, C., Cornford, D., Opper, M., and Shawe-Taylor, J. 2007. Gaussian process approximations of stochastic differential equations. Pages 1–16 of: Lawrence, N. D., Schwaighofer, A., and Quiñonero Candela, J. (eds.), *Gaussian Processes in Practice*. Proceedings of Machine Learning Research, vol. 1. Bletchley Park, UK: PMLR. (Cited on pages 272, 273, and 274.)

Åström, K. J. and Wittenmark, B. 1997. *Computer-Controlled Systems: Theory and Design*. Third edn. Upper Saddle River, NJ: Prentice Hall. (Cited on page 9.)

Axelsson, P. and Gustafsson, F. 2015. Discrete-time solutions to the continuous-time differential Lyapunov equation with applications to Kalman filtering. *IEEE Transactions on Automatic Control*, **60**(3), 632–643. (Cited on pages 81, 83, and 84.)

Bar-Shalom, Y., Li, X.-R., and Kirubarajan, T. 2001. *Estimation with Applications to Tracking and Navigation*. New York, NY: Wiley. (Cited on pages 29 and 82.)

Bell, B. M. 1994. The iterated Kalman smoother as a Gauss–Newton method. *SIAM Journal on Optimization*, **4**(3), 626–636. (Cited on pages 221 and 229.)

Bell, B. M. and Cathey, F. W. 1993. The iterated Kalman filter update as a Gauss–Newton method. *IEEE Transactions on Automatic Control*, **38**(2), 294–297. (Cited on page 221.)

Beneš, V. E. 1981. Exact finite-dimensional filters for certain diffusions with nonlinear drift. *Stochastics*, **5**(1–2), 65–92. (Cited on page 207.)

Beskos, A. and Roberts, G. O. 2005. Exact simulation of diffusions. *Annals of Applied Probability*, **15**(4), 2422–2444. (Cited on pages 157 and 160.)

Beskos, A., Papaspiliopoulos, O., Roberts, G., and Fearnhead, P. 2006a. Exact and computationally efficient likelihood-based estimation for discretely observed diffusion processes (with discussion). *Journal of the Royal Statistical Society: Series B (Statistical Methodology)*, **68**(3), 333–382. (Cited on pages 111, 157, and 160.)

Beskos, A., Papaspiliopoulos, O., and Roberts, G. O. 2006b. Retrospective exact simulation of diffusion sample paths with applications. *Bernoulli*, **12**(6), 1077–1098. (Cited on pages 157 and 160.)

Beskos, A., Papaspiliopoulos, O., and Roberts, G. O. 2008. A factorisation of diffusion measure and finite sample path. *Methodology and Computing in Applied Probability*, **10**(1), 85–104. (Cited on pages 157 and 160.)

Bishop, C. M. 2006. *Pattern Recognition and Machine Learning*. New York, NY: Springer. (Cited on page 251.)

Brandt, M. W. and Santa-Clara, P. 2002. Simulated likelihood estimation of diffusions with an application to exchange rate dynamics in incomplete markets. *Journal of Financial Economics*, **63**(2), 161–210. (Cited on page 192.)

Brenner, S. and Scott, R. 2002. *The Mathematical Theory of Finite Element Methods*. Second edn. Texts in Applied Mathematics, vol. 15. New York, NY: Springer. (Cited on pages 185 and 189.)

Brigo, D., Hanzon, B., and Le Gland, F. 1999. Approximate nonlinear filtering by projection on exponential manifolds of densities. *Bernoulli*, **5**(3), 495–534. (Cited on page 173.)

Brooks, S., Gelman, A., Jones, G. L., and Meng, X.-L. 2011. *Handbook of Markov Chain Monte Carlo*. Boca Raton, FL: Chapman & Hall/CRC. (Cited on pages 237, 238, and 242.)

Bucy, R. S. 1965. Nonlinear filtering theory. *IEEE Transactions on Automatic Control*, **10**(2), 198–198. (Cited on page 206.)

Burrage, K., Burrage, P., Higham, D. J., Kloeden, P. E., and Platen, E. 2006. Comment on "Numerical methods for stochastic differential equations". *Physical Review E*, **74**(6), 068701. (Cited on page 145.)

Burrage, K., Lenane, I., and Lythe, G. 2007. Numerical methods for second-order stochastic differential equations. *SIAM Journal on Scientific Computing*, **29**(1), 245–264. (Cited on page 155.)

Cameron, R. H. and Martin, W. T. 1947. The orthogonal development of non-linear functionals in series of Fourier–Hermite functionals. *Annals of Mathematics*, **48**(2), 385–392. (Cited on page 194.)

Cappé, O., Moulines, E., and Rydén, T. 2005. *Inference in Hidden Markov Models*. Springer Series in Statistics. New York, NY: Springer-Verlag. (Cited on pages 234 and 247.)

Chaichian, M. and Demichev, A. 2001a. *Path Integrals in Physics, Volume 1: Stochastic Processes and Quantum Mechanics*. Bristol, UK: IOP Publishing. (Cited on pages 101, 105, and 117.)

Chaichian, M. and Demichev, A. 2001b. *Path Integrals in Physics, Volume 2: Quantum Field Theory, Statistical Physics & Other Modern Applications*. Bristol, UK: IOP Publishing. (Cited on page 105.)

Chow, C. C. and Buice, M. A. 2015. Path integral methods for stochastic differential equations. *Journal of Mathematical Neuroscience*, **5**(1), 8. (Cited on page 118.)

Chow, P.-L. 2007. *Stochastic Partial Differential Equations*. Chapman & Hall/CRC Applied Mathematics and Nonlinear Science Series, vol. 11. Boca Raton, FL, and London: Chapman & Hall/CRC Press. (Cited on page 268.)

Cover, T. M. and Thomas, J. A. 2006. *Elements of Information Theory*. Second edn. Hoboken, NJ: John Wiley & Sons. (Cited on pages 73 and 246.)

Crisan, D. and Rozovskiĭ, B. L. (eds.). 2011. *The Oxford Handbook of Nonlinear Filtering*. Oxford and New York, NY: Oxford University Press. (Cited on pages 111, 197, and 209.)

Da Prato, G. and Zabczyk, J. 2014. *Stochastic Equations in Infinite Dimensions*. Second edn. Encyclopedia of Mathematics and its Applications, vol. 152. Cambridge: Cambridge University Press. (Cited on pages 253, 267, and 279.)

Daum, F. E. 1984. Exact finite-dimensional nonlinear filters for continuous time processes with discrete time measurements. Pages 16–22 of: *Proceedings of the 23rd Conference on Decision and Control*. Piscataway, NJ: IEEE. (Cited on page 215.)

Doucet, A. 2010. *A Note on Efficient Conditional Simulation of Gaussian Distributions*. Technical note. Departments of Computer Science and Statistics, University of British Columbia, Canada. (Cited on page 266.)

Dutra, D. A., Teixeira, B. O. S., and Aguirre, L. A. 2014. Maximum a posteriori state path estimation: Discretization limits and their interpretation. *Automatica*, **50**(5), 1360–1368. (Cited on pages 204 and 205.)

Einstein, A. 1905. Über die von molekularkinetischen Theorie der Wärme geforderte Bewegung von in ruhenden Flüssigkeiten suspendierten Teilchen. *Annalen der Physik*, **17**, 549–560. (Cited on page 23.)

Feynman, R. P. and Hibbs, A. R. 2010. *Quantum Mechanics and Path Integrals*. Emended by Daniel F. Styer edn. New York, NY: Dover. (Cited on page 117.)

Frigola, R., Lindsten, F., Schön, T. B., and Rasmussen, C. E. 2013. Bayesian inference and learning in Gaussian process state-space models with particle MCMC. Pages 3156–3164 of: Burges, C. J. C., Bottou, L., Welling, M., Ghahramani, Z., and Weinberger, K. Q. (eds.), *Advances in Neural Information Processing Systems 26*. Red Hook, NY: Curran Associates, Inc. (Cited on page 269.)

García-Fernández, Á. F., Svensson, L., Morelande, M. R., and Särkkä, S. 2015. Posterior linearization filter: principles and implementation using sigma points. *IEEE Transactions on Signal Processing*, **63**(20), 5561–5573. (Cited on pages 174 and 221.)

García-Fernández, Á. F., Svensson, L., and Särkkä, S. 2017. Iterated posterior linearization smoother. *IEEE Transactions on Automatic Control*, **62**(4), 2056–2063. (Cited on pages 174, 221, 229, and 245.)

Gardiner, C. W. 2004. *Handbook of Stochastic Methods*. Third edn. Berlin: Springer. (Cited on pages 51 and 64.)

Gelb, A. (ed.). 1974. *Applied Optimal Estimation*. Cambridge, MA: MIT Press. (Cited on pages 167, 209, and 222.)

Gelman, A., Carlin, J. B., Stern, H. S., Dunson, D. B., Vehtari, A., and Rubin, D. B. 2013. *Bayesian Data Analysis*. Third edn. Boca Raton, FL: Chapman and Hall/CRC. (Cited on pages 237 and 242.)

Gilsing, H. and Shardlow, T. 2007. SDELab: A package for solving stochastic differential equations in MATLAB. *Journal of Computational and Applied Mathematics*, **205**(2), 1002–1018. (Cited on page 148.)

Girolami, M. and Calderhead, B. 2011. Riemann manifold Langevin and Hamiltonian Monte Carlo methods. *Journal of the Royal Statistical Society: Series B (Statistical Methodology)*, **73**(2), 123–214. (Cited on page 75.)

Girsanov, I. V. 1960. On transforming a certain class of stochastic processes by absolutely continuous substitution of measures. *Theory of Probability and Its Applications*, **5**(3), 285–301. (Cited on pages 104 and 107.)

Grewal, M. S. and Andrews, A. P. 2001. *Kalman Filtering, Theory and Practice Using MATLAB*. New York, NY: Wiley. (Cited on pages 29, 77, 79, 82, and 83.)

Hairer, E., Nørsett, S. P., and Wanner, G. 2008. *Solving Ordinary Differential Equations I: Nonstiff Problems*. Springer Series in Computational Mathematics, vol. 1. Berlin: Springer Science & Business. (Cited on page 4.)

Hartikainen, J. and Särkkä, S. 2010. Kalman filtering and smoothing solutions to temporal Gaussian process regression models. Pages 379–384 of: *Proceedings of the IEEE International Workshop on Machine Learning for Signal Processing (MLSP)*. Piscataway, NJ: IEEE. (Cited on pages 251, 260, 261, and 266.)

Hartikainen, J. and Särkkä, S. 2011. Sequential inference for latent force models. Pages 311–318 of: *Proceedings of the 27th Conference on Uncertainty in Artificial Intelligence (UAI)*. Corvallis, OR: AUAI Press. (Cited on pages 251, 266, 270, and 271.)

Hartikainen, J., Seppänen, M., and Särkkä, S. 2012. State-space inference for non-linear latent force models with application to satellite orbit prediction. Pages 903–910 of: Langford, J. and Pineau, J. (eds.), *Proceedings of the 29th International Conference on Machine Learning (ICML)*. New York, NY: Omnipress. (Cited on pages 270 and 271.)

Henderson, D. and Plaschko, P. 2006. *Stochastic Differential Equations in Science and Engineering*. Singapore: World Scientific Publishing Company. (Cited on page 54.)

Higham, N. J. 2008. *Functions of Matrices: Theory and Computation*. Philadelphia, PA: Society for Industrial and Applied Mathematics. (Cited on page 265.)

Iacus, S. M. 2008. *Simulation and Inference for Stochastic Differential Equations*. New York, NY: Springer. (Cited on pages 54, 98, 157, 175, 181, 183, 192, 234, and 248.)

Ikeda, N. and Watanabe, S. 1981. *Stochastic Differential Equations and Diffusion Processes*. Amsterdam: North-Holland Publishing Company. (Cited on pages 54 and 204.)

Itô, K. 1951. Multiple Wiener integral. *Journal of the Mathematical Society of Japan*, **3**(1), 157–169. (Cited on page 136.)

Ito, K. and Xiong, K. 2000. Gaussian filters for nonlinear filtering problems. *IEEE Transactions on Automatic Control*, **45**(5), 910–927. (Cited on page 169.)

Jazwinski, A. H. 1970. *Stochastic Processes and Filtering Theory*. New York, NY: Academic Press. (Cited on pages 29, 69, 168, 170, 197, 205, 209, 219, 222, and 247.)

Jeisman, J. 2005. *Estimation of the Parameters of Stochastic Differential Equations*. Ph.D. thesis, Queensland University of Technology, Brisbane, Australia. (Cited on page 234.)

Julier, S. J. and Uhlmann, J. K. 2004. Unscented filtering and nonlinear estimation. *Proceedings of the IEEE*, **92**(3), 401–422. (Cited on page 169.)

Julier, S. J., Uhlmann, J. K., and Durrant-Whyte, H. F. 1995. A new approach for filtering nonlinear systems. Pages 1628–1632 of: *Proceedings of the 1995 American Control Conference*. Piscataway, NJ: IEEE. (Cited on page 169.)

Kalman, R. E. 1960. A new approach to linear filtering and prediction problems. *Transactions of the ASME, Journal of Basic Engineering*, **82**(3), 35–45. (Cited on page 216.)

Kalman, R. E. and Bucy, R. S. 1961. New results in linear filtering and prediction theory. *Transactions of the ASME, Journal of Basic Engineering*, **83**(3), 95–108. (Cited on page 208.)

Karatzas, I. and Shreve, S. E. 1991. *Brownian Motion and Stochastic Calculus*. New York, NY: Springer-Verlag. (Cited on pages ix, 42, 46, 47, 50, 60, 100, 101, and 104.)

Kessler, M. 1997. Estimation of an ergodic diffusion from discrete observations. *Scandinavian Journal of Statistics*, **24**(2), 211–229. (Cited on pages 179 and 181.)

Kloeden, P. E. and Platen, E. 1999. *Numerical Solution to Stochastic Differential Equations*. Applications of mathematics, vol. 23. New York, NY: Springer. (Cited on pages 51, 52, 53, 54, 129, 132, 133, 137, 138, 145, 150, 151, and 153.)

Kloeden, P. E., Platen, E., and Schurz, H. 1994. *Numerical Solution of SDE Through Computer Experiments*. Berlin: Springer. (Cited on pages 129, 145, and 150.)

Kokkala, J., Solin, A., and Särkkä, S. 2016. Sigma-point filtering and smoothing based parameter estimation in nonlinear dynamic systems. *Journal of Advances in Information Fusion*, **11**(1), 15–30. (Cited on page 249.)

Kreyszig, E. 1993. *Advanced Engineering Mathematics*. New York, NY: John Wiley & Sons, Inc. (Cited on pages 4, 13, and 185.)

Kushner, H. J. 1964. On the differential equations satisfied by conditional probability densities of Markov processes, with applications. *Journal of the Society for Industrial and Applied Mathematics, Series A: Control*, **2**(1), 106–119. (Cited on page 206.)

Kushner, H. J. 1967. Approximations to optimal nonlinear filters. *IEEE Transactions on Automatic Control*, **12**(5), 546–556. (Cited on page 166.)

Langevin, P. 1908. Sur la théorie du mouvement brownien (Engl. On the Theory of Brownian Motion). *Comptes-rendus de l'Académie des sciences (Paris)*, **146**, 530–533. (Cited on page 26.)

Leondes, C. T., Peller, J. B., and Stear, E. B. 1970. Nonlinear smoothing theory. *IEEE Transactions on Systems Science and Cybernetics*, **6**(1), 63–71. (Cited on pages 225 and 229.)

Liu, J. S. 2001. *Monte Carlo Strategies in Scientific Computing*. Springer Series in Statistics. New York, NY: Springer. (Cited on page 237.)

Loève, M. 1963. *Probability Theory II*. Springer-Verlag. (Cited on page 103.)

Luenberger, D. G. and Ye, Y. 2008. *Linear and Nonlinear Programming*. Third edn. New York, NY: Springer. (Cited on pages 237, 241, and 242.)

Luo, W. 2006. *Wiener Chaos Expansion and Numerical Solutions of Stochastic Partial Differential Equations*. Ph.D. thesis, California Institute of Technology, Pasadena, CA. (Cited on pages 104, 193, and 194.)

Lyons, S. M. J., Särkkä, S., and Storkey, A. J. 2014. Series expansion approximations of Brownian motion for non-linear Kalman filtering of diffusion processes. *IEEE Transactions on Signal Processing*, **62**(6), 1514–1524. (Cited on page 194.)

Lyons, S. M., Storkey, A. J., and Särkkä, S. 2012. The coloured noise expansion and parameter estimation of diffusion processes. Pages 1952–1960 of: Pereira, F., Burges, C. J. C., Bottou, L., and Weinberger, K. Q. (eds.), *Advances in Neural Information Processing Systems 25*. Red Hook, NY: Curran Associates, Inc. (Cited on page 194.)

MacKay, D. J. 1998. Introduction to Gaussian processes. Pages 133–166 of: Bishop, C. M. (ed.), *Neural Networks and Machine Learning*. NATO ASI Series F Computer and Systems Sciences, vol. 168. Berlin: Springer. (Cited on page 251.)

Maybeck, P. S. 1979. *Stochastic Models, Estimation and Control*. Vol. 1. New York, NY: Academic Press. (Cited on page 197.)

Maybeck, P. S. 1982a. *Stochastic Models, Estimation and Control*. Vol. 2. New York, NY: Academic Press. (Cited on pages 168, 174, 197, 219, and 222.)

Maybeck, P. S. 1982b. *Stochastic Models, Estimation and Control*. Vol. 3. New York, NY: Academic Press. (Cited on page 279.)

Mbalawata, I. S. 2014. *Adaptive Markov Chain Monte Carlo and Bayesian Filtering for State Space Models*. Doctoral dissertation, Lappeenranta University of Technology, Lappeenranta, Finland. (Cited on page 234.)

Mbalawata, I. S., Särkkä, S., and Haario, H. 2013. Parameter estimation in stochastic differential equations with Markov chain Monte Carlo and non-linear Kalman filtering. *Computational Statistics*, **28**(3), 1195–1223. (Cited on pages 241, 247, 248, and 250.)

McNamee, J. and Stenger, F. 1967. Construction of fully symmetric numerical integration formulas. *Numerische Mathematik*, **10**(4), 327–344. (Cited on page 169.)

Mercer, J. 1909. Functions of positive and negative type, and their connection with the theory of integral equations. *Philosophical Transactions of the Royal Society of London A: Mathematical, Physical and Engineering Sciences*, **209**(441–458), 415–446. (Cited on page 103.)

Møller, J. K. and Madsen, H. 2010. *From State Dependent Diffusion to Constant Diffusion in Stochastic Differential Equations by the Lamperti Transform*. Tech. rept. Technical University of Denmark, Lyngby, Denmark. (Cited on pages 98 and 99.)

Nickisch, H., Solin, A., and Grigorevskiy, A. 2018. State space Gaussian processes with non-Gaussian likelihood. Pages 3789–3798 of: Dy, J. and Krause, A. (eds.), *Proceedings of the 35th International Conference on Machine Learning (ICML)*. Proceedings of Machine Learning Research, vol. 80. Stockholm, Sweden: PMLR. (Cited on page 251.)

Nielsen, J. N., Madsen, H., and Young, P. C. 2000. Parameter estimation in stochastic differential equations: an overview. *Annual Reviews in Control*, **24**, 83–94. (Cited on page 234.)

Nualart, D. 2006. *The Malliavin Calculus and Related Topics*. Probability and Its Applications. Berlin and New York, NY: Springer. (Cited on pages 42, 272, and 279.)

O'Hagan, A. 1978. Curve fitting and optimal design for prediction (with discussion). *Journal of the Royal Statistical Society. Series B (Methodological)*, **40**(1), 1–42. (Cited on page 251.)

Øksendal, B. 2003. *Stochastic Differential Equations: An Introduction with Applications*. Sixth edn. New York, NY: Springer. (Cited on pages ix, 29, 42, 46, 47, 55, 56, 59, 60, 68, 104, 108, 118, 121, and 279.)

Øksendal, B. and Sulem, A. 2007. *Applied Stochastic Control of Jump Diffusions*. Second edn. Berlin: Springer. (Cited on page 279.)

Ozaki, T. 1992. A bridge between nonlinear time series models and nonlinear stochastic dynamical systems: a local linearization approach. *Statistica Sinica*, **2**(1), 113–135. (Cited on pages 173, 175, and 177.)

Ozaki, T. 1993. A local linearization approach to nonlinear filtering. *International Journal of Control*, **57**(1), 75–96. (Cited on pages 173, 175, and 177.)

Papoulis, A. 1984. *Probability, Random Variables, and Stochastic Processes*. Singapore: McGraw-Hill. (Cited on page 167.)

Pardoux, E. and Răşcanu, A. 2014. *Stochastic Differential Equations, Backward SDEs, Partial Differential Equations*. Stochastic Modelling and Applied Probability, vol. 69. Berlin: Springer. (Cited on pages 118, 120, 121, and 122.)

Pedersen, A. R. 1995. New approach to maximum likelihood estimation for stochastic differential equations based on discrete observations. *Scandinavian Journal of Statistics*, **22**(1), 55–71. (Cited on page 192.)

Piiroinen, P. and Simon, M. 2016. From Feynman–Kac formulae to numerical stochastic homogenization in electrical impedance tomography. *Annals of Applied Probability*, **26**(5), 3001–3043. (Cited on pages 118, 120, and 121.)

Platen, E. and Bruti-Liberati, N. 2010. *Numerical Solution of Stochastic Differential Equations with Jumps in Finance*. Stochastic Modelling and Applied Probability, vol. 64. Berlin: Springer. (Cited on page 279.)

Protter, P. E. 2013. *Stochastic Integration and Differential Equations*. Second edn. Berlin: Springer. (Cited on pages 55 and 56.)

Rao, B. L. S. P. 1999. *Statistical Inference for Diffusion Type Processes*. Kendall's Library of Statistics, vol. 8. Chichester: Wiley. (Cited on pages 175, 181, 183, 234, 248, and 249.)

Rasmussen, C. E. and Williams, C. K. 2006. *Gaussian Processes for Machine Learning*. Cambridge, MA: MIT Press. (Cited on pages 251, 252, 254, 259, 264, and 265.)

Rogers, L. C. G. and Williams, D. 2000a. *Diffusions, Markov Processes and Martingales, Volume 1: Foundations*. Cambridge Mathematical Library. Cambridge: Cambridge University Press. (Cited on pages 59, 60, 101, 104, and 113.)

Rogers, L. C. G. and Williams, D. 2000b. *Diffusions, Markov Processes and Martingales, Volume 2: Itô Calculus*. Cambridge Mathematical Library. Cambridge: Cambridge University Press. (Cited on pages 59, 60, 113, and 116.)

Rößler, A. 2006. Runge–Kutta methods for Itô stochastic differential equations with scalar noise. *BIT Numerical Mathematics*, **46**(1), 97–110. (Cited on page 150.)

Rößler, A. 2009. Second order Runge–Kutta methods for Itô stochastic differential equations. *SIAM Journal on Numerical Analysis*, **47**(3), 1713–1738. (Cited on pages 151 and 153.)

Rößler, A. 2010. Runge–Kutta methods for the strong approximation of solutions of stochastic differential equations. *SIAM Journal on Numerical Analysis*, **48**(3), 922–952. (Cited on pages 146 and 148.)

Ruttor, A., Batz, P., and Opper, M. 2013. Approximate Gaussian process inference for the drift function in stochastic differential equations. Pages 2040–2048 of: Burges, C. J. C., Bottou, L., Welling, M., Ghahramani, Z., and Weinberger, K. Q. (eds.), *Advances in Neural Information Processing Systems 26*. Red Hook, NY: Curran Associates, Inc. (Cited on pages 249, 268, and 269.)

Särkkä, S. 2006. *Recursive Bayesian Inference on Stochastic Differential Equations*. Doctoral dissertation, Helsinki University of Technology, Espoo, Finland. (Cited on pages 79, 82, 83, 84, 197, 219, 222, and 224.)

Särkkä, S. 2007. On unscented Kalman filtering for state estimation of continuous-time nonlinear systems. *IEEE Transactions on Automatic Control*, **52**(9), 1631–1641. (Cited on pages 209, 211, and 222.)

Särkkä, S. 2011. Linear operators and stochastic partial differential equations in Gaussian process regression. Pages 151–158 of: Honkela, T., Duch, W., Girolami, M., and Kaski, S. (eds.), *Artificial Neural Networks and Machine Learning – ICANN 2011*. Lecture Notes in Computer Science, vol. 6792. Berlin and Heidelberg: Springer. (Cited on page 256.)

Särkkä, S. 2013. *Bayesian Filtering and Smoothing*. Institute of Mathematical Statistics Textbooks, vol. 3. Cambridge: Cambridge University Press. (Cited on pages 68, 79, 82, 83, 168, 169, 197, 202, 203, 206, 212, 213, 216, 219, 221, 222, 224, 228, 229, 232, 234, 237, 247, 248, and 249.)

Särkkä, S. and Hartikainen, J. 2012. Infinite-dimensional Kalman filtering approach to spatio-temporal Gaussian process regression. Pages 993–1001 of: Lawrence, N. D. and Girolami, M. (eds.), *Proceedings of the Fifteenth International Conference on Artificial Intelligence and Statistics*. Proceedings of Machine Learning Research, vol. 22. La Palma, Canary Islands: PMLR. (Cited on pages 251 and 266.)

Särkkä, S. and Piché, R. 2014. On convergence and accuracy of state-space approximations of squared exponential covariance functions. Pages 1–6 of: *Proceedings of the*

IEEE International Workshop on Machine Learning for Signal Processing (MLSP). Piscataway, NJ: IEEE. (Cited on page 261.)

Särkkä, S. and Sarmavuori, J. 2013. Gaussian filtering and smoothing for continuous-discrete dynamic systems. *Signal Processing*, **93**(2), 500–510. (Cited on pages 166, 167, 168, 170, 209, 210, 221, 222, and 229.)

Särkkä, S. and Solin, A. 2012. On continuous-discrete cubature Kalman filtering. Pages 1221–1226 of: *16th IFAC Symposium on System Identification (SYSID)*. IFAC Proceedings Volumes, vol. 45. Elsevier. (Cited on pages 135, 169, and 175.)

Särkkä, S. and Sottinen, T. 2008. Application of Girsanov theorem to particle filtering of discretely observed continuous-time non-linear systems. *Bayesian Analysis*, **3**(3), 555–584. (Cited on pages 104, 105, 106, 111, and 222.)

Särkkä, S., Solin, A., Nummenmaa, A., Vehtari, A., Auranen, T., Vanni, S., and Lin, F.-H. 2012. Dynamic retrospective filtering of physiological noise in BOLD fMRI: DRIFTER. *NeuroImage*, **60**(2), 1517–1527. (Cited on page 32.)

Särkkä, S., Solin, A., and Hartikainen, J. 2013. Spatiotemporal learning via infinite-dimensional Bayesian filtering and smoothing. *IEEE Signal Processing Magazine*, **30**(4), 51–61. (Cited on pages 251, 256, 261, 266, 267, 268, and 275.)

Särkkä, S., Tolvanen, V., Kannala, J., and Rahtu, E. 2015a. Adaptive Kalman filtering and smoothing for gravitation tracking in mobile systems. Pages 1–7 of: *Proceedings of Indoor Positioning and Indoor Navigation (IPIN)*. Piscataway, NJ: IEEE. (Cited on page 31.)

Särkkä, S., Hartikainen, J., Mbalawata, I. S., and Haario, H. 2015b. Posterior inference on parameters of stochastic differential equations via non-linear Gaussian filtering and adaptive MCMC. *Statistics and Computing*, **25**(2), 427–437. (Cited on pages 234, 247, 248, and 266.)

Schweppe, F. 1965. Evaluation of likelihood functions for Gaussian signals. *IEEE Transactions on Information Theory*, **11**(1), 61–70. (Cited on page 247.)

Shoji, I. and Ozaki, T. 1998. Estimation for nonlinear stochastic differential equations by a local linearization method. *Stochastic Analysis and Applications*, **16**(4), 733–752. (Cited on pages 173, 175, and 177.)

Shubin, M. A. 1987. *Pseudodifferential Operators and Spectral Theory*. Springer Series in Soviet Mathematics. Berlin: Springer. (Cited on page 268.)

Shumway, R. H. and Stoffer, D. S. 1982. An approach to time series smoothing and forecasting using the EM algorithm. *Journal of Time Series Analysis*, **3**(4), 253–264. (Cited on page 249.)

Simon, M. 2015. *Anomaly Detection in Random Heterogeneous Media*. Wiesbaden: Springer Spektrum. (Cited on pages 118, 120, and 121.)

Singer, H. 2002. Parameter estimation of nonlinear stochastic differential equations: simulated maximum likelihood versus extended Kalman filter and Itô-Taylor expansion. *Journal of Computational and Graphical Statistics*, **11**(4), 972–995. (Cited on page 247.)

Šmídl, V. and Quinn, A. 2006. *The Variational Bayes Method in Signal Processing*. Berlin: Springer. (Cited on page 249.)

Socha, L. 2008. *Linearization Methods for Stochastic Dynamic Systems*. Lecture Notes in Physics, vol. 730. Berlin: Springer. (Cited on pages 72, 73, and 167.)

Solin, A. 2016. *Stochastic Differential Equation Methods for Spatio-Temporal Gaussian Process Regression*. Doctoral dissertation, Aalto University, Helsinki, Finland. (Cited on pages 251, 259, 266, and 268.)

Solin, A. and Särkkä, S. 2014a. Explicit link between periodic covariance functions and state space models. Pages 904–912 of: Kaski, S. and Corander, J. (eds.), *Proceedings of the Seventeenth International Conference on Artificial Intelligence and Statistics*. Proceedings of Machine Learning Research, vol. 33. Reykjavik, Iceland: PMLR. (Cited on pages 262, 263, and 265.)

Solin, A. and Särkkä, S. 2014b. Gaussian quadratures for state space approximation of scale mixtures of squared exponential covariance functions. Pages 1–6 of: *Proceedings of the IEEE International Workshop on Machine Learning for Signal Processing (MLSP)*. Piscataway, NJ: IEEE. (Cited on page 262.)

Sørensen, H. 2004. Parametric inference for diffusion processes observed at discrete points in time: a survey. *International Statistical Review*, **72**(3), 337–354. (Cited on page 234.)

Stengel, R. F. 1994. *Optimal Control and Estimation*. New York, NY: Dover. (Cited on pages 83 and 279.)

Stratonovich, R. L. 1968. *Conditional Markov Processes and Their Application to the Theory of Optimal Control*. New York, NY: American Elsevier. (Cited on pages 55, 56, 197, and 206.)

Svensson, A., Solin, A., Särkkä, S., and Schön, T. 2016. Computationally efficient Bayesian learning of Gaussian process state space models. Pages 213–221 of: Gretton, A. and Robert, C. C. (eds.), *Proceedings of the 19th International Conference on Artificial Intelligence and Statistics*. Proceedings of Machine Learning Research, vol. 51. Cadiz, Spain: PMLR. (Cited on page 269.)

Tenenbaum, M. and Pollard, H. 1985. *Ordinary Differential Equations*. New York, NY: Dover. (Cited on pages 4 and 19.)

Tronarp, F., García-Fernández, Á. F., and Särkkä, S. 2018. Iterative filtering and smoothing in nonlinear and non-Gaussian systems using conditional moments. *IEEE Signal Processing Letters*, **25**(3), 408–412. (Cited on pages 221, 229, and 245.)

Turner, R., Deisenroth, M., and Rasmussen, C. 2010. State-space inference and learning with Gaussian processes. Pages 868–875 of: Teh, Y. W. and Titterington, M. (eds.), *Proceedings of the Thirteenth International Conference on Artificial Intelligence and Statistics*. Proceedings of Machine Learning Research, vol. 9. Sardinia, Italy: PMLR. (Cited on page 269.)

Van Trees, H. L. 1968. *Detection, Estimation, and Modulation Theory, Part I: Detection, Estimation, and Linear Modulation Theory*. New York, NY: John Wiley & Sons. (Cited on page 95.)

Wan, E. A. and van der Merwe, R. 2001. The unscented Kalman filter. Chapter 7 of: Haykin, S. (ed.), *Kalman Filtering and Neural Networks*. New York, NY: Wiley. (Cited on page 169.)

Wiener, N. 1923. Differential-space. *Journal of Mathematics and Physics*, **2**(1–4), 131–174. (Cited on page 101.)

Wiktorsson, M. 2001. Joint characteristic function and simultaneous simulation of iterated Itô integrals for multiple independent Brownian motions. *Annals of Applied Probability*, **11**(2), 470–487. (Cited on page 148.)

Wilkie, J. 2004. Numerical methods for stochastic differential equations. *Physical Review E*, **70**(1), 017701. (Cited on page 145.)

Wong, E. and Zakai, M. 1965. On the convergence of ordinary integrals to stochastic integrals. *Annals of Mathematical Statistics*, **36**(5), 1560–1564. (Cited on pages 55 and 193.)

Wu, Y., Hu, D., Wu, M., and Hu, X. 2006. A numerical-integration perspective on Gaussian filters. *IEEE Transactions on Signal Processing*, **54**(8), 2910–2921. (Cited on pages 169 and 170.)

Zakai, M. 1969. On the optimal filtering of diffusion processes. *Probability Theory and Related Fields*, **11**(3), 230–243. (Cited on page 207.)

Zeitouni, O. 1989. On the Onsager–Machlup functional of diffusion processes around non C^2 curves. *Annals of Probability*, **17**(3), 1037–1054. (Cited on page 205.)

Zinn-Justin, J. 2002. *Quantum Field Theory and Critical Phenomena*. Fourth edn. International Series of Monographs on Physics, vol. 113. Oxford: Clarendon Press. (Cited on page 117.)

Symbols and Abbreviations

Notational Conventions

$\bullet$	Arbitrary argument
$[\bullet]$	Integer part of the argument
$a, b, c, \ldots$	Scalars
$\alpha, \beta, \gamma, \ldots$	Scalars
$\mathbf{a}, \mathbf{b}, \mathbf{c}, \ldots$	Vectors
$\boldsymbol{\alpha}, \boldsymbol{\beta}, \boldsymbol{\gamma}, \ldots$	Vectors
$\mathbf{A}, \mathbf{B}, \mathbf{C}, \ldots$	Matrices
$[\mathbf{a}]_i$	ith element of vector $\mathbf{a}$
$[\mathbf{a}]_\times$	Cross-product matrix such that $\mathbf{a} \times \mathbf{b} = [\mathbf{a}]_\times \mathbf{b}$
$a \leftrightarrow b$	a and b are the same objects or operations in a different basis
$\mathbf{A}^\top$	Transpose of matrix $\mathbf{A}$
$\mathbf{A}^{-1}$	Inverse of matrix $\mathbf{A}$
$\mathbf{A}^{-\top}$	Inverse of transpose of matrix $\mathbf{A}$
$\text{tr}\,\mathbf{A}$	Trace of matrix $\mathbf{A}$
$|\mathbf{A}|$	Determinant of matrix $\mathbf{A}$
$[\mathbf{A}]_i$	ith column of matrix $\mathbf{A}$
$[\mathbf{A}]_{ij}$	Element at ith row and jth column of matrix $\mathbf{A}$
$\mathbf{A} \otimes \mathbf{B}$	Kronecker product of matrices $\mathbf{A}$ and $\mathbf{B}$
$\mathcal{A}^*$	Adjoint of operator $\mathcal{A}$
$\text{Cov}[\mathbf{x}]$	Covariance of $\mathbf{x}$
$\text{Cov}[\mathbf{x} \mid \mathbf{y}]$	Conditional covariance of $\mathbf{x}$ given $\mathbf{y}$
$\text{E}[\mathbf{x}]$	Expectation of $\mathbf{x}$
$\text{E}[\mathbf{x} \mid \mathbf{y}]$	Conditional expectation of $\mathbf{x}$ given $\mathbf{y}$
$\text{E}_\text{N}[\bullet]$	Expectation with respect to moment-matched Gaussian
$\exp(\mathbf{F})$	Matrix exponential $\exp(\mathbf{F}) = \mathbf{I} + \mathbf{F} + \frac{1}{2!}\mathbf{F}^2 + \cdots$
$\exp(\mathcal{A})$	Operator exponential of operator $\mathcal{A}$
$\langle f, g \rangle$	Inner product of functions f and g

$\lim_{s\downarrow 0} f(s)$	Limit of $f(s)$ then s approaches zero from positive side
$\int_{t_0}^t f(\tau)\, d\tau$	Lebesgue integral of $f(t)$ over the range $[t_0, t]$
$\int f(\mathbf{x})\, d\mathbf{x}$	Lebesgue integral of $f(\mathbf{x})$ over the space $\mathbb{R}^D$ (assuming that $\mathbf{x} \in \mathbb{R}^D$)
$\int_{t_0}^t f(\tau)\, d\beta(\tau)$	Itô integral of $f(t)$ over Brownian motion $\beta(t)$ on the range $[t_0, t]$
$g'(x)$	Derivative $g'(x) = dg(x)/dx$
$\frac{\partial g(\mathbf{x},t)}{\partial t}$	Partial derivative of g with respect to t
$\frac{\partial g(\mathbf{x},t)}{\partial x_i}$	Partial derivative of g with respect to the ith component of $\mathbf{x}$
$\frac{\partial^2 g(\mathbf{x})}{\partial x_i\, \partial x_j}$	Second partial derivative of g with respect to the ith and jth components of $\mathbf{x}$
∇g	Gradient (column vector of partial derivatives) of scalar function $g(\mathbf{x})$ with input $\mathbf{x} \in \mathbb{R}^D$
$\nabla\nabla^{\mathsf{T}} g$	Hessian matrix (matrix of second derivatives) of scalar function $g(\mathbf{x})$ with input $\mathbf{x} \in \mathbb{R}^D$
$\int_{t_0}^t \mathbf{L}(\tau)\, d\boldsymbol{\beta}(\tau)$	Itô integral of $\mathbf{L}(t)$ over vector of Brownian motions $\boldsymbol{\beta}(t)$ on the range $[t_0, t]$
$\int_{t_0}^t \mathbf{L}(\tau) \circ d\boldsymbol{\beta}(\tau)$	Stratonovich integral of $\mathbf{L}(t)$ over vector of Brownian motions $\boldsymbol{\beta}(t)$ on the range $[t_0, t]$
$\mathbf{m}(t \mid s)$	Mean of the transition density $p(\mathbf{x}(t) \mid \mathbf{x}(s))$
$n!$	Factorial of n, that is, $n! = 1 \times 2 \times \cdots \times (n-1) \times n$
$\mathrm{N}(x \mid m, P)$	Univariate Gaussian probability density with mean m and variance P
$\mathrm{N}(\mathbf{x} \mid \mathbf{m}, \mathbf{P})$	Multivariate Gaussian probability density with mean $\mathbf{m}$ and covariance $\mathbf{P}$
$p(\mathbf{x}(t))$	Probability density of $\mathbf{x}(t)$
$p(\mathbf{x}, t)$	Probability density of $\mathbf{x}(t)$
$p(\mathbf{x}, t) \triangleq p(\mathbf{x}(t))$	$p(\mathbf{x}, t)$ is defined to be equal to $p(\mathbf{x}(t))$ (i.e., it is just an alternative notation)
$p(\mathbf{x}(t) \mid \mathbf{x}(s))$	Conditional (transition) probability density of $\mathbf{x}(t)$ given $\mathbf{x}(s)$
$p(\mathbf{x}, t \mid \mathbf{y}, s)$	Conditional (transition) probability density of $\mathbf{x}(t)$ given $\mathbf{x}(s) = \mathbf{y}(s)$
$\mathbf{P}(t \mid s)$	Covariance of the transition density $p(\mathbf{x}(t) \mid \mathbf{x}(s))$
$\sqrt{\mathbf{P}}$	Matrix such that $\mathbf{P} = \sqrt{\mathbf{P}}\,\sqrt{\mathbf{P}}^{\mathsf{T}}$
$u \sim \mathrm{U}(0, 1)$	Random variable u has a uniform distribution on range $[0, 1]$
$\frac{d\mathbf{x}}{dt}$	Time derivative of $\mathbf{x}(t)$

$\dfrac{d^2\mathbf{x}}{dt^2}$	Second time derivative of $\mathbf{x}(t)$
$x \sim \mathrm{N}(m, P)$	Random variable x has univariate Gaussian distribution with mean m and variance P
$\mathbf{x} \sim \mathrm{N}(\mathbf{m}, \mathbf{P})$	Random variable $\mathbf{x}$ has a multivariate Gaussian distribution with mean $\mathbf{m}$ and covariance $\mathbf{P}$
$\dot{\mathbf{x}}$	Time derivative of $\mathbf{x}(t)$
$\ddot{\mathbf{x}}$	Second time derivative of $\mathbf{x}(t)$
$\mathbf{y}_{1:k}$	Collection of vectors $\{\mathbf{y}_1, \mathbf{y}_2, \dots, \mathbf{y}_k\}$
$\partial\Omega$	Boundary of domain $\Omega \subset \mathbb{R}^D$

Symbols

α	Scalar constant, weak order of convergence
α_i	Scalar constant ($i = 1, 2, \dots$), acceptance probability
$\boldsymbol{\alpha}$	Constant vector in SRK, multi-index
β	Brownian motion (with diffusion coefficient q)
$\tilde{\beta}$	Physical Brownian motion (in Langevin's model)
$\check{\beta}$	Brownian motion of a weak solution of SDE, biased Brownian motion
$\boldsymbol{\beta}$	Vector of Brownian motions (with diffusion matrix $\mathbf{Q}$)
γ	Damping parameter, strong order of convergence
$\boldsymbol{\gamma}$	A constant vector in SRK
Γ	Gamma function
Γ_{ij}	Element at row i and column j of matrix $\boldsymbol{\Gamma}$
$\boldsymbol{\Gamma}$	Matrix of operator-transformed basis function evaluations or inner products
δ	Dirac's delta function, positive scalar
δ_{ij}	Kronecker's delta
Δ	Spatial displacement
Δt	Length of a time interval
Δt_k	Length of the time interval $\Delta t_k = t_{k+1} - t_k$
$\Delta\beta_k$	Increment of scalar Brownian motion $\Delta\beta_k = \beta(t_{k+1}) - \beta(t_k)$
$\Delta\boldsymbol{\beta}_k$	Increment of vector of Brownian motions $\Delta\boldsymbol{\beta}_k = \boldsymbol{\beta}(t_{k+1}) - \boldsymbol{\beta}(t_k)$
$\Delta\beta_k^{(i)}$	Element $\Delta\beta_k^{(i)} \triangleq [\Delta\boldsymbol{\beta}_k]_i$
$\Delta\hat{\boldsymbol{\beta}}_k$	Random variable in weak SDE solution method
$\Delta\hat{\beta}_k^{(i)}$	Element $\Delta\hat{\beta}_k^{(i)} \triangleq [\Delta\hat{\boldsymbol{\beta}}_k]_i$
$\Delta\boldsymbol{\zeta}_k$	Gaussian increment in Itô–Taylor method
$\Delta\hat{\boldsymbol{\zeta}}_k$	Random variable in weak SDE solution method

$\Delta\boldsymbol{\chi}_{j,k}$	Vector of iterated Itô integrals
$\Delta\chi_{j,k}^{(i)}$	Element $\Delta\chi_{j,k}^{(i)} \triangleq [\Delta\boldsymbol{\chi}_{j,k}]_i$
ε	Positive constant
ε_k	Gaussian noise
$\boldsymbol{\varepsilon}$	White noise in measurement noise
$\boldsymbol{\varepsilon}_k$	Gaussian noise vector
ζ	An arbitrary scalar function, a complex Gaussian process
η	Viscosity of liquid, a damping parameter
$\boldsymbol{\eta}$	Brownian motion in measurement model
θ	Angle, a scalar constant
θ_n	Scalar constant, angle, random variable, parameter ($n = 1, 2, \ldots$)
$\boldsymbol{\theta}$	Arbitrary Itô process, vector of parameters
$\boldsymbol{\theta}_{\mathrm{ML}}$	Maximum likelihood estimate of $\boldsymbol{\theta}$
$\boldsymbol{\theta}_{\mathrm{MAP}}$	Maximum a posteriori estimate of $\boldsymbol{\theta}$
$\boldsymbol{\theta}^*$	Proposal sample
$\boldsymbol{\theta}^{(i)}$	MCMC sample from round i
$\boldsymbol{\Theta}$	Arbitrary Itô process
κ	Parameter of unscented transform
λ	Positive constant, parameter of unscented transform
λ_n	nth eigenvalue
$\boldsymbol{\lambda}$	Lagrange multiplier
μ	Drift constant, mean of mean-reverting process, sufficient statistic
$\boldsymbol{\mu}$	Mean function
$\boldsymbol{\mu}_k$	Measurement mean
ν	Angular velocity parameter of spring model, parameter of Matérn covariance function
ξ	Arbitrary scalar, a random variable, noise-free solution
$\boldsymbol{\xi}$	Gaussian process input
$\boldsymbol{\xi}'$	Other Gaussian process input
$\boldsymbol{\xi}_i$	Unit sigma point in a sigma-point method ($i = 0, 1, \ldots$)
π	The constant $\pi \approx 3.14159265359$
ρ	Order of integration method
σ	Volatility constant, standard deviation, scale parameter
Σ	Scalar discrete-time variance
Σ_k	Scalar discrete-time variance at time step k
$\boldsymbol{\Sigma}$	Discretization covariance of LTI SDE, covariance matrix
$\boldsymbol{\Sigma}_k$	Step k discretization covariance of linear SDE
τ	Time, time interval, dummy integration variable
ϕ	Probability density function of displacements, a scalar function

ϕ_n	nth eigen/basis function
$\boldsymbol{\phi}$	Test function in Onsager–Machlup functional
Φ	Poisson process
Φ_{ij}	Element at row i and column j of matrix $\boldsymbol{\Phi}$
$\boldsymbol{\Phi}$	Matrix exponential in matrix fraction decomposition, matrix of basis function evaluations or inner products in PDE solver
φ	Arbitrary scalar function
φ_n	nth orthonormal basis function, other basis function
$\boldsymbol{\varphi}$	Other function in Onsager–Machlup functional
$\boldsymbol{\varphi}_n$	Approximating function in the Picard iteration ($n = 0, 1, \ldots$)
ψ	Scalar function, scalar transition matrix, potential function
Ψ	Boundary value
$\boldsymbol{\Psi}$	Transition matrix of a linear differential equation (ODE or SDE), Lagrange multiplier
ω	Angular velocity (e.g., in Fourier transform), event space element, scalar parameter
ω_0	Period length
$\boldsymbol{\omega}$	Angular velocity vector
Ω	Integration domain $\Omega \subset \mathbb{R}^D$, domain of PDE
a	Scalar constant, scalar discretization coefficient, scalar function
a_k	Scalar constant at step k of with index k
A	Scalar constant
$\mathbf{a}$	Acceleration vector
$\mathbf{a}_k$	Function in Itô–Taylor method
$\mathbf{A}$	Discretization matrix of LTI SDE, feedback matrix in variational approximation
$\mathbf{A}_k$	Step k discretization matrix in linear SDE
$\mathcal{A}$	Generator of diffusion
$\mathcal{A}_t$	Generalized generator of diffusion
$\mathcal{A}^*$	Fokker–Planck–Kolmogorov operator
b	Scalar constant, a scalar function
$\mathbf{b}$	Bias vector in variational approximation
$\mathbf{b}_{j,k}$	Function in Itô–Taylor method
B	Scalar constant, ε-tube
$\mathbf{B}$	Input coefficient matrix
$\mathcal{B}$	Space of Brownian paths
c	Scalar constant
c_n	Scalar constant ($n = 1, 2, \ldots$)
$c_{\boldsymbol{\alpha}}$	Scalar constant with multi-index $\boldsymbol{\alpha}$
C	Capacitance, covariance function (scalar), a scalar constant

$\mathbf{C}$	Covariance function (matrix), a matrix in the matrix fraction decomposition
$\mathbf{C}_\Sigma$	Matrix in the LTI matrix fraction decomposition
$\mathbf{C_w}$	Covariance function of $\mathbf{w}$
$\mathbf{C_x}$	Covariance function of $\mathbf{x}$
$\mathbf{C}_k$	Cross-covariance matrix
d	Scalar constant, GP input dimension
D	Scalar constant, dimensionality of state
$\mathbf{D}$	Matrix in the matrix fraction decomposition
$\mathbf{D}_\Sigma$	Matrix in the LTI matrix fraction decomposition
$\mathbf{e}_k$	Gaussian random variable
f	Arbitrary function, a scalar drift function, frequency
f_i	The ith element of $\mathbf{f}$ $(i = 1, 2, \dots)$
$\mathbf{f}$	State feedback function (in ODE), the drift function (in SDE), or a force vector
$\tilde{f}_i$	ith element of $\tilde{\mathbf{f}}$ $(i = 1, 2, \dots)$
$\tilde{\mathbf{f}}$	Modified drift function (in Itô–Stratonovich conversion)
F	Laplace transform of function f, force in Langevin model, functional, coefficient in LTI system
$\mathbf{F}$	Feedback matrix in linear differential equation (ODE or SDE)
$\mathbf{F_x}$	Jacobian matrix of function $\mathbf{f}$
$\mathfrak{F}$	Fourier transform operator
$\mathfrak{F}^{-1}$	Inverse Fourier transform operator
g	Arbitrary time-dependent function, gravitational acceleration, a constant, Gaussian process input function
$\mathbf{g}$	Local gravitation vector, arbitrary vector function, a drift function
G	Fourier transform of function g, a scalar function
$\mathbf{G}$	Matrix in the matrix fraction decomposition
$\mathbf{G}_k$	Smoother gain
$\mathscr{F}$	Feedback operator in spatiotemporal model
h	Impulse response of linear system, change-of-variable function, function in Doob's h-transform, bias distribution, discretization step length
$\mathbf{h}$	Measurement model function, change of variables function
$\mathbf{h}_k$	Measurement model function at step k
H	Fourier or Laplace transform of h, the transfer function of a linear system with the impulse response h, MAP path functional, scalar measurement model parameter
H_i	ith Hermite polynomial $(i = 0, 1, \dots)$

H_α	Multivariate Hermite polynomial with multi-index $\boldsymbol{\alpha}$
$\mathbf{H}$	Measurement model matrix
$\mathbf{H_x}$	Jacobian matrix of function $\mathbf{h}$
$\mathcal{H}$	Measurement operator
i	Nonnegative integer, generic index
i	Imaginary unit with the property $\mathrm{i}^2 = -1$
I	Binary random variable
I_α	Modified Bessel function of the first kind
$\mathbf{I}$	Identity matrix
j	Nonnegative integer, generic index
J	Nonnegative integer, Onsager–Machlup functional
J_e	Minimum energy functional
k	Time step number, nonnegative integer
K	Constant, number of time steps
K_ν	Modified Bessel function of second kind
$\mathbf{K}_k$	Kalman gain at step k
ℓ	Negative log-likelihood, length scale
$\hat{\ell}$	Approximation of negative log-likelihood
ℓ_p	Unnormalized negative log-posterior
ℓ_i	Length scale of index i
L	Dispersion coefficient of scalar SDE, (half) length of interval
L_{ij}	Element at row i and column j in matrix $\mathbf{L}$ ($i, j = 1, 2, \ldots$)
L	Laguerre polynomial
$\mathbf{L}$	Input feedback matrix (in ODE) or dispersion matrix (in SDE)
$\mathbf{L}_i$	ith column of matrix L (in stochastic Runge–Kutta methods)
$\mathcal{L}$	Linear operator in Taylor series for ODEs or dispersion operator
$\mathcal{L}_t$	Linear operator in Itô–Taylor series for SDEs
$\mathcal{L}_{\beta,\nu}$	Linear operator in Itô–Taylor series for SDEs
$\mathfrak{L}$	Laplace transform operator
$\mathfrak{L}^{-1}$	Inverse Laplace transform operator
m	Mass of a particle or car, scalar mean, nonnegative integer
m_i	ith element of vector $\mathbf{m}$ ($i = 1, 2, \ldots$)
m_0	Initial mean
m_k	Updated mean or sufficient statistic at time step k
m_k^-	Predicted mean or sufficient statistic at time step k
$\mathbf{m}$	Mean vector
$\mathbf{m}^s$	Smoother mean vector
$\mathbf{m}_0$	Initial mean vector
$\mathbf{m}_k$	Updated mean vector at time step k

$\mathbf{m}_k^-$	Predicted mean vector at time step k
$\mathbf{m}_k^s$	Smoother mean at time step k
M	Constant, number of subintervals, measurement dimensionality
n	Nonnegative integer with many meanings
n_i	ith component of $\mathbf{n}$ $(i = 1, 2, \ldots)$
$\mathbf{n}$	Normal of a boundary
N	Avogardo constant, number of series terms, number of particles
q	Spectral density of white noise, diffusion constant/coefficient of Brownian motion, probability density, unnormalized probability density
q_k	Discrete-time Gaussian noise, probability distribution
$\mathbf{q}_k$	Discrete-time Gaussian noise
$\mathbf{Q}$	Spectral density (matrix) of white noise, diffusion matrix of Brownian motion
p	Probability density function, probability distribution
$\hat{p}$	Approximation of p
P	Discrete probability distribution
$\mathbf{p}$	Vector of values of $p(\mathbf{x}_i, t)$ where $i = 1, 2, \ldots$
P	Variance of an SDE solution, probability measure, sufficient statistic
P_{ij}	Element at row i and column j in matrix $\mathbf{P}$ $(i, j = 1, 2, \ldots)$
P_0	Initial variance
P_k	Updated variance or sufficient statistic at time step k
P_k^-	Predicted variance or sufficient statistic at time step k
$\mathbf{P}$	Covariance matrix
$\mathbf{P}^s$	Smoother covariance matrix
$\mathbf{P}_0$	Initial covariance matrix
$\mathbf{P}_k$	Updated covariance matrix at time step k
$\mathbf{P}_k^-$	Predicted covariance matrix at time step k
$\mathbf{P}_k^s$	Smoother covariance matrix at time step k
$\mathbf{P}_\infty$	Stationary stage covariance matrix
r	Diameter of particle, a constant, diffusion constant/coefficient of Brownian motion
$\mathbf{r}$	Remainder term, space location
$\mathbf{r}_k$	Discrete-time measurement noise
$\mathbb{R}$	Space of real numbers
$\mathbb{R}^n$	n-dimensional space of real vectors
R	Gas constant, resistance, variance
$\mathbf{R}$	Spectral density or diffusion matrix of measurement noise
$\mathbf{R}_k$	Measurement noise covariance

s	Laplace transform input variable, dummy integration variable, time, number of RK stages, smoothing density, discrete-time system variance
s_k	Sum of k random variables
S	Dimensionality of (white noise) input or Brownian motion, area of domain
S_n	Process defined as via a sum of n random variables
S_x	Spectral density of x
$\mathbf{S}$	Spectral density (matrix)
$\mathbf{S_w}$	Spectral density (matrix) of $\mathbf{w}$ (which should equal to $\mathbf{Q}$)
$\mathbf{S_x}$	Spectral density (matrix) of $\mathbf{x}$
$\mathbf{S}_k$	Kalman filter innovation covariance
t	Time
t'	Time (typically $t \neq t'$)
t_0	Initial time
t_k	Time of the step or measurement k
t_k^*	Time in the interval $[t_k, t_{k+1}]$
T	End of time interval, number of time steps, number of measurements
T_e	Exit time
u	Heaviside step function, number of particles per volume, dummy integration variable, an arbitrary function, dummy index, uniform random variable
$\mathbf{u}$	Input function
v	Velocity of a particle, voltage, an arbitrary function, potential function, dummy index
$\mathbf{v}_k$	Kalman filter innovation
w	Scalar-valued time-dependent differential equation input, white noise
w_i	Laguerre weight
W	Fourier or Laplace transform of w, sigma-point weight
$\mathbf{w}$	Differential equation input, vector of white noises
$\mathbf{W}$	The matrix $\mathbf{W} = \mathbf{L}\,\mathbf{Q}\,\mathbf{L}^\mathsf{T}$
x	Differential equation solution (scalar), scalar Itô process
x'	Arbitrary (scalar) state
x_0	Initial value of x
x_i	ith element of vector $\mathbf{x}$
$\mathbf{x}$	Differential equation solution (vector), state of a first-order differential equation (ODE or SDE), vector-valued Itô process
$\mathbf{x}'$	Arbitrary (vector) state

$\mathbf{x}_0$	Initial value of $\mathbf{x}$
$\mathbf{x}_n$	nth subcomponent of state ($n = 1, 2, \ldots$)
$\hat{\mathbf{x}}$	Approximation of $\mathbf{x}$
$\tilde{\mathbf{x}}$	Weak solution of SDE
X	Fourier or Laplace transform of x
$\mathbf{X}$	Fourier or Laplace transform of $\mathbf{x}$
$\mathcal{X}_t$	History of $\mathbf{x}$ up to time t
y	Differential equation solution (scalar), transformed SDE solution
$\mathbf{y}$	State of an SDE, continuous-time measurement, Gaussian process measurement
$\mathbf{y}_k$	Discrete-time measurement
Y	Laplace transform of y
$\mathcal{Y}_t$	History of $\mathbf{y}$ up to time t
z	Displacement of particle, a real number
z_n	Gaussian random variable, element of pseudonormalized increment
$\mathbf{z}$	Pseudonormalized increment, integrated measurement
Z	Likelihood ratio
Z_k	Discretized likelihood ratio, normalization constant
$\mathcal{Z}_t$	History of $\mathbf{z}$ up to time t
∞	Infinity

Abbreviations

ARE	Algebraic Riccati equation
BOLD	Blood-oxygen-level dependent
CIR	Cox–Ingersoll–Ross
EA	Exact algorithm
EKBF	Extended Kalman–Bucy filter
EKF	Extended Kalman filter
FEM	Finite-element method
fMRI	Functional magnetic resonance imaging
FPK	Fokker–Planck–Kolmogorov
GP	Gaussian process
GPSS	Gaussian process state-space
KL	Kullback–Leibler
KS	Kushner–Stratonovich
LFM	Latent force model
LTI	Linear time-invariant
MAP	Maximum a posteriori
MCMC	Markov chain Monte Carlo

MH	Metropolis–Hastings
ML	Maximum likelihood
MLP	Multi-layer perceptron
ODE	Ordinary differential equation
PDE	Partial differential equation
PLL	Phase locked loop
RBF	Radial basis function
RC	Resistor–capacitor
RK	Runge–Kutta
RK4	Fourth-order Runge–Kutta
RQ	Rational quadratic
RTS	Rauch–Tung–Striebel
SDE	Stochastic differential equation
SPDE	Stochastic partial differential equation
STD	Standard deviation
ZOH	Zeroth-order-hold

List of Examples

List of Algorithms

Index

Printed in the United States
By Bookmasters